T0136076

About Island Press

Since 1984, the nonprofit Island Press has been stimulating, shaping, and communicating the ideas that are essential for solving environmental problems worldwide. With more than 800 titles in print and some 40 new releases each year, we are the nation's leading publisher on environmental issues. We identify innovative thinkers and emerging trends in the environmental field. We work with world-renowned experts and authors to develop cross-disciplinary solutions to environmental challenges.

Island Press designs and implements coordinated book publication campaigns in order to communicate our critical messages in print, in person, and online using the latest technologies, programs, and the media. Our goal: to reach targeted audiences—scientists, policymakers, environmental advocates, the media, and concerned citizens—who can and will take action to protect the plants and animals that enrich our world, the ecosystems we need to survive, the water we drink, and the air we breathe.

Island Press gratefully acknowledges the support of its work by the Agua Fund, Inc., The Margaret A. Cargill Foundation, Betsy and Jesse Fink Foundation, The William and Flora Hewlett Foundation, The Kresge Foundation, The Forrest and Frances Lattner Foundation, The Andrew W. Mellon Foundation, The Curtis and Edith Munson Foundation, The Overbrook Foundation, The David and Lucile Packard Foundation, The Summit Foundation, Trust for Architectural Easements, The Winslow Foundation, and other generous donors.

The opinions expressed in this book are those of the author(s) and do not necessarily reflect the views of our donors.

SEASONALLY DRY TROPICAL FORESTS

Seasonally Dry Tropical Forests

ECOLOGY AND CONSERVATION

EDITED BY

Rodolfo Dirzo, Hillary S. Young,
Harold A. Mooney, and Gerardo Ceballos

Washington | Covelo | London

Library of Congress Cataloging-in-Publication Data

Seasonally dry tropical forests: ecology and conservation / edited by Rodolfo Dirzo
... [et al.].
p. cm.
ISBN-13: 978-1-59726-703-8 (hardcover)
ISBN-10: 1-59726-703-1 (hardcover)
ISBN-13: 978-1-59726-704-5 (pbk.)
ISBN-10: 1-59726-704-X (pbk.)
1. Forest ecology–Tropics. 2. Forest biodiversity–Tropics. 3. Forest conservation–Tropics.
I. Dirzo, Rodolfo.
QH84.5.S43 2010
577.30913–dc22
2010026222

Printed on recycled, acid-free paper

Manufactured in the United States of America

10 9 8 7 6 5 4 3 2 1

Key Words: Dry forests, seasonally dry tropical forests, Latin America, Neotropics, conservation, biodiversity, ecosystem functioning, ecosystem services, biogeography, global change, anthropogenic change, habitat loss, sustainability.

CONTENTS

INTRODUCTION

RODOLFO DIRZO, HILLARY S. YOUNG, HAROLD A. MOONEY, AND GERARDO CEBALLOS

The usual perception that the term *tropical forests* refers to evergreen tropical rain or moist forests is inaccurate. The tropical forest biome is, in reality, a mosaic of different vegetation entities including, at mid elevations of the tropics, the patchy and biogeographically restricted tropical cloud forests and, in the lowlands, the rain forest per se and the seasonally dry tropical forests (SDTFs). At least part of the biased perception of the term *tropical forest* stems from the fact that, by far, tropical rain forests are the most studied and, indeed, most popularized among the general public. SDTFs, in contrast, have been seriously neglected. For example, only 14 percent of articles published on tropical environments between 1950 and 2005 focus on dry forests (Sánchez-Azofeifa et al. 2005). Such scientific bias, however, determines that our understanding of the planet's biodiversity, the ecosystem services it provides, and the anthropogenic threats to it in general, and to the tropical forest biome in particular, is in turn biased and will remain grossly incomplete if we do not pay attention to the SDTFs still present in the different parts of the world. The present volume is an attempt to fill part of this lacuna in our knowledge on tropical ecology by analyzing the ecology and conservation of SDTFs in Latin America. This volume represents, also, a sequel to the first and only other global synthesis, (Bullock et al. 1995) and provides a complement to some recent efforts conducted at a more local level (e.g., Ceballos et al. 2010).

SDTFs are forests with a mean annual temperature typically greater than 17 degrees Celsius, rainfall ranging from 250 to 2000 millimeters annually, and an annual ratio of potential evapotranspiration to precipitation of less than 1.0. However, by far the most distinctive character of this forest type is its seasonality, with 4 to 6 dry months (rainfall less than 100 millimeters), which in turn determines the distinctive phenology of the plants and the forest as a whole: an alternating deciduousness during the dry season, followed by an evergreen physiognomy during the rainy season. Such environmental seasonality represents a unique combination of challenges for the living biota contained within SDTFs and, accordingly, results in a series of special morphological, physiological, and behavioral adaptations of plants (chap. 8), animals (chap. 5, 6), fungi and soil organisms (chap. 4), and probably microorganisms. The climatic seasonality and the coupled seasonality of organisms and their ecological roles determine in turn the ecosystem processes (productivity, water and nutrient cycling, etc.) that characterize SDTFs (chap. 7–10).

Beyond their phenology and seasonality, three "macroscopic" features define the importance of SDTFs. The first is their wide coverage, encompassing 42 percent of tropical ecosystems worldwide, globally representing the second largest type of tropical forest (Miles et al. 2006; chap. 3). Second is their high biological diversity, which, although not comparable with the species richness of tropical rain forests, is nevertheless considerable (chap. 1, 4, 5, 6, 12). SDTF biodiversity includes other facets of great significance, in particular SDTFs' remarkable concentration of endemic species, their diversity of life-forms and functional groups of plants and animals (Dirzo and Raven 2003), and their incomparable beta diversity, or spatial species turnover (chap. 1). SDTF beta diversity is underscored by the high plant species dissimilarity (or floristic distance values) among sites, both within a relatively restricted region (e.g., Mexico, with a mean dissimilarity of 91 percent among all possible pair-wise comparisons of 20 study sites; Trejo and Dirzo 2002) and among the 21 major geographic nuclei that encompass the SDTF in the Latin American region (with 203 out of 253 possible pairs having dissimilarity values of more than 70 percent; chap. 1). The unusual SDTF beta diversity, combined with the impressive concentration of endemic taxa (e.g., 60 percent of plant species in Mexican SDTF), is an aspect that has important biogeographic (chap. 2) and conservation (e.g., chap. 12–16) implications. Finally, a third distinctive feature of SDTFs is their uneven distribution across the tropical regions of the world. Such forests have a greater distribution in the Neotropical and Caribbean region, encompassing approximately 700,000 square kilometers of land covered by SDTFs, representing 67 percent of the global coverage of this ecosystem.

On the other hand, among the region's tropical forests, SDTFs are regarded as the most threatened, with an estimated conversion of at least 48 percent of its extent into other land uses (Miles et al. 2006), an estimate similar to that of chapter 3, suggesting that only 44 percent of SDTF remains in the region (see also chap. 12, which cites an estimate of only 30 percent). Furthermore, a significant proportion of the remaining area of SDTF is fragmented to a varying degree, with important negative consequences on species and genetic diversity (chap. 11), as well as on several ecological processes, including species interactions crucial to plant reproduction, plant recruitment, and forest regeneration (chap. 11). In addition to land use change, other global environmental changes, in particular climatic change (chap. 16), have the potential to affect the structure, diversity, and functioning of SDTFs as well as the delivery of crucial ecosystem services they provide to human societies (chap. 15).

Given the dramatic magnitude of forest conversion and the persisting high rate of SDTF deforestation, coupled with the fact that protected areas including SDTF are extremely limited (e.g., only about 6 percent of SDTF in Central America has protected-area status; Miles et al. 2006), conservation of such vegetation and its biodiversity, ecosystem processes, and services will depend on how much SDTF biodiversity can be preserved in the mosaic of forest remnants and human-occupied areas—the agroscape (chap. 12). Conservation of SDTFs into the future will depend also on the extent to which such landscape mosaics can be used as biotic sources for restoration of degraded areas (chap. 12, 13) and the extent to which such agroscape can be valued for its biodiversity and maintenance of ecosystem services (chap. 15). Recent research suggests that SDTF biodiversity conservation in the agroscape, although quite challenging, holds high promise (chap. 12). Such hope is enhanced by isolated examples that show that the useful flora of seasonally dry Neotropical forests is of considerable cultural and economic importance (chap. 14). This combination of facts coupled with an appreciation of the traditional knowledge of the rural inhabitants of SDTF areas suggest that forest management, involving local communities, has potential to become sustainable (chap. 14).

The exuberant biodiversity of SDTFs and the ecosystem processes that characterize them represent an ecological resource we are just beginning to learn how to interpret. This is a task we urgently need to confront. We hope this volume will contribute to such an endeavor.

We are grateful to the Center for Latin American Studies of Stanford University for the support to hold a conference on the ecology and conservation of SDTFs, from which the present volume is derived. We also thank Fundación Telmex and Fundación Telcel, from Mexico, for partly sponsoring the production of this volume.

PART I

Seasonally Dry Tropical Forests as a Natural System

Chapter 1

Neotropical Seasonally Dry Forests: Diversity, Endemism, and Biogeography of Woody Plants

Reynaldo Linares-Palomino,
Ary T. Oliveira-Filho, and R. Toby Pennington

Neotropical seasonally dry forests are found from northwestern Mexico to northern Argentina and southwestern Brazil in separate areas of varying size (fig. 1-1). Their different variants have not always been considered the same vegetation type (e.g., Hueck 1978) or biogeographic unit (e.g., Cabrera and Willink 1980), but recent work has helped to define the extent, distribution, and phytogeography of seasonally dry tropical forest (SDTF) as a coherent biome with a wide Neotropical distribution (Prado and Gibbs 1993; Pennington et al. 2000; Pennington, Lewis et al. 2006). This unified interpretation is important both for biogeographic inference and for setting conservation priorities in Neotropical SDTF, which is the most threatened tropical forest type in the world (Miles et al. 2006).

Pleistocene climatic changes have been proposed as a possible force influencing the overall distribution of SDTF in the Neotropics (Prado and Gibbs 1993) and in driving evolution in SDTF plants (Pennington et al. 2000). Prado and Gibbs (1993) and Pennington et al. (2000) proposed a hypothesis in which during glacial times of cooler and drier climate, SDTFs were much more extensive than at present, perhaps forming contiguous forests across wide areas of the Neotropics. This view of current more-restricted areas of SDTF as "refugia" has been challenged by palynological studies that suggest the rain forests of Amazonia occupied hardly any less area than today and that the SDTFs of the Bolivian Chiquitano have been assembled recently (reviewed by Mayle 2004, 2006).

If there have been connections between some or all of the seasonal forests in the Neotropics during recent geological time, we would expect to find high floristic similarity among them. Prado and Gibbs (1993) and Pennington et al. (2000) highlighted a number of unrelated SDTF tree species that are widespread and found in several of the disjunct areas of Neotropical SDTF. They argued that these repeated distribution patterns were evidence of a once more widespread and perhaps continuous seasonal forest formation. These authors failed, however, to place these widespread species in the context of the entire woody flora of these areas, and no analyses of overall floristic similarity were presented.

In this chapter, we present a quantitative analysis of floristic similarity of the flora of the major areas of seasonal forests (SFs) in the Neotropics, including those of the floristically and ecologically unrelated but geographically adjacent vegetation of the Cerrados (savannas) and Chaco woodlands (fig. 1-1). This is the first quantitative analysis of the floras of these forests since Sarmiento (1975), who considered genera, and not species. Our species-level analysis provides a more fine-grained view of floristic variation. We use an ordination approach to analyze inventory data of woody plants from sites throughout Neotropical SDTF and examine the implications of the results for (1) patterns of diversity and endemism, (2) patterns of floristic relationships, (3) beta diversity, (4) biogeographic history, and (5) conservation prioritization.

Quantitative Analyses of Neotropical Seasonally Dry Tropical Floristic Nuclei

We define SDTFs following the broad concepts of Beard (1955) and Murphy and Lugo (1995), including tall evergreen SFs on moister sites, at one extreme of the series, to thorn woodland and cactus scrub at the other. We delimited 21 floristic nuclei of Neotropical SDTF, plus the Cerrado and Chaco areas (fig. 1-1). When nuclei are geographically isolated, this definition was straightforward, and in other cases we used previous phytogeographical studies that have revealed high affinities between some areas (e.g., Gentry 1995) for the equatorial Pacific SDTF in Peru and Ecuador. Published and unpublished but reliable tree inventory data from plots and sites for each of these regions were then aggregated to produce an initial species list for each of the floristic nuclei. Each nucleus' species list was enriched,

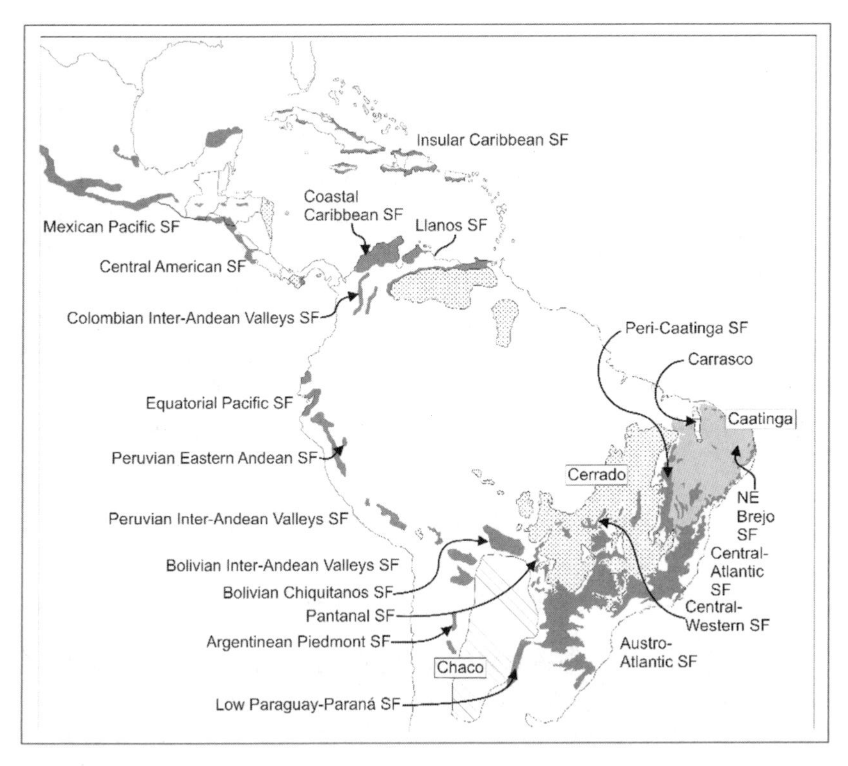

FIGURE 1-1. Floristic nuclei of Neotropical seasonally dry vegetation used in the analyses (SF = seasonal forests).

whenever possible, with additional information of plants reported for the area (e.g., herbarium collections, checklists, and our own field experience). We considered plants that are woody and reach at least 3 meters during some stage of their life cycle, excluding woody lianas and climbers. Main sources of data were Ratter et al. (2003), Linares-Palomino et al. (2003), and Oliveira-Filho (unpublished data). The data were homogenized using relevant taxonomic literature and online databases (W3Tropicos, IPNI, ILDIS) by checking for synonyms and misspellings. Doubtful identifications and records were excluded. The taxonomic treatment of families follows the Angiosperm Phylogeny Group II classification (APG 2003).

The final dataset included 3839 species from 806 floristic lists (table 1-1). Classification (UPGMA using group average and TWINSPAN) and ordination (nonmetric multidimensional scaling, MDS) analyses using standard settings in PC-ORD (McCune and Mefford 1999) were performed on a subset of 1901 species present in two or more floristic nuclei.

The MDS ordination and the UPGMA cluster analysis were performed using the Sørensen distance. The same index was used to assess beta diversity among floristic nuclei, allowing comparison of our results with beta diversity studies of the Cerrado (Bridgewater et al. 2004).

Each species found in 10 or more floristic nuclei (i.e., widespread species) was then annotated as ecologically versatile if it occurred in several forest types, including SDTF (e.g., *Maclura tinctoria*, *Trema micrantha*; table 1-2). We also annotated SDTF specialists (e.g., *Anadenanthera colubrina*, *Sideroxylon obtusifolium*) and SDTF generalists—species that generally grow in SDTF but are occasionally found in other vegetation (e.g., *Guazuma ulmifolia*, *Tabebuia impetiginosa*). Annotation was based on bibliographic sources (e.g., Flora Neotropica Monographs) and our own field experience.

Patterns of Plant Species Diversity

Diversity and Endemism

The number of floristic lists per nucleus ranged from 2 to 376 (table 1-1). While this does not represent even geographic coverage of inventories, we do not believe our results are excessively biased, because nuclei covered by few studies often have many species and vice versa. For example, the Peruvian Eastern Andean SF nucleus has 101 species from just 2 lists, whereas 358 species are recorded from 376 lists in the Cerrado. This pattern of nuclei with more lists but lower overall species numbers (e.g., Coastal Caribbean SF, 19 lists, 135 species) and nuclei with a low number of lists but high numbers of species (e.g., Bolivian Chiquitanos SF) may reflect several historical and ecological factors, including the relative size of the nuclei and different rainfall regimes.

Species numbers ranged from 45 to 1602 per nucleus (table 1-1). The percentage of unique species present in each nucleus ranged from 1.9 percent in the Paraguay-Paraná SF to 77.5 percent in the Insular Caribbean SF (table 1-1). While these unique species are not strictly endemics (they may be present in other areas outside our nuclei), their numbers offer a reasonable proxy for levels of endemism.

Of the 3839 species, 457 were present in 5 or more nuclei, and only 55 (1.43 percent of the total; table 1-2, fig. 1-2A) have been recorded in 10 or more nuclei. Of the latter, 24 are ecologically versatile species, 22 are SDTF generalists, and only 9 are SDTF specialists (table 1-2).

The uneven geographic coverage and heterogeneous nature (from plots,

TABLE 1-1. Geography and diversity of Neotropical seasonally dry floristic nuclei

Floristic nucleus	*Geographical location*	*No. lists*	*No. species*			*Sources*
			Orig.	*≥ 2*	*Uniq. %*	
Insular Caribbean (InsuCari)	Jamaica, Cuba, Haiti, Dominican Republic, Puerto Rico	8	298	67	77.5	L-P
Mexican Pacific (MexiPaci)	Mexican Pacific coast (Yucatán not included)	8	200	69	65.5	L-P
Central American (AmCePaci)	Pacific coast, from Guatemala to Panama	17	202	121	40.1	L-P
Coastal Caribbean (CoVeCari)	Colombian and Venezuelan Caribbean coast	19	134	92	31.3	L-P
Llanos (VeneLlan)	Northern fringes of the Venezuelan Llanos domain	3	45	32	28.9	L-P
Colombian inter-Andean valleys (ColoAndV)	Magdalena and Cauca valleys in central Colombia	5	70	48	31.4	L-P
Equatorial Pacific (EcPePaci)	Western Andes in Ecuador and Peru	54	308	163	47.1	L-P
Peruvian inter-Andean valleys (PeruAnVa)	Marañon, Mantaro, and Apurimac valleys in Peru	6	222	119	46.4	L-P
Peruvian eastern Andean (PeruEAnd)	Tarapoto region in northeastern Peru	2	101	64	36.6	L-P
Bolivian inter-Andean valleys (BoliAnVa)	Bolivian eastern Andean valleys	4	61	51	16.4	L-P
Bolivian Chiquitanos (BoliChiq)	Santa Cruz province, eastern Bolivia	4	307	271	11.7	L-P+O-F
Argentinean piedmont (ArgePied)	Northwestern Argentina	4	169	144	14.8	L-P+O-F
Low Paraguay-Parana (PargParn)	Low course of the Paraguay and Paraná rivers in Argentina	6	213	209	1.9	L-P+O-F
Pantanal (Pantanal)	Forest patches in the Pantanal region, western Brazil	16	383	360	6.0	O-F
Central-western (CentWest)	Forest patches in the Brazilian Cerrado domain	61	1079	975	9.6	O-F

(table continues)

TABLE 1-1. *(continued)*

Floristic nucleus	*Geographical location*	*No. lists*	*No. species*			*Sources*
			Orig.	*≥ 2*	*Uniq. %*	
Austro-Atlantic (AustAtla)	Middle Paraná River basin in Paraguay, Argentina, and Brazil	72	1187	1000	15.8	O-F
Central Atlantic (CentAtla)	Minas Gerais and Bahia, east of the Espinhaço Range in Brazil	56	1602	1179	26.4	O-F
Peri-Caatinga (PeriCaat)	Western fringes of the Caatinga Domain in Central Brazil	20	480	467	2.7	O-F
Northeastern Brejo (BrejoFor)	Montane Forests in the Brazilian Caatinga Domain	16	933	796	14.7	O-F
Chaco (Chaco)	Northern Argentina and Paraguay and Southern Bolivia	7	151	100	33.8	L-P + O-F
Cerrado (Cerrado)	Woody Savannas of Central-Western Brazil	376	370	341	7.8	R
Caatinga (Caatinga)	Sclerophyllous Woodlands of Northeastern Brazil	38	272	194	28.7	O-F
Carrasco (Carrasco)	Ceará and Piauí states, Northeastern Brazil	4	183	171	6.6	O-F

Abbreviations in parentheses are those used in the ordination and classification analyses.
Sources: L-P = Linares-Palomino et al. 2003, O-F = Oliveira-Filho (unpublished data), R = Ratter et al. 2003.

transects, and floristic lists) of the basic data limit us from objectively comparing alpha diversity levels and total species numbers in the SDTF nuclei. Our data perhaps are more robust for analyzing patterns of endemism because some nuclei for which we have few inventories show high numbers of unicates (e.g., Caribbean, Mexico), and others for which we have sampled far more thoroughly show low numbers (e.g., Brejo and Peri-Caatinga). Though the percentage of unicate species varies widely from 1.9 to 77.5 percent, in general it is high, with 12 of 23 nuclei showing greater than 20 percent unicates, suggesting high endemism. While such high numbers of en-

TABLE 1-2. Ecological characteristics of widespread species in Neotropical seasonally dry forests

Species	*Family*	*SDTF specialist*	*SDTF generalist*	*Ecological generalist*
Acacia farnesiana (L.) Willd.	Fabaceae Mimosoideae			
Acacia polyphylla DC.	Fabaceae Mimosoideae		x	
Acacia riparia Kunth	Fabaceae Mimosoideae		x	
Acacia tenuifolia (L.) Willd.	Fabaceae Mimosoideae			x
Acrocomia aculeata (Jacq.) Lodd. ex Mart.	Arecaceae		x	
Agonandra brasiliensis Miers ex Benth. & Hook.	Opiliaceae			x
Allophylus edulis (A.St.-Hil. et al.) Radlk.	Sapindaceae		x	
Aloysia virgata (Ruiz & Pav.) A.Juss.	Verbenaceae	x		
Amburana cearensis (Allemão) A.C.Sm.	Fabaceae Faboideae		x	
Anadenanthera colubrina (Vell.) Brenan	Fabaceae Mimosoideae	x		
Apeiba tibourbou Aubl.	Malvaceae		x	
Aspidosperma cuspa (Kunth) S.F.Blake ex Pittier	Apocynaceae		x	
Aspidosperma pyrifolium Mart.	Apocynaceae	x		
Astronium fraxinifolium Schott ex Spreng.	Anacardiaceae		x	
Brosimum gaudichaudii Trécul	Moraceae			x
Capparis flexuosa (L.) L.	Brassicaceae		x	
Casearia sylvestris Sw.	Salicaceae			x
Celtis brasiliensis (Gardner) Planch.	Cannabaceae		x	
Celtis ehrenbergiana (Klotzsch) Liebm.	Cannabaceae		x	
Celtis iguanaea (Jacq.) Sarg.	Cannabaceae	x		
Cochlospermum vitifolium (Willd.) Spreng.	Bixaceae			x
Copaifera langsdorffii Desf.	Fabaceae Caesalpinioideae			x
Cordia trichotoma (Vell.) Arrab. ex Steud.	Boraginaceae		x	
Genipa americana L.	Rubiaceae			x
Guapira noxia (Netto) Lundell	Nyctaginaceae			x
Guazuma ulmifolia Lam.	Malvaceae		x	
Hymenaea courbaril L.	Fabaceae Caesalpinioideae			x
Luehea candicans Mart. & Zucc.	Malvaceae		x	
Luehea grandiflora Mart. & Zucc.	Malvaceae		x	
Luehea paniculata Mart. & Zucc.	Malvaceae		x	

(table continues)

TABLE 1-2. *(continued)*

Species	*Family*	*SDTF specialist*	*SDTF generalist*	*Ecological generalist*
Machaerium acutifolium Vogel	Fabaceae Faboideae			x
Maclura tinctoria (L.) Steud.	Moraceae			x
Myracrodruon urundeuva Allemão	Anacardiaceae	x		
Prockia crucis P.Browne ex L.	Salicaceae			x
Pterocarpus rohri Vahl	Fabaceae Faboideae		x	
Pterogyne nitens Tul.	Fabaceae Caesalpinioideae		x	
Randia armata (Sw.) DC.	Rubiaceae			x
Sapindus saponaria L.	Sapindaceae			x
Sapium glandulosum (L.) Morong	Euphorbiaceae			x
Sebastiania brasiliensis Spreng.	Euphorbiaceae			x
Senna spectabilis (DC.) H.S.Irwin & Barneby	Fabaceae Caesalpinioideae			x
Sideroxylon obtusifolium (Roem. & Schult.) T.D.Penn.	Sapotaceae	x		
Spondias mombin L.	Anacardiaceae			x
Tabebuia impetiginosa (Mart. ex DC.) Standl.	Bignoniaceae		x	
Tabebuia ochracea (Cham.) Standl.	Bignoniaceae			x
Tabebuia roseo-alba (Ridl.) Sandwith	Bignoniaceae		x	
Talisia esculenta (A.St.-Hil.) Radlk.	Sapindaceae			x
Trema micrantha (L.) Blume	Cannabaceae			x
Trichilia elegans A.Juss.	Meliaceae			x
Trichilia hirta L.	Meliaceae		x	
Trichilia pallida Sw.	Meliaceae			x
Urera baccifera (L.) Gaudich. ex Wedd.	Urticaceae	x		
Ximenia americana L.	Olacaceae	x		
Zanthoxylum fagara (L.) Sarg.	Rutaceae	x		
Zanthoxylum rhoifolium Lam.	Rutaceae		x	
TOTAL SPECIES		**9**	**22**	**24**

demic species might be produced by recent, rapid evolution, it seems more likely that in many SDTF nuclei they represent the result of the considerable age of the biome, prolonged isolation, and limited arrival of immigrant lineages by dispersal (Pennington et al. 2009). This view is partly derived from the fossil record, which shows evidence for SDTF in the Ecuadorean Andes

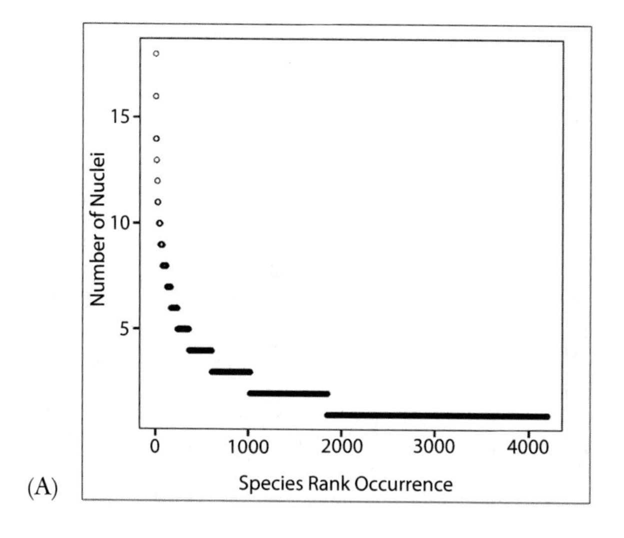

(A)

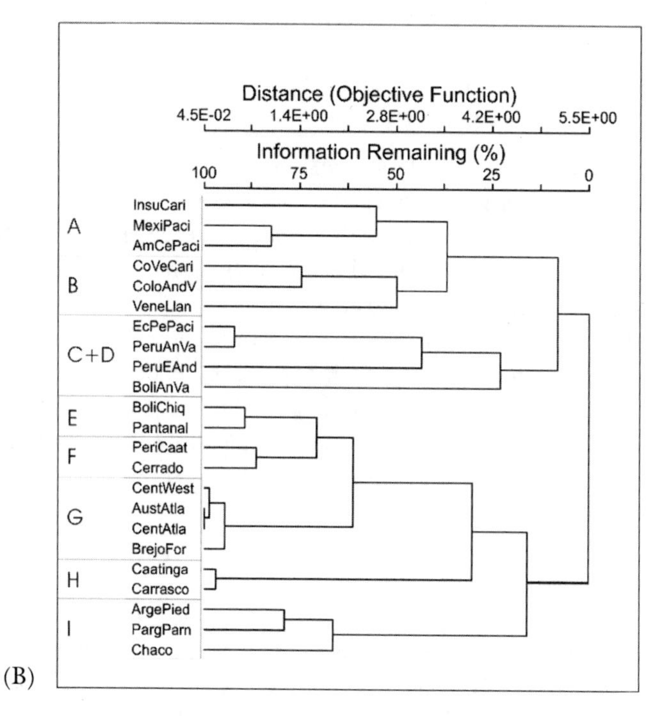

(B)

FIGURE 1-2. (A) Number of nuclei in which each of the 3839 woody species occurs; more than 50 percent of species (1938) are unicates, occurring in only one nucleus. (B) UPGMA dendrogram.

(figure continues)

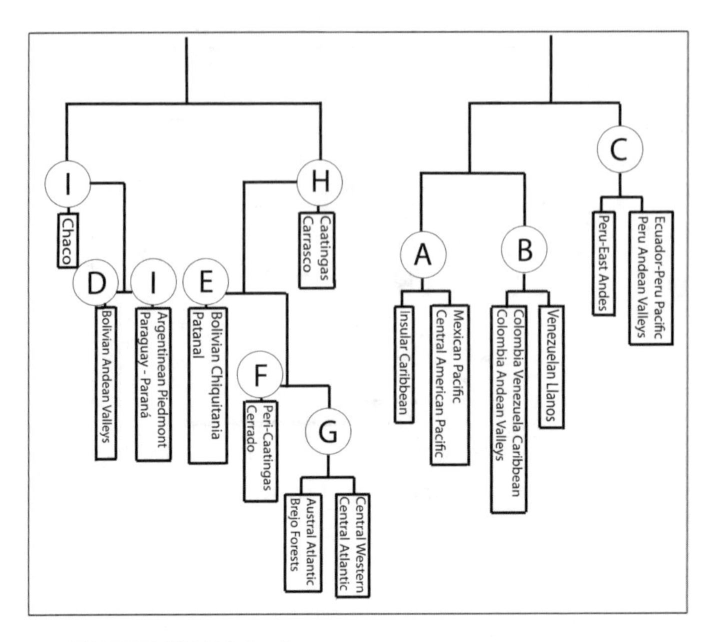

FIGURE 1-2. (C) TWINSPAN dendrogram.

in the late Miocene, 10 to 12 million years ago (e.g., Burnham and Carranco 2004). Dated molecular phylogenies in general show patterns of speciation that predate the Pleistocene and high phylogenetic geographic structure where closely related species occupy the same geographic area (see Pennington, Lewis et al. 2006 and Pennington, Richardson et al. 2006 for reviews). This view of limited dispersal is corroborated by the contribution of Caetano and Naciri in this volume (chap. 2). Their population genetics approach to investigating the widespread distributions of two SDTF tree species, *Astronium urundeuva* and *Geoffroea spinosa*, shows high population genetic structure that is consistent with limited gene flow between major SDTF nuclei.

Floristic Relationships

The three quantitative analyses we applied to the data consistently identified four major SDTF regions (fig. 1-2B–D): Caribbean/Mesoamerican, Andean (not including Bolivian Andes), Southern South American, and Brazilian. The position of the Bolivian Andes nucleus is intermediate, with affinities to both neighboring Andean and adjacent lowland sites.

The SDTFs in Mesoamerica have been considered a distinct phyto-

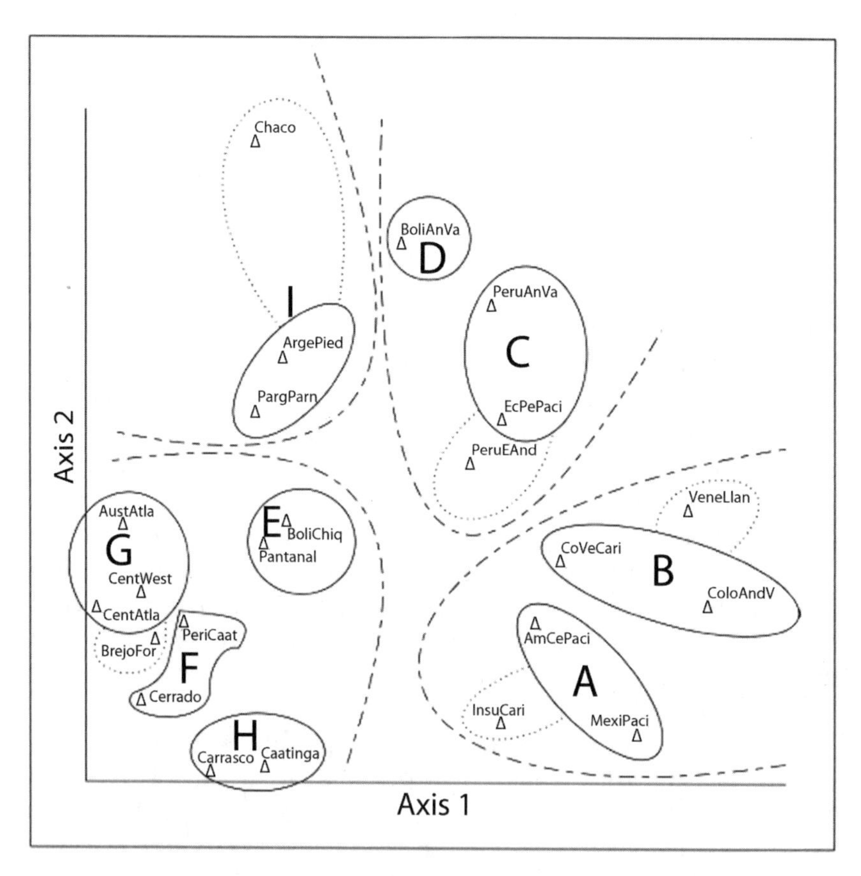

FIGURE 1-2. (D) MDS ordination of the 23 Neotropical seasonally dry floristic nuclei. Abbreviations follow those in table 1-1. SDTF groups discussed in the text are indicated by uppercase letters and circled by continuous lines. Closely related nuclei are indicated by stippled lines. Major SDTF regions are separated by broken lines.

geographic unit since the studies of Rzedowski (1978) and recent floristic data (e.g., Trejo and Dirzo 2002) have confirmed their remarkable plant diversity. Likewise, the Caribbean islands are also interesting because of their high endemism of vascular plant species (with more than 50 percent of approximately 12,000 species endemic) (Santiago-Valentin and Olmstead 2004). This fact is also reflected in the SDTF flora by high Sørensen distance values with the adjacent Mesoamerican SF (65 to 81 percent) and the highest unicates percentage (77.5 percent) in our analyses (table 1-1). There are, however, surprisingly few studies evaluating large-scale phytogeographic relationships in the entire Mesoamerican-Caribbean region (Santiago-Valentin and Olmstead 2004), apart from research on the floristic affinities between the vascular floras of the Yucatán peninsula and the greater Caribbean islands, particularly Cuba (Chiappy-Jhones et al. 2001). The only wide-ranging study evaluating the affinities of the SDTF floras of the region remains that of Gentry (1995), which was based on a rather small sample of transect inventories. Our analyses show strong evidence for a floristic connection between the Insular Caribbean SDTF (including the Greater and Lesser Antilles) and the Mexican and Central American SDTFs (highest Sørensen similarity was with the latter).

Gentry (1982c) noted a strong relationship between SDTF in northern Colombia-Venezuela and the Central American Pacific SDTF, suggesting that the wet Chocó forests in Colombia, which probably constituted a major rain forest refuge during glacial dry periods, had functioned as a barrier to the drier forests north and south of it (see also Simpson and Neff 1985). The Chocó has been suggested to have been a low and swampy area even before the Andean orogeny (Haffer 1970) and so a barrier to SDTF species. Our data support Gentry's view and also the high dissimilarity of Central American SDTF and the SDTF in coastal Ecuador that he discussed (Gentry 1982c).

Our analyses, placing all Brazilian sites (plus the Argentinean, Paraguayan, and Bolivian Chiquitano area) in the first major UPGMA and TWINSPAN divisions, support the floristic relationship of these areas within the "Pleistocenic Arc" of seasonal vegetation, as proposed by Prado and Gibbs (1993). They also suggested that the SDTF in the dry inter-Andean valleys of Peru might constitute remnants of a previously much wider expansion of this arc, but the Peruvian Andean areas are resolved separately in our analyses, just as Prado (2000) anticipated. Prado (2003) proposed several complex migration routes for the floristic elements that formed the caatinga, such as the Caribbean route (see also Sarmiento 1975) and the Andean route (see also Weberbauer 1914). Nevertheless, few species are disjunct between the caatingas and these areas, and our analyses show the caatinga to be firmly embedded in the Brazilian group.

Despite being situated adjacent to two major South American seasonal woodland ecosystems (the Chaco and Cerrado), the Chiquitano SF is unrelated floristically to either. Killeen et al. (1998) suggested that the Chiquitano SF had more in common with the semideciduous forests of the Andean piedmont of northwestern Argentina and the Misiones region of eastern Paraguay and northeastern Argentina, as well as with the Caatinga region of northeastern Brazil. More recently, Killeen et al. (2006)

showed the transitional, albeit distinct nature of these forests. Our data, showing high Sørensen similarity values with the adjacent Pantanal, Argentinean Piedmont, and Paraguay-Paraná SF, provide evidence of strong floristic relationships. The low-level unicate species in these forests (table 1-1) provide some support for the view that the Chiquitano forests have been assembled recently (Mayle 2006). López (2003) argued that the Bolivian inter-Andean dry valley flora was more related to the Chaco and other Argentinean SFs. Of the 1156 species he reported for the Bolivian Andean dry valleys, more than half had their northernmost distribution in central Bolivia and parts of southern Brazil. More recently, Wood (2006) showed that the biogeographical relationships of the dry areas in the Bolivian Andes were variable and highly dependent on which family was studied: Labiatae showed an essentially Andean distribution, suggesting a pattern between SDTF areas similar to that shown by our UPGMA analysis. Asclepiadaceae and Acanthaceae instead showed stronger links with the lowland vegetation in southern South America (Argentinean SF, Chaco, and Cerrado), a pattern suggested by our TWINSPAN analysis. It seems that the Inter-Andean Bolivian SFs, due to their geologic and climatic history, as well as to their unique position close to major distinct biomes, are composed of plant species of variable biogeographic affinity, making generalizations difficult.

Weberbauer (1945) proposed that the xerophytic floras of Peru, Bolivia, and the Argentinean Chaco are remnants of a formerly homogeneous flora fragmented by Andean uplift. More recently, Sarmiento (1975) proposed the existence of a major disjunction, located somewhere in the Andes of southern Peru and northern Bolivia, separating the dry floras from northern Peru to Venezuela from those south and east of Bolivia to Argentina and Brazil. Evidence in support of this comes from Kessler and Helme (1999) and López (2003), who were unable to find strong connections between the Bolivian inter-Andean dry floras and southern Peru. Unfortunately, the unstable position of the Bolivian inter-Andean valleys in our analyses does not confirm or reject any of these theories.

Prado (2000) was able to find clear floristic differences between the Chaco (and closely related Chaquenian formations), the Cerrado, and South American SDTFs by quantitative floristic analysis. Our results agree that the Chaco vegetation has a peripheral position with respect to other seasonal forest formations in southern South America. The Cerrado woodlands, in contrast, seem to have closer floristic relationships with the adjacent Brazilian SF, particularly with those of the Peri-Caatingas, demonstrating the importance of ecological generalists shared between these forests. Several species characteristic of mesotrophic soils in the cerrados are also found in adjacent SDTF formations, such as *Dilodendron bipinnatum* (Sapindaceae) and *Callisthene fasciculata* (Vochysiaceae).

Beta Diversity Levels in SDTFs and Savannas in the Neotropics

As expected, highest distance (or beta diversity) values were found between the Chaco and other seasonal forests (table 1-3). Highest similarity (or lowest beta diversity) was found between the Central-Atlantic and Austro-Atlantic SDTFs, the Central-Western and Brejo SFs, and the Central-Western and Austral-Atlantic SDTFs. There were 61 pairs of nuclei, out of 253 possible pairs, that had 90 percent or more dissimilarity, and 203 pairs had dissimilarity of more than 70 percent. In contrast, only 2 pairs of nuclei showed a similarity higher than 70 percent (table 1-3). Beta diversity estimates for vegetation units over large geographical areas are rare. One such study for the Brazilian Cerrado biome, an area covering some 2 million square kilometers (Bridgewater et al. 2004), compared floristic nuclei defined in a similar manner with those in this chapter. Sørensen distance values among 6 Cerrado nuclei were 0.38 and 0.74, indicating that the Cerrado flora is heterogeneous. Our data present higher distance values (table 1-3). Eighty percent of the pairwise comparisons had distances over 0.70. This high level of heterogeneity reflects both the continental scale of the study area and that SDTF exists as scattered areas in comparison to the continuous Brazilian Cerrado. However, in SDTF, it is clear that floristic similarity can be very low between geographically close areas of SDTF, even within some of the nuclei. For example, the similarity between the Marañon and Mantaro inter-Andean dry valleys in Peru, separated by only about 400 kilometers, is only 14 percent, with only 16 species shared from a total of nearly 200 woody species, and Trejo and Dirzo (2002) showed the average Sørensen similarity between 20 Mexican SDTF sites (sampled using 0.1-hectare plots) to be only 0.09. The generally low floristic similarity argues for lack of recent floristic links and dispersal between isolated SDTF areas, as discussed under Biogeographic History below.

Sørensen distance values of the same magnitude as those found between floristic nuclei in the Cerrado (less than 75 percent, table 1-3) are largely confined to comparisons between nuclei resolved in the southern South American and Brazilian groups (fig. 1-2B–D). This may reflect the relatively high continuity of SDTF in these areas in comparison with the more isolated nuclei elsewhere in the Neotropics (fig. 1-1). However, across these areas, SDTF is still scattered compared with the continuous area of distribution of the Cerrado, which occupies a similar overall area. We therefore believe that the pattern of beta diversity uncovered by our analyses lends some support to the contention that SDTF may have been more widespread and continuous in dry glacial times in these areas (Prado and Gibbs 1993; see Biogeographic History section below).

Bridgewater et al. (2004) also identified a suite of frequently occurring species that were dominant (contributing to both high species richness and importance value index) in all Cerrado nuclei. This is a similar scenario to that found in the rain forests of western Amazonia by Pitman et al. (1999), who suggested that most tree species in the region are habitat generalists occurring over large areas of the Amazonian lowlands at low densities but large absolute population sizes. The presence of such a ubiquitous "oligarchy" of species argues for their free dispersal over large areas. There is little evidence for any such oligarchy in Neotropical SDTF as a whole, probably reflecting limited dispersal between isolated areas. If such an oligarchy exists, it should be sought in individual SDTF nuclei, or in pairs or suites of nuclei shown to be closely related in our analyses.

Biogeographic History

We find little support for a wide-ranging Pleistocene SDTF formation throughout the Neotropics or South America, because few species are widespread (fig. 1-2A). Moreover, of the 55 most widespread species in our data set, 24 are ecologically versatile, while only 9 are SDTF specialists (table 1-2). It seems likely that the wide distributions of SDTF specialist and generalist species must reflect, at least in part, long-distance dispersal events, as has been proposed for similarly widespread rain forest tree species (Dick et al. 2003). For SDTF specialist species, these long-distance dispersal events must have traversed expanses of non-SDTF vegetation. There is a precedent for this role for long-distance dispersal: at higher taxonomic levels, transoceanic dispersal events rather than plate tectonics have been implicated as the cause of intercontinental distributions in several families represented in our data set (e.g., Leguminosae, Rhamnaceae, Annonaceae; see Pennington, Lavin et al. 2004, Richardson et al. 2006, and Renner 2005 for reviews).

It is only within some of the four major SDTF regions identified by the quantitative analyses (fig. 1-2B–D) that levels of floristic similarity may be high enough to suggest more widespread SDTF vegetation in the past.

TABLE 1-3. Beta diversity (Sørensen distance) among Neotropical seasonally dry forest nuclei

	InsuCari	*MexiPaci*	*AmCePaci*	*CoVeCari*	*VeneLlan*	*ColoAndV*	*EcPePaci*	*PeruAnVa*	*PeruEAnd*	*BoliAnVa*
MexiPaci	0.81									
AmCePaci	0.66	0.49								
CoVeCari	0.83	0.66	0.66							
VeneLlan	0.88	0.82	0.72	0.54						
ColoAndV	0.89	0.86	0.74	0.52	0.77					
EcPePaci	0.78	0.85	0.77	0.72	0.78	0.77				
PeruAnVa	0.92	0.92	0.88	0.85	0.84	0.94	0.43			
PeruEAnd	0.86	0.95	0.80	0.83	0.91	0.91	0.75	0.72		
BoliAnVa	0.91	0.97	0.89	0.89	0.92	0.96	0.74	0.83	0.82	
BoliChiq	0.82	0.90	0.83	0.84	0.84	0.90	0.85	0.85	0.70	0.69
ArgePied	0.88	0.94	0.90	0.92	0.89	0.96	0.88	0.85	0.89	0.72
PargParn	0.88	0.93	0.92	0.87	0.89	0.94	0.86	0.87	0.85	0.77
Pantanal	0.76	0.90	0.81	0.84	0.85	0.88	0.84	0.88	0.75	0.73
CentWest	0.73	0.90	0.80	0.85	0.82	0.85	0.85	0.86	0.71	0.73
AustAtla	0.75	0.89	0.82	0.85	0.85	0.86	0.85	0.87	0.76	0.72
CentAtla	0.75	0.90	0.83	0.85	0.84	0.87	0.85	0.89	0.74	0.76
PeriCaat	0.82	0.92	0.90	0.88	0.88	0.94	0.90	0.88	0.78	0.77
BrejoFor	0.75	0.89	0.82	0.84	0.79	0.88	0.87	0.90	0.77	0.80
Chaco	0.98	0.96	0.96	0.94	0.98	1.00	0.90	0.91	0.97	0.87
Cerrado	0.91	0.95	0.90	0.92	0.91	0.95	0.95	0.95	0.81	0.93
Caatinga	0.93	0.93	0.95	0.88	0.89	0.95	0.93	0.92	0.91	0.91
Carrasco	0.90	0.97	0.93	0.92	0.94	0.99	0.94	0.96	0.89	0.91

Abbreviations follow those in table 1.1. Highlighted, underlined values show similarities below 10% (i.e., distances above 90%), and values in bold show similarities above 70% (i.e., distances below 30%).

For example, our analyses do not contradict the idea of a "Pleistocenic Arc" of SDTF vegetation (Prado and Gibbs 1993) stretching from the Caatingas south through Brazil to Paraguay and Argentina, because these areas

BoliChiq	*ArgePied*	*PargParn*	*Pantanal*	*CentWest*	*AustAtla*	*CentAtla*	*PeriCaat*	*BrejoFor*	*Chaco*	*Cerrado*	*Caatinga*
0.63											
0.67	0.56										
0.46	0.65	0.61									
0.54	0.71	0.66	0.46								
0.62	0.64	0.54	0.56	0.32							
0.65	0.76	0.73	0.62	0.33	**0.24**						
0.61	0.78	0.77	0.58	0.40	0.57	0.51					
0.68	0.78	0.79	0.64	0.43	0.53	**0.29**	0.50				
0.75	0.58	0.54	0.71	0.88	0.84	0.92	0.91	0.92			
0.70	0.88	0.86	0.62	0.38	0.58	0.60	0.52	0.60	0.96		
0.81	0.89	0.89	0.82	0.73	0.86	0.75	0.53	0.63	0.91	0.88	
0.76	0.90	0.89	0.74	0.64	0.77	0.69	0.58	0.58	0.96	0.75	0.43

emerge as closely related in our analyses, and floristic similarity amongst them is relatively high.

The strongly supported Mesoamerican-Caribbean group that emerges in all analyses probably reflects a Laurasian evolutionary history of the taxonomic groups in the SDTF vegetation of these areas. This is supported by molecular phylogenies of several SDTF Laurasian plant taxa that are diverse in these northern areas, such as *Bursera* (Weeks et al. 2005), *Leucaena* (Hughes et al. 2003), and *Vigna* (Delgado-Salinas et al. 2006). In contrast, other SDTF genera such as *Coursetia* and *Ruprechtia* have been shown to have South American origin, with any Central American species more recently derived (Pennington, Richardson et al. 2004).

The close relationship of the northern South American nuclei with the Mesoamerican and Caribbean area may reflect enhanced opportunities for overland dispersal via the Isthmus of Panama since its closure 3.5 million years ago as well as stepping-stone dispersal via islands and putative land bridges such as GAARlandia, which is hypothesized to have joined northern South America with the Greater Antilles 33 to 35 million years ago (Iturralde-Vinent and MacPhee 1999). Dated phylogenies of SDTF groups from these areas might distinguish these possibilities.

The grouping of the Peruvian and Ecuadorean sites with the northern sites is intriguing as there is no obvious biogeographic scenario to explain it. Similarity values between these sites, however, are low and range from 4 to 23 percent (the highest values being between the Ecuadorean-Peruvian coastal SDTF and the Central American/Colombian-Venezuelan Caribbean sites) and seem to be a reflection of a few locally common species distributed from the central Andes northwards (e.g., *Stemmadenia obovata*, *Tabebuia billbergii*, or *Bursera graveolens*).

Conservation Implications

Recently, Miles et al. (2006) analyzed the conservation status of extant tropical dry forest ecosystems in the world (see also chap. 3). They indicated that (1) the two most extensive contiguous SDTF areas are located in South America (northeastern Brazil and southeastern Bolivia, Paraguay, and northern Argentina) while most other SDTF areas are fragmented and scattered, and (2) more than half (54.2 percent) of the remaining SDTFs in the world are in South America. Notably, another 12.5 percent are located in North and Central America, making the Neotropics the most important SDTF region in terms of extension (66.7 percent of the world's SDTF). They also renewed Janzen's (1988c) statement that dry forests are the most threatened major tropical forest type, based on rates of deforestation but also shown by the degree of threat by forest fragmentation, climate change, conversion to agriculture, human population density, and the low level of protected areas cover. Our results complement and refine this information by showing which areas of SDTF within the Neotropics might deserve conservation attention based on floristic distinctness. From the perspective of conservation, endemism may be more important than diversity (Gentry 1995), and our data indicate potentially high endemism for many areas.

The 2006 World Database on Protected Areas (www.unep-wcmc.org/wdpa/index.htm) lists 153 protected areas (IUCN categories I–IV) in the 23 SDTF nuclei defined here. Miles et al. (2006) reported percentages of protected SDTF of 5.7 percent for North and Central America and 37.8 percent for South America. The low values for North and Central America are not surprising since several studies have reported the rapid rates of deforestation of this ecosystem (e.g., Janzen 1988c; Trejo and Dirzo 2000; chap. 3). The high percentage of protected areas reported for South America is, however, misleading. For one part, as they state, Miles et al. (2006) identified grid cells containing protected areas that also contain SDTF, rather than identifying the precise area of SDTF that is protected. For another, most of this percentage is probably contributed by large protected areas in the extensive SDTF nuclei in the Caatingas, Coastal Caribbean, Bolivian Chiquitanos, and the several Atlantic SDTFs in Brazil. The other South American SDTF nuclei have much smaller total areas and are also very heavily impacted. Extreme cases are the Inter-Andean SDTFs in Colombia and Peru, both of which show high endemism (30 to 46 percent unicate species; table 1-1) but are not covered by any protected areas. Conservation measures are urgently needed there, as well as in Mesoamerica.

Acknowledgments

We thank the Darwin Initiative for a Scholarship to Reynaldo Linares-Palomino, the Royal Society of Edinburgh International Exchange Programme for a travel bursary to Ary Oliveira-Filho, and the Leverhulme Trust for a Study Abroad Fellowship to Toby Pennington.

Chapter 2

The Biogeography of Seasonally Dry Tropical Forests in South America

Sofia Caetano and Yamama Naciri

In the previous chapter much attention was given to the study of the present status of the seasonally dry tropical forest (SDTF) in the Neotropics. Nevertheless, a broader picture of this ecosystem requires consideration of the historical events governing it. Herein, we address the major biogeographical hypotheses concerning the colonization of the SDTF in South America.

Biogeography, being the study of the natural distributions of living organisms, addresses the contemporary ecological processes, together with the historical changes in landscape and environmental features (Schaal et al. 1998). The purpose of this chapter is to understand how SDTF colonized the continent, especially during the last glacial maximum. We highlight the specificity of South America when compared with other continents.

State of Knowledge

Quaternary climatic fluctuations had genetic consequences in boreal, temperate, and tropical zones (Hewitt 2000). For several reasons, including easier sampling and the availability of pollen cores, Europe and North America have been more studied than other parts of the world. The northern continents were characterized by cycles of south-north recolonizations during the warm periods, often associated with rapid range expansions, followed by north-south retreats during the cold periods. In Europe, numerous pollen cores indicate that species now inhabiting the boreal and temperate regions had their ice age refugia south of the permafrost, in the Iberian Peninsula, Italy, the Balkans, and Turkey/Greece (Demesure et al. 1996; Petit et al. 1997; Heuertz et al. 2006). In North America, refugia were also located south of the permafrost with peculiarities due to the Appalachians running north-south and to the land mass existing farther south where the Mediterranean stands at similar latitudes in Europe. In each case, populations at the northern limit of the glacial refugia are suspected to have expanded to large, open zones cleared by ice and permafrost thaw, which allowed for demographic and/or range expansions. The cold periods had the reverse effect, with demographic and/or range contractions associated with genetic drift (Jesus et al. 2006). Rapid range expansions are expected to have reduced the level of genetic heterozygosity with a concomitant reduction in allelic diversity. This colonization pattern should therefore lead to large genetically homogenous patches, as suggested by theoretical and simulation studies (Ibrahim et al. 1996; Le Corre et al. 1997; Klopfstein et al. 2006). This was confirmed by numerous empirical studies that show a latitudinal loss of genetic diversity (Demesure et al. 1996; Sewell et al. 1996; Petit et al. 1997; Hewitt 1999, 2000; Heuertz et al. 2006). Still, other studies show more complex patterns due to the shaping of modern genetic structure over multiple glacial-interglacial cycles (Jesus et al. 2006; Magri et al. 2006) or to the existence of cryptic glacial refugia in small and localized northern spots that remained unglaciated (Provan and Bennett 2008). The expectation of a higher diversity in the southern refugia has also been sometimes refuted, for instance when lineages originating from different refugia came into contact during the recolonization process and then produced a higher diversity in the north compared with the south (Comps et al. 2001). The basic expansion-contraction model initially stated is therefore growing in complexity, as studies accumulate on different taxa.

Colonization Hypotheses in South America

The climatic changes during the Pleistocene were less severe in South America than in the northern continents, with the ice sheets and the permafrost never reaching the lowlands. The consequence of this particularity was that instead of recolonization of empty territories, as documented in Europe and North America, substitutions between the vegetation types (e.g., rain forest and drier vegetation) seem to have occurred in South America. From this perspective, the fragmentation-isolation of the Amazonian rain forests into pocket areas during the drier/cooler periods of the Pleistocene (van der Hammen and Absy 1974; Connor 1986) would coincide with the expansion of some sort of drier vegetation (Haffer 1982; Prance 1982).

The Pleistocenic Arc theory recently proposed that the present-day SDTF distribution patterns represent relics of a more contiguous formation having reached its maximum extension during periods of humid forest contraction. The more continuous expanse of SDTF stretched from the Brazilian Caatinga to the peripheral areas of the Chaco domain, possibly also reaching the dry inter-Andean valleys of Bolivia, Peru, and Ecuador (Prado and Gibbs 1993). Pennington et al. (2000) went even further, suggesting that SDTF species may have also penetrated into the Amazon basin during the glacial periods.

The validity of the Pleistocenic Arc theory has been increasingly debated over the last years (Linares-Palomino et al. 2003; Bonaccorso et al. 2006; Naciri et al. 2006). Herein, it is treated as a biogeographical colonization hypothesis of the SDTF in South America—that the currently scattered SDTF nuclei may be the result of vicariance (i.e., fragmentation) of a more widespread formation that reached its maximum expansion during the last glacial maximum (Pennington et al. 2000; Prado 2000).

Pennington, Richardson et al. (2004) failed at demonstrating common patterns in area relationships among SDTF genera that would support speciation by vicariance. Therefore, in analogy with temperate tree species for which rare long-distance dispersal events during the late glacial period and Holocene have been recognized (e.g., Davis 1983; Petit et al. 1997), it can be hypothesized that SDTF species could have also spread by long-distance dispersal, in the search for suitable climatic/edaphic conditions (Gentry 1982a; Mayle 2006; Naciri et al. 2006).

In the present chapter, the vicariance versus dispersal hypotheses will be tested and discussed by examination of the geographical distribution of the genetic variability within two tree species: *Astronium urundeuva* (Allemão) Engler (Anacardiaceae, considered by some authors as *Myracrodruon urundeuva* F.F. and M.F. Allemão) and *Geoffroea spinosa* Jacq. (Leguminosae).

The Model System Species

Astronium urundeuva is a deciduous tree species well represented in northwestern Argentina, Paraguay, Bolivia, and central and northeastern Brazil, growing at up to 1800 meters altitude in the Andes (Vargas-Salazar 1993). It can be found from humid, dense formations to drier and open areas, but it is largely confined to SDTF and can be considered an ecological specialist of this vegetation. It is an important member of the caatinga woodland and is frequently considered an indicator species of mesotrophic soil conditions within the Cerrado region. *A. urundeuva* is dioecious and insect pollinated, and its small seeds can easily be dispersed by wind (Allen 1991). Occasional events of asexual formation of seeds without fertilization may also occur. Two congeneric species, *A. fraxinifolium* Schott and *A. balansae* Engl., are sympatric with *A. urundeuva* in some regions, but they are quite distinct on the basis of the last taxonomic revision (Muñoz 1990). This was further confirmed by a recent study that used a Bayesian clustering analysis with microsatellite data (Caetano et al. 2005; Caetano, Nusbaumer and Naciri 2008) on the three species. The authors revealed that confusion may however, exist at the chloroplast level, between *A. urundeuva* and *A. balansae*, for which a recent divergence and/or introgression is suspected.

Geoffroea spinosa is a well-defined, deciduous, hermaphrodite tree species whose morphological variation is related to habitat specificities. Spiny trees with small leaflets were reported in dry habitats, whereas the spineless specimens with large leaflets were mainly found near water. This species is also specific to SDTF and occurs in five disjunct areas: (1) northeastern Brazil; (2) northeastern Argentina, Paraguay, and Bolivia; (3) Ecuador and northern Peru; (4) Galápagos (Floreana and Española islands); and (5) Colombia, Venezuela, and the Antilles (Ireland and Pennington 1999). To what extent *G. spinosa* disperses is hard to evaluate at this point, because direct observations are lacking. The almond-type fruits with a fleshy outer wall, however, appear to be suited to vertebrate dispersal. Also, dispersal by water cannot be rejected, as the species is frequently present on flooded ground or next to water. The fact that the seed is well protected by an indehiscent, woody inner fruit wall further supports this dispersal mode.

The collection sites for *A. urundeuva* included 53 natural populations used in a previous study (Caetano, Prado et al. 2008), from Argentina (9), Bolivia (6), Brazil (22), and Paraguay (16), and two additional populations in eastern Brazil. For *G. spinosa*, a total of 38 natural populations were sampled in Argentina (5), Bolivia (2), Brazil (11), Galápagos (2: Floreana and Española islands), Paraguay (14), and Peru (4). Each population was georeferenced directly by GPS or indirectly by correspondence to a nearby town (table 2-1).

Molecular and Statistical Tools

The use of noncoding regions of the chloroplast genome (Taberlet et al. 1991; Hamilton 1999) has been shown to provide convenient information to trace back recolonization events in various plants (e.g., Petit et al. 1997; Cavers et al. 2003; Dutech et al. 2003). The chloroplast genome presents several characteristics very useful in biogeographical research, as it allows us to study the processes of migration and colonization over long periods of time (Ennos et al. 1999).

Two chloroplast spacers were selected for this study: HA (*trn*H-*psb*A) and SG (*trn*S-*trn*G; Hamilton 1999). The amplification and sequencing procedures have been recently published (Caetano, Prado et al. 2008), and the haplotype sequences have been deposited in GenBank. Nucleotide sequences were subsequently aligned (BioEdit; Hall 1999) and revised manually. Within-population diversities, measured as the number of haplotypes, nucleotide diversity (π; Tajima 1983), and gene diversity (h; Nei 1987), were computed. A haplotype network was constructed (Network; Bandelt et al. 1999) in order to visualize the way the different lineages were genealogically related.

The geographical patterns of the chloroplast diversity for both species were analyzed using different approaches:

- Haplotype maps were constructed using ArcMap GIS (Environmental Systems Research Institute, Inc., Redlands, CA, USA).
- A coefficient of correlation (r^M) between genetic and geographical distances was estimated with a Mantel test (Arlequin; Excoffier, Laval, and Schneider 2005).
- An analysis of molecular variance (AMOVA; Excoffier, Smouse, and Quattro 1992) was performed in order to identify the genetic structure at three different geographic levels: (1) continental (the entire species range), (2) regional (defined as domains), and (3) local (populations). The definition of the regional level requires the designation a priori of groups of populations. This was quite simple for *G. spinosa* because of its patchy distribution: (1) the populations from Peru and Galápagos Islands (the Pacific domain), (2) the Argentinean, Bolivian, and Paraguayan populations (Chaco domain), and (3) the northeastern Brazilian populations (Caatinga domain). On the contrary, defining groups of populations within *A. urundeuva* was problematic and was therefore based on the clustering results obtained with nuclear microsatellites (Caetano, Prado et al. 2008): (1) the 10 northeasternmost Brazilian popu-

TABLE 2-1. Locations of the *Astronium urundeuva* (55 populations) and *Geoffroea spinosa* (38 populations) plants used in this study

Astronium urundeuva				
Population	*Code*	*Country*	*Longitude*	*Latitude*
Las Lajitas	Ag_LLj	Argentina	–64.278	–24.758
Ceibalito	Ag_Cei	Argentina	–64.443	–24.927
El Rey	Ag_ERy	Argentina	–64.595	–25.029
San Pedro	Ag_SPd	Argentina	–64.971	–24.365
Ledesma	Ag_Led	Argentina	–64.816	–23.910
Calilegua	Ag_Cal	Argentina	–64.860	–23.762
Oran	Ag_Orn	Argentina	–64.291	–23.232
Tartagal	Ag_Ttg	Argentina	–63.827	–22.508
Embarcación	Ag_Emb	Argentina	–63.906	–23.016
San Luis 1	Pa_SL1	Paraguay	–57.439	–22.624
San Luis 2	Pa_SL2	Paraguay	–57.534	–22.657
San Luis 3	Pa_SL3	Paraguay	–57.556	–22.886
Cordillera	Pa_Cor	Paraguay	–56.420	–25.117
Mbaracayú 1	Pa_Mb1	Paraguay	–55.561	–24.120
Mbaracayú 2	Pa_Mb2	Paraguay	–55.651	–24.194
Mbaracayú 3	Pa_Mb3	Paraguay	–55.673	–24.271
Punta del Este	Pa_PuE	Paraguay	–58.776	–23.825
Ruta Asunción-Mariscal	Pa_AsM	Paraguay	–59.200	–23.417
Riochito	Pa_Ric	Paraguay	–58.572	–23.996
Altos	Pa_Alt	Paraguay	–57.238	–25.257
Cerro Mbatovi	Pa_CMb	Paraguay	–57.181	–25.593
Pozo Colorado	Pa_PCo	Paraguay	–58.383	–23.533
Madrejón 1	Pa_Mj1	Paraguay	–59.837	–20.558
Madrejón 2	Pa_Mj2	Paraguay	–59.704	–19.988
Cerro Léon	Pa_CLe	Paraguay	–60.317	–20.432
Puerto Quijarro	Bo_PQu	Bolivia	–57.128	–17.783
Cuevas de Motacucito	Bo_CMt	Bolivia	–57.948	–17.783
Tacuaral	Bo_Tac	Bolivia	–57.723	–17.783

Astronium urundeuva				
Population	*Code*	*Country*	*Longitude*	*Latitude*
San José 1	Bo_SJ1	Bolivia	–60.930	–17.850
San José 2	Bo_SJ2	Bolivia	–60.615	–17.850
San José 3	Bo_SJ3	Bolivia	–60.750	–17.310
Alvorada do Norte	Br_AvN	Brazil	–46.901	–14.898
Iaciara	Br_Iac	Brazil	–46.648	–14.181
Parque Terra Ronca	Br_PTR	Brazil	–46.395	–13.496
Iguatama	Br_Igu	Brazil	–45.711	–20.174
Arcos	Br_Arc	Brazil	–45.539	–20.282
Barreiras	Br_Bar	Brazil	–45.133	–12.128
Javi	Br_Jav	Brazil	–43.899	–12.490
Jaborandi	Br_Jab	Brazil	–44.438	–13.563
Monte Rey	Br_MtR	Brazil	–44.223	–14.512
Porteirinha	Br_Por	Brazil	–42.937	–15.621
Boquira	Br_Boq	Brazil	–42.781	–12.608
Medina	Br_Med	Brazil	–40.220	–14.097
Paraopeba	Br_Peb	Brazil	–44.355	–19.345
Bocaiuva	Br_Uva	Brazil	–43.827	–17.096
Junco	Br_Jun	Brazil	–40.082	–11.352
Jequié	Br_Jeq	Brazil	–40.154	–13.916
Riachão Bacamarte	Br_RBa	Brazil	–35.542	–7.204
Pombal	Br_Pom	Brazil	–37.927	–6.737
Milagres	Br_Mil	Brazil	–39.104	–7.256
Picos	Br_Pic	Brazil	–41.555	–7.035
Presidente Dutra	Br_Pre	Brazil	–44.468	–5.367
Cratéus	Br_Cra	Brazil	–40.530	–5.253
Limoeiro do Norte	Br_Lim	Brazil	–38.157	–5.148
Santa Cruz	Br_SCr	Brazil	–36.074	–6.197

(table continues)

TABLE 2-1. *(continued)*

Geoffroea spinosa				
Population	*Code*	*Country*	*Longitude*	*Latitude*
Isla del Cerrito	Ag_ICt	Argentina	–58.634	–27.305
Antequera	Ag_Ant	Argentina	–58.876	–27.432
Cerrito-Antequera	Ag_ICA	Argentina	–58.770	–27.333
Estancia Mongay	Ag_EMg	Argentina	–58.674	–27.046
San Hilario	Ag_SHi	Argentina	–58.272	–26.269
Trans-Chaco	Pa_TCh	Paraguay	–58.043	–24.583
Mariscal Estigarribia	Pa_MEg	Paraguay	–60.620	–22.028
La Patria	Pa_LPa	Paraguay	–61.376	–21.467
Fortin Garrapatal	Pa_FGp	Paraguay	–61.492	–21.451
Loma Plata 1	Pa_LP1	Paraguay	–59.889	–22.352
Loma Plata 2	Pa_LP2	Paraguay	–59.915	–22.592
Gesudi	Pa_Ges	Paraguay	–59.947	–21.892
Madrejón	Pa_Mjo	Paraguay	–59.863	–20.592
Agua Dulce	Pa_ADu	Paraguay	–59.703	–19.988
Cerro Léon 1	Pa_CL1	Paraguay	–60.326	–20.442
Cerro Léon 2	Pa_CL2	Paraguay	–60.220	–20.529
Concépcion 1	Pa_CC1	Paraguay	–58.199	–23.513
Concépcion 2	Pa_CC2	Paraguay	–57.793	–23.525
Destacamiento Caballero	Pa_DCb	Paraguay	–58.784	–24.655
Comunidad Tetarembey	Bo_CTt	Bolivia	–61.079	–18.612
Campo Tita	Bo_CTi	Bolivia	–61.313	–18.579
Barra	Br_Baa	Brazil	–43.028	–11.100
Riachão Bacamarte	Br_RBa	Brazil	–35.542	–7.204
Pombal	Br_Pom	Brazil	–35.602	–7.235
Milagres	Br_Mil	Brazil	–38.995	–7.317
Picos	Br_Pic	Brazil	–41.483	–7.078
Boa Vista	Br_BVi	Brazil	–39.689	–5.051
Maranguape	Br_Mar	Brazil	–38.688	–3.932

Geoffroea spinosa				
Population	*Code*	*Country*	*Longitude*	*Latitude*
Aracati	Br_Art	Brazil	–37.770	–4.579
Jaguarabira	Br_Jag	Brazil	–38.429	–5.472
Jucurutu	Br_Juc	Brazil	–37.027	–6.031
Guanambi	Br_Gua	Brazil	–35.489	–6.791
Inter-Andean 1	Pe_IA1	Peru	–79.694	–6.057
Inter-Andean 2	Pe_IA2	Peru	–78.583	–5.802
Pacific Coast 1	Pe_PC1	Peru	–80.517	–3.800
Pacific Coast 2	Pe_PC2	Peru	–79.975	–4.099
Floreana	Gp_Flo	Galápagos	–90.417	–1.300
Española	Gp_Esp	Galápagos	–89.564	–1.344

lations (Caatinga domain), (2) the central Brazilian populations (Cerrado domain), and (3) the Argentinean, Bolivian, and Paraguayan populations (Chaco domain).

It is useful to clarify the terms *Pacific*, *Chaco*, *Cerrado*, and *Caatinga domains*. The Caatinga and Pacific domains are well recognized as SDTF nuclei. However, the Cerrado and Chaco domains of SDTF may be erroneously confounded with savanna (termed *cerrado* in Brazilian Portuguese) and chaco vegetations, which are distinct from SDTF. Small enclaves of SDTF vegetation exist within these vegetations where soil conditions are favorable. The Chaco domain, as used here, also includes the well-defined Misiones, Chiquitano, and Piedmont SDTF nuclei.

Patterns within *Astronium urundeuva*

A total of 389 individuals were used for the analyses. Three haplotypes were detected for each locus separately (table 2-2), and the two loci considered together resulted in five combined haplotypes. The length of the consensus sequences was of 1249 base pairs, for which the first letter corresponded to the HA haplotype and the second to the SG's. Out of the 55 populations, 10 were polymorphic (table 2-3), most of them being located

TABLE 2-2. Characterization of haplotypes found for *trn*H-*psb*A and *trn*S-*trn*G spacers in *Astronium urundeuva*

	*trn*H-*psb*A (593 pb)			*trn*S-*trn*G (656 pb)							
Haplotype	*Gene bank*	*ID 36*	*S 341*	*Gene bank*	*S 108*	*ID 253*	*ID 254*	*S 480*	*S 573*	*ID 576*	*S 601*
A	EF513743	—	T	EF513746	C	A	—	G	A	—	G
B	EF513744	C	T	EF513747	C	A	A	G	A	—	G
C	EF513745	C	G	EF513748	A	—	—	A	C	G	T

S = substitution and ID = indel. Mutation, including indels of one base pair, and indel positions are numbered from the end of the *trn*H and *trn*S primers, respectively.

in central Paraguay. A significant positive correlation between genetic and geographical distances (Mantel test: $r^M = 0.497$) was found.

A simple network was obtained (fig. 2-1), with AA (36 percent of the total sample) occupying the central position and AC (25 percent) being separated from it by six mutational steps. These haplotypes showed opposite distributions, both being found over a range of approximately 2000 kilometers (fig. 2-2). A small overlapping zone was detected within the Cerrado domain, where two polymorphic populations were identified. Accordingly, these populations were found among the ones with the highest nucleotide and gene diversities (Br_Igu: $h = 0.0032$ plus or minus 0.0028, $\pi = 0.667$ plus or minus 0.314; and Br_Arc: $h = 0.0026$ plus or minus 0.0018, $\pi = 0.533$ plus or minus 0.172).

The distribution of the divergent haplotypes AA and AC seems to indicate that secondary contact between these lineages may have occurred. Given the more or less even distribution of the SDTF from the Caatinga through the Cerrado to the Chaco domains (though, in some cases, as scattered patches), there is no reason to believe that the Caatinga-Cerrado transition would represent more of a present-day boundary than the Chaco-Cerrado transition. Instead, this pattern suggests at least one first vicariance event that gave rise to two ancestral lineages, which have progressively diverged due to respective isolation in the Chaco and the Caatinga domains. The highest pairwise F_{CT} value between the Caatinga and Chaco domains (0.891; table 2-4) supports this picture. After the vicariance event, the secondary contact seems to have occurred through rapid range expansions, probably when the suitable conditions for the species' spread were gathered. Microsatellites support this model of colonization (Caetano et al. 2008a).

TABLE 2-3. Distribution of the combined *trn*H-*psb*A and *trn*S-*trn*G haplotypes for *Astronium urundeuva* populations

Population	*Code*	*Haplotypes*					*N*	*n*	*h*	π
		AA	*BA*	*CA*	*AB*	*AC*				
Las Lajitas	Ag_LLj	0	0	0	6	0	6	1	—	—
Ceibalito	Ag_Cei	0	0	0	7	0	7	1	—	—
El Rey	Ag_ERy	0	0	0	9	0	9	1	—	—
San Pedro	Ag_SPd	0	0	0	8	0	8	1	—	—
Ledesma	Ag_Led	0	0	0	5	0	5	1	—	—
Calilegua	Ag_Cal	0	0	0	5	0	5	1	—	—
Oran	Ag_Orn	0	0	0	7	0	7	1	—	—
Tartagal	Ag_Ttg	0	0	0	8	0	8	1	—	—
Embarcación	Ag_Emb	0	0	0	8	0	8	1	—	—
San Luis 1	Pa_SL1	14	19	2	0	0	35	3	0.558±0.045	0.0005±0.0004
San Luis 2	Pa_SL2	2	7	0	0	0	9	2	0.389±0.164	0.0003±0.0004
San Luis 3	Pa_SL3	1	10	0	0	0	11	2	0.182±0.144	0.0002±0.0002
Cordillera	Pa_Cor	0	11	0	0	0	11	1	—	—
Mbaracayú 1	Pa_Mb1	8	15	0	0	0	23	2	0.474±0.067	0.0004±0.0004
Mbaracayú 2	Pa_Mb2	3	17	0	0	0	20	2	0.268±0.113	0.0002±0.0003
Mbaracayú 3	Pa_Mb3	13	0	0	0	0	13	1	—	—
Punta del Este	Pa_PuE	9	0	0	0	0	9	1	—	—
Ruta Asunción-Mariscal	Pa_AsM	4	1	0	0	0	5	2	0.400±0.237	0.0003±0.0004
Riochito	Pa_Ric	6	8	0	0	0	14	2	0.528±0.064	0.0004±0.0004
Altos	Pa_Alt	0	8	0	0	0	8	1	—	—
Cerro Mbatovi	Pa_CMb	0	6	0	0	0	6	1	—	—
Pozo Colorado	Pa_PCo	5	2	0	0	0	7	2	0.476±0.171	0.0004±0.0004
Madrejón 1	Pa_Mj1	5	0	0	0	0	5	1	—	—
Madrejón 2	Pa_Mj2	9	0	0	0	0	9	1	—	—
Cerro Léon	Pa_CLe	11	0	0	0	0	11	1	—	—
Puerto Quijarro	Bo_PQu	3	0	0	0	0	3	1	—	—
Cuevas de Motacucito	Bo_CMt	3	0	0	0	0	3	1	—	—
Tacuaral	Bo_Tac	4	0	0	0	0	4	1	—	—
San José 1	Bo_SJ1	3	0	0	0	0	3	1	—	—
San José 2	Bo_SJ2	5	0	0	0	0	5	1	—	—

(table continues)

TABLE 2-3. *(continued)*

Population	*Code*	*Haplotypes*					*N*	*n*	*h*	*π*
		AA	*BA*	*CA*	*AB*	*AC*				
San José 3	Bo_SJ3	4	0	0	0	0	4	1	—	—
Alvorada do Norte	Br_AvN	5	0	0	0	0	5	1	—	—
Iaciara	Br_Iac	4	0	0	0	0	4	1	—	—
Parque Terra Ronca	Br_PTR	4	0	0	0	0	4	1	—	—
Iguatama	Br_Igu	1	0	0	0	2	3	2	0.667±0.314	0.0032±0.0028
Arcos	Br_Arc	4	0	0	0	2	6	2	0.533±0.172	0.0026±0.0018
Barreiras	Br_Bar	0	0	0	0	5	5	1	—	—
Javi	Br_Jav	0	0	0	0	5	5	1	—	—
Jaborandi	Br_Jab	0	0	0	0	4	4	1	—	—
Monte Rey	Br_MtR	0	0	0	0	5	5	1	—	—
Porteirinha	Br_Por	0	0	0	0	5	5	1	—	—
Boquira	Br_Boq	0	0	0	0	3	3	1	—	—
Medina	Br_Med	0	0	0	0	4	4	1	—	—
Paraopeba	Br_Peb	0	0	0	0	5	5	1	—	—
Bocaiuva	Br_Uva	0	0	0	0	5	5	1	—	—
Junco	Br_Jun	0	0	0	0	5	5	1	—	—
Jequié	Br_Jeq	0	0	0	0	5	5	1	—	—
Riachão Bacamarte	Br_RBa	0	0	0	0	4	4	1	—	—
Pombal	Br_Pom	0	0	0	0	3	3	1	—	—
Milagres	Br_Mil	0	0	0	0	4	4	1	—	—
Picos	Br_Pic	0	0	0	0	5	5	1	—	—
Presidente Dutra	Br_Pre	0	0	0	0	5	5	1	—	—
Cratéus	Br_Cra	0	0	0	0	5	5	1	—	—
Limoeiro do Norte	Br_Lim	0	0	0	0	4	4	1	—	—
Santa Cruz	Br_SCr	0	0	0	0	5	5	1	—	—

Based on 55 *Astronium urundeuva* populations. The number of individuals analyzed (*N*), the total number of haplotypes in each population (*n*), the nucleotide (*π*), and the gene (*h*) diversities are also given.

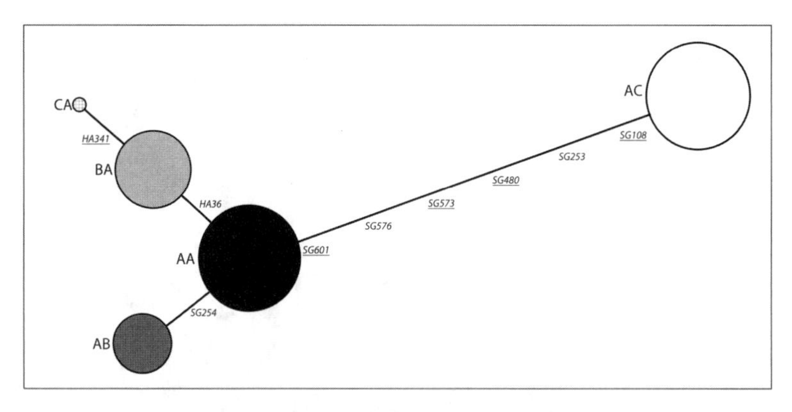

FIGURE 2-1. Median joining network of *Astronium urundeuva* haplotypes considering all mutations. The radius of each circle is proportional to the number of individuals displaying the haplotype. Mutation names starting with "HA" and "SG" referred to mutations observed in *trn*H-*psb*A and *trn*S-*trn*G, respectively. Indels are in italics, and substitutions are underlined.

The haplotypes that have been recognized in the Chaco domain (fig. 2-1)—AA, BA (20 percent), AB (18 percent), and CA (1 percent)—are roughly arranged in a starlike way (fig. 2-2), which could be consistent with a past demographic expansion pattern. Accordingly, the microsatellites also reported the highest diversities within this domain (Caetano et al. 2008a).

Despite the few sampling gaps, especially in the southern limit of the Cerrado domain, the phylogeographical pattern obtained for *A. urundeuva* is well assessed. Indeed, given the location of AA haplotype northwards of the sampling gap, a more intense sampling is not expected to dramatically change the results and the subsequent interpretations.

Patterns within *Geoffroea spinosa*

The 208 individuals analyzed revealed a total of eight combined haplotypes (three for HA and eight for SG; table 2-5). The consensus sequence was 1075 base pairs long. A strong geographical pattern was reported, as evidenced by the highly significant correlation between genetic and geographic distances (Mantel test: $r^M = 0.713$).

A more complicated network than that of *A. urundeuva* was obtained (fig. 2-3). Nineteen mutations separated the most differentiated haplotypes, KK (21 percent) and MM (15 percent), and a single indel differenti-

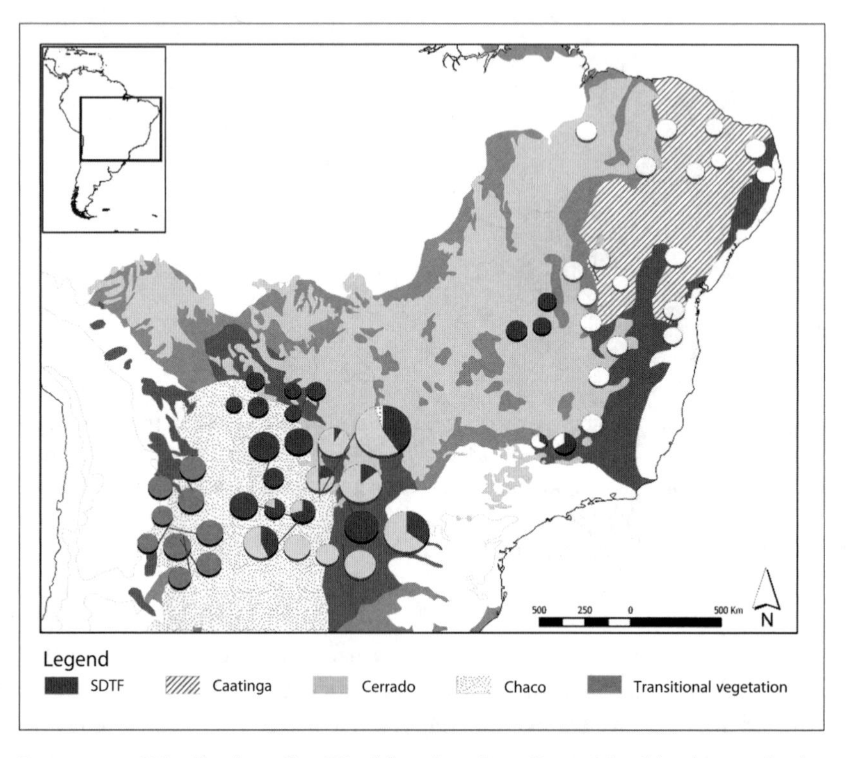

FIGURE 2-2. Distribution of *trn*H-*psb*A and *trn*S-*trn*G combined haplotypes in the 55 populations of *Astronium urundeuva*. Haplotypes were colored according to Figure 2-1, and each circle represents the relative percentage of each haplotype in the population. Their sizes are functions of the number of individuals sequenced.

ated haplotype KK from KR (3 percent). Both KK and KR were found only in the Pacific domain (fig. 2-4), KR being restricted to Floreana Island (Galápagos), which had the only polymorphic population (h = 0.0004 plus or minus 0.0004 and π = 0.429 plus or minus 0.089; table 2-6).

The differentiation that characterized the Pacific domain (pairwise F_{CT} of 0.882 and 0.973 with Chaco and Caatinga domains, respectively; table 2-7) seems to be the outcome of an ancient isolation event, as already argued by Naciri-Graven et al. (2005). Two major natural barriers exist within this domain: the Andes and the oversea distance separating the Galápagos Islands from the continent. In the particular case of *G. spinosa*, neither one of them seemed to be acting as barrier to gene flow, as reflected by the sharing of the same haplotype among populations east and west of the Andes and in the Galápagos Islands.

TABLE 2-4. Pairwise F_{CT} among *Astronium urundeuva* pairs of groups, estimated on merged *trn*H-*psb*A and *trn*S-*trn*G haplotypes

	Chaco	*Cerrado*	*Caatinga*
Chaco	—		
Cerrado	0.731***	—	
Caatinga	0.891***	0.195 (NS)	—

***significant values with $\alpha = 0.001$

Three haplotypes were exclusive of the Caatinga domain: LN (2 percent), LP (13 percent), and LQ (5 percent). A maximum of two mutational steps separated them, and LN occupied the central position of the network (fig. 2-3). The Chaco domain was also characterized by three haplotypes: MM, LL (37 percent), and LO (4 percent). LO was restricted to the two westernmost Paraguayan populations, LL was present in Bolivia and northern Paraguay, and MM was shared among the two southernmost populations from Paraguay and the ones from Argentina. When the analyses were performed without the nucleotide repeats, LO, LP, and LQ haplotypes merged together into LN, which became the shared haplotype between the whole Caatinga domain and the two westernmost populations from the Chaco. This suggests a recent colonization of the area in between the Chaco and the Caatinga domains, followed by a fragmentation that led to today's species distribution. This was further sustained by the smallest pairwise F_{CT} reported between these domains (0.500).

The Chaco was identified as the domain with the highest diversity, as shown by the number of mutations differentiating the haplotypes present in this region (four to seven). Moreover, their geographical distributions seem to argue for an earlier differentiation between the northern and southern parts of the Chaco domain. The history of *G. spinosa* is still incomplete, due to the lack of populations from the Caribbean coast in our analyses. Therefore, the interpretations given here remain preliminary, and the proposed scenario may need future readjustment.

Vicariance or Dispersal: Inferences on the History of the SDTF

Although the two species chosen as models for the present study illustrate individual evolutionary histories, some of the observed patterns helped to illustrate the SDTF biogeography. Herein, a phylogeographic approach

TABLE 2-5. Characterization of haplotypes found for *trn*H-*psb*A and *trn*S-*trn*G spacers in *Geoffroea spinosa*

		*trn*H-*psb*A (273 pb)				
Haplotype	*Gene bank*	*S 23*	*S 68*	*S 87*	*ID 194*	*S 212*
K	EF564430	C	T	T	A	C
L	EF564431	A	C	T	—	A
M	EF564429	A	C	G	—	A

		*trn*S-*trn*G (802 pb)						
Haplotype	*Gene bank*	*ID 94*	*ID 96*	*S 114*	*S 156*	*Inv 163*	*S 200*	*ID 213*
K	EF564432	—	—	T	T	TTA	T	—
L	EF564433	T	—	T	T	AAT	C	—
M	EF564434	T	—	G	C	AAT	C	—
N	EF564435	T	—	T	T	AAT	C	—
O	EF564436	T	T	T	T	AAT	C	—
P	EF564437	T	T	T	T	AAT	C	—
Q	EF564438	T	T	T	T	AAT	C	A
R	EF564439	—	—	T	T	TTA	T	—

S = substitution, ID = indel, and Inv = inversion. Mutation, including indels of one base pair, and indel positions are numbered from the end of the *trn*H and *trn*S primers, respectively.

(Avise 1994; Bermingham and Moritz 1998) was for the first time applied to SDTF species. Previous studies have centered on the evolution of Neotropical widespread species from rain forests (e.g., Caron et al. 2000; Richardson et al. 2001; Dutech et al. 2003; Salgueiro et al. 2004) and the savanna (e.g., Lacerda et al. 2001; Collevatti et al. 2003; Martins et al. 2006; Ramos et al. 2007). Therefore, because our approach lacks comparative data from other SDTF species, the following inferences should be treated with caution. Moreover, the paucity of paleobotanical data from the dry forest areas, together with the lack of a suitable molecular clock to date the different events, makes our interpretations preliminary.

The first topic to be addressed in this section concerns the probable pre-Pleistocene age of both *A. urundeuva* and *G. spinosa*. This is in agreement with

*trn*S-*trn*G (802 pb)									
ID 258	*ID 381*	*ID 382*	*ID 383*	*S 472*	*S 588*	*ID 652*	*ID 672*	*S 693*	*S 702*
T	—	—	—	A	G	—	—	A	T
—	A	A	A	A	A	C	A	G	C
—	A	A	A	C	A	C	—	G	C
—	A	A	—	A	A	C	—	G	C
—	A	—	—	A	A	C	—	G	C
—	A	A	—	A	A	C	—	G	C
—	A	A	—	A	A	C	—	G	C
T	A	—	—	A	G	—	—	A	T

the findings of Pennington et al. (2004b), indicating that a certain number of endemic SDTF species appeared during the late Miocene and Pliocene. Our results point towards an older age and subsequent deeper colonization history for *G. spinosa*, when compared with *A. urundeuva*. Their respective ranges, especially the disjunct presence of *G. spinosa* in the Pacific and Caribbean coasts, seem indeed to reflect more ancient vicariance events for this species.

The Pacific SDTF nucleus seems to have remained isolated since its colonization, accounting for the huge differentiation that characterises *G. spinosa*'s populations within this region. Still, whether this vicariance-isolation event only concerned this nucleus or if it also affected the Caribbean one is still an open question. Getting populations from the Caribbean domain is, therefore, extremely important, as they would improve insights on

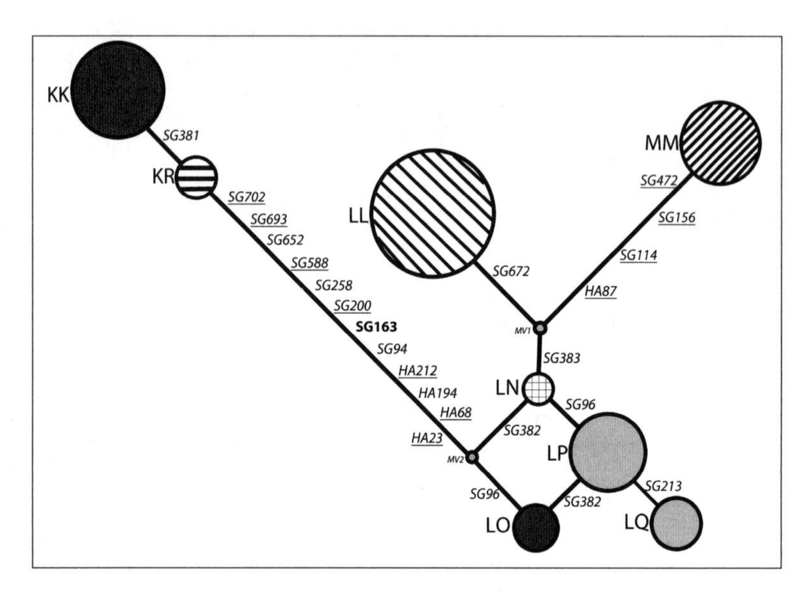

FIGURE 2-3. Median joining network of *Geoffroea spinosa trn*H-*psb*A and *trn*S-*trn*G combined haplotypes considering all mutations. The radius of each circle is proportional to the number of individuals displaying the haplotype. Mutation names starting with "HA" and "SG" referred to mutations observed in *trn*H-*psb*A and *trn*S-*trn*G, respectively. Indels are in italics, substitutions are underlined, and the single inversion is in bold.

the history of *G. spinosa*. Two different results may be expected: (1) a close relationship between Caribbean and Pacific populations and (2) a small differentiation of the Caribbean populations from the Caatinga and Chaco domains. The first result would imply one single ancient event that fragmented populations into two groups, which evolved independently and without contact northwest and southeast of the Amazon basin. The second result may instead support the expansion of the SDTF into the Amazon basin sometime in the recent past.

Our data do not test the suggestion that the SDTF species penetrated the Amazon basin (Pennington et al. 2000), but they clearly indicate that the Pleistocenic Arc did not stretch northwards up to the Andes. This is also congruent with Linares-Palomino et al.'s analyses (chap. 1) that show low levels of floristic similarity between the SDTFs of eastern South America and those in Peru and Ecuador.

Another issue that also deserves attention when considering the common geographical range of the two species concerns the historical relation-

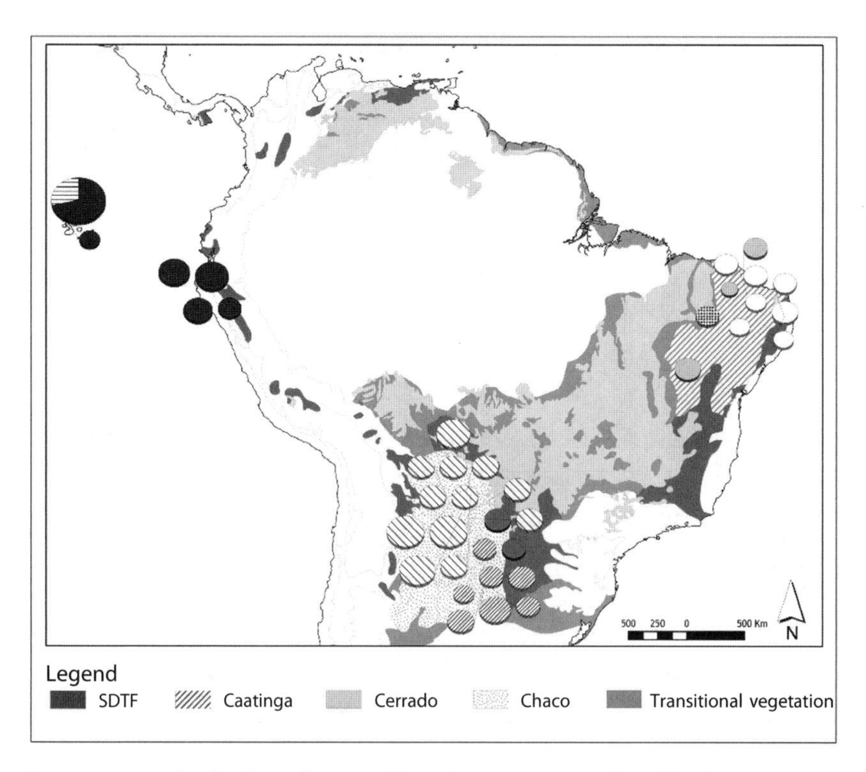

FIGURE 2-4. Distribution of *trn*H-*psb*A and *trn*S-*trn*G combined haplotypes in the 38 populations of *Geoffroea spinosa*. Haplotypes were colored according to Figure 2-3, and each circle represents the relative percentage of each haplotype in the population. Their sizes are functions of the number of individuals sequenced.

ships between the remaining SDTF nuclei, namely the Caatinga and the Chaco domains. It is obvious from the previous sections that each species presents a unique colonization history within this region. The differentiation levels reported between the Chaco and Caatinga domains for the two species (0.891 for *A. urundeuva* and 0.500 for *G. spinosa*) indeed reflect the two different scenarios that were proposed. For *A. urundeuva* an ancient vicariance and divergence between the Chaco and Caatinga domains is suggested, followed by rapid range expansion and secondary contact between them. For *G. spinosa* a recent fragmentation following the colonization of the area between the Chaco and Caatinga domains is hypothesized.

Despite the lack of a molecular clock that would allow dating of these different events, it seems reasonable to assume that the suggested presence of both species in the region between the Chaco and Caatinga domains occurred

TABLE 2-6. Distribution of the combined *trn*H-*psb*A and *trn*S-*trn*G haplotypes for *Geoffroea spinosa* populations

Code	*Haplotypes*								N	n	h	π
	KK	*KR*	*LL*	*LN*	*LO*	*LP*	*LQ*	*MM*				
Ag_ICt	0	0	0	0	0	0	0	4	4	1	—	—
Ag_Ant	0	0	0	0	0	0	0	5	5	1	—	—
Ag_ICA	0	0	0	0	0	0	0	7	7	1	—	—
Ag_EMg	0	0	0	0	0	0	0	3	3	1	—	—
Ag_SHi	0	0	0	0	0	0	0	4	4	1	—	—
Pa_TCh	0	0	0	0	0	0	0	5	5	1	—	—
Pa_MEg	0	0	9	0	0	0	0	0	9	1	—	—
Pa_LPa	0	0	5	0	0	0	0	0	5	1	—	—
Pa_FGp	0	0	10	0	0	0	0	0	10	1	—	—
Pa_LP1	0	0	5	0	0	0	0	0	5	1	—	—
Pa_LP2	0	0	5	0	0	0	0	0	5	1	—	—
Pa_Ges	0	0	5	0	0	0	0	0	5	1	—	—
Pa_Mjo	0	0	10	0	0	0	0	0	10	1	—	—
Pa_ADu	0	0	5	0	0	0	0	0	5	1	—	—
Pa_CL1	0	0	5	0	0	0	0	0	5	1	—	—
Pa_CL2	0	0	5	0	0	0	0	0	5	1	—	—
Pa_CC1	0	0	0	0	4	0	0	0	4	1	—	—
Pa_CC2	0	0	0	0	5	0	0	0	5	1	—	—
Pa_DCb	0	0	0	0	0	0	0	4	4	1	—	—
Bo_CTt	0	0	8	0	0	0	0	0	8	1	—	—
Bo_CTi	0	0	5	0	0	0	0	0	5	1	—	—
Br_Baa	0	0	0	0	0	0	5	0	5	1	—	—
Br_RBa	0	0	0	0	0	4	0	0	4	1	—	—
Br_Pom	0	0	0	0	0	3	0	0	3	1	—	—
Br_Mil	0	0	0	0	0	3	0	0	3	1	—	—
Br_Pic	0	0	0	4	0	0	0	0	4	1	—	—
Br_BVi	0	0	0	0	0	0	2	0	2	1	—	—
Br_Mar	0	0	0	0	0	0	4	0	4	1	—	—
Br_Art	0	0	0	0	0	4	0	0	4	1	—	—
Br_Jag	0	0	0	0	0	4	0	0	4	1	—	—

TABLE 2-6. *(continued)*

Code	*Haplotypes*								*N*	*n*	*h*	π
	KK	*KR*	*LL*	*LN*	*LO*	*LP*	*LQ*	*MM*				
Br_Juc	0	0	0	0	0	3	0	0	3	1	—	—
Br_Gua	0	0	0	0	0	5	0	0	5	1	—	—
Pe_IA1	6	0	0	0	0	0	0	0	6	1	—	—
Pe_IA2	4	0	0	0	0	0	0	0	4	1	—	—
Pe_PC1	7	0	0	0	0	0	0	0	7	1	—	—
Pe_PC2	8	0	0	0	0	0	0	0	8	1	—	—
Gp_Flo	15	6	0	0	0	0	0	0	21	2	0.427±0.089	0.0004±0.0004
Gp_Esp	3	0	0	0	0	0	0	0	3	1	—	—

38 *Geoffroea spinosa* populations. The number of individuals analyzed (*N*), the total number of haplotypes in each population (*n*), the nucleotide (π), and the gene (*h*) diversities are also given.

TABLE 2-7. Pairwise F_{CT} among *Geoffroea spinosa* pairs of groups, estimated on merged *trn*H-*psb*A and *trn*S-*trn*G haplotypes

	Pacific	*Chaco*	*Caatinga*
Pacific	—		
Chaco	882***	—	
Caatinga	0.973***	0.500***	—

***significant values with α = 0.001

more or less simultaneously. We believe that this corresponded to a precise moment in history, possibly the last glacial maximum, during which the climate conditions allowed the existence of a more extensive and contiguous SDTF formation that connected the Chaco and the Caatinga domains. Therefore, our study seems to agree with the Pleistocenic Arc theory in eastern South America (Prado and Gibbs 1993), which is also well supported by the levels of floristic similarity between the SDTF in this area (chap. 1).

The way the three SDTF nuclei within the Chaco domain, namely the Misiones, the Chiquitano, and the Piedmont, have been historically related is more complex to assess, given the difficulty of allocating populations to one or the other of the nuclei. However, the clear assignment of the different haplotypes to particular populations suggests that some degree of isolation might effectively be or have been at work within the Chaco domain. But more important than the possible isolation between these SDTF nuclei is actually the role that the whole domain seems to have had in the maintenance of the genetic variability for both *A. urundeuva* and *G. spinosa*.

The two species agree in identifying the Chaco domain as the region with the highest variability. The Chaco plain itself is characterized by saline soils and a highly seasonal climate, which effectively represent a hostile environment for the presence of the SDTF species. The *cerros*, hills, cordilleras, and net of gallery forests in between the chaco matrix (Spichiger et al. 1991), together with the peripheral SDTF nuclei, however, offer the necessary conditions for the growth of SDTF species (Spichiger et al. 2004). Hence, and in agreement with Spichiger et al. (1995), who defined the Chaco as "a huge ecotone where floristic elements of distinct origins are converging or diverging, according to the climate fluctuations," the whole domain may be considered a center of genetic diversity for SDTF species.

It is difficult at this point to place the ancestral populations in space and time, and impossible to date precisely the different colonization events proposed above. Yet, this study favors the vicariance hypothesis for the SDTF colonization in eastern South America. If the colonization had occurred by long-distance dispersal, it is indeed odd to find such homogeneity in the geographical patterns, especially between the Chaco and the Caatinga domains. Hence, our results support the Pleistocenic Arc as proposed by Prado and Gibbs (1993). Nevertheless, dating (with pollen cores or with a suitable molecular clock) is necessary to confirm that the expansion of the SDTF between the Chaco and the Caatinga domains, as suggested by both species, overlaps with the Pleistocene period.

Acknowledgments

We thank Professor R. Spichiger for his support; T. Pennington, D. Prado, A. Teixeira-Filho, S. Beck, and F. Méreles for their commitment in the project; P. Silveira, K. Elizeche, L. Oakley, R. Santos, A. Daza, and E. Rosero for their help with sampling; D. de Carvalho for the lab facilities at the Lavras University (Brazil); L. Schneider for technical help at the CJBG lab; and Dr. N. Wyler for the maps. This work was supported by the Swiss NSF (grants n°3100A0/100806-1 and 2), the CJBG, and the three societies—Société Académique de Genève, Société de Physique et d'Histoire Naturelle, and the Swiss Zoological Society—that contributed to travel expenses and lab consumables.

Chapter 3

Extent and Drivers of Change of Neotropical Seasonally Dry Tropical Forests

G. Arturo Sánchez-Azofeifa and Carlos Portillo-Quintero

Seasonally dry tropical forests (SDTFs) are considered one of the most endangered tropical ecosystems (Janzen 1988c). High degrees of degradation are reported, not only for the Neotropics, but also in the old tropics (Miles et al. 2006). Causes and consequences of such degradation are known on a limited basis, and much needs to be learned in terms of gaining a full understanding of what controls environmental deterioration trends in these ecosystems and their impact on ecosystem services. Furthermore, current knowledge on the extent and degree of fragmentation of tropical dry forests is constrained because of the low priority for conservation within governmental and nongovernmental funding agencies. In general, the perception that tropical forests do not exist outside of the Amazon basin, or that high priority should be given to tropical rain forests, has limited the current body of scientific literature (Sánchez-Azofeifa et al. 2005).

The current imbalance in scientific knowledge is reflected in terms of the estimation of the current extent of tropical dry forests and their degree of degradation and fragmentation. Major world efforts to understand the process of land use/cover change in the tropics have concentrated on humid regions (Aldhous 1993; Skole and Tucker 1993; Chomentowski et al. 1994). The NASA Pathfinder study is one of those cases in which little or no effort has been placed to better understand the extent of tropical dry forests. Furthermore, in general, large mapping efforts tend to confuse tropical dry forests with woody savannas and agricultural fields, contributing to a misinformed estimation of their extent and degree of fragmentation (Pfaff et al. 2000; Kalacska et al. 2008). In general, tropical dry forests have received much less attention from the scientific community and management planners (Masera et al. 1997), although they have a higher degree of endemism when compared with tropical rain forests (Trejo and Dirzo 2002).

Understanding of the causes of land use/cover change in tropical dry forests is also limited. Tropical dry forests are, in general, located in areas with good to excellent conditions for agricultural and cattle development (Fajardo et al. 2005) and, most recently, for large megatourism projects. It is estimated that more than 50 percent of all tropical dry forests in Costa Rica were cut to promote the cattle industry in Costa Rica between the late 1950s and the early 1970s (Calvo-Alvarado et al. 2009). Similar observations, but with different time frames, have been documented for Mexico, Venezuela, Brazil, Bolivia, and Paraguay. A recent analysis calculated that only 15 percent of the original SDTFs remain in Venezuela (Rodriguez et al. 2008). The fundamental problem in documenting those changes is that tropical dry forests are, in general, not the last but the first frontier of agriculture development, some of them so early that the quantification of rates of change and the dynamics of land use/cover change are compounded by a lack of complete remote-sensing records that could be used to construct comprehensive deforestation and secondary-growth rates.

In this chapter, we aim to document two fundamental issues that can be considered critical to the conservation of tropical dry forests. The first one deals with the challenges associated with mapping the extent of tropical dry forests. We explore the different methodologies used to extract their extent and discuss their major limitations and problems. We also compare current estimates of tropical dry forests among the main sources of information currently available. The second section of this chapter deals with an analysis of the forces of change currently present in tropical dry forest environments. We conduct this analysis for the Neotropics based on information derived from the World Wildlife Fund's Global 200 priority ecoregions assessment (Olson and Dinerstein 2002). In this section, we divide the analysis into three subsections: all tropical dry forests (continental and insular), continental level, and insular level. Finally, we complement our analysis with a specific case study of the role of fire as a tool for land use/cover change in the area of the Chamela-Cuixmala Biosphere Reserve in Mexico.

Extent of Neotropical Dry Forests

Regional mapping efforts of Neotropical dry forests have been limited. Recent efforts conducted by Muchoney et al. (2000), Bartholome et al. (2002), Mas et al. (2002), Giri and Jenkins (2005), and Miles et al. (2006) have been comprehensive but still not conclusive in terms of defining the current extent of tropical dry forests globally. Nonetheless, their work has been a valuable approach to assessing the actual extent and geographical distribution of Neotropical dry forest. One of the most important land-cover mapping efforts, the Global Land Cover 2000 (GLC2000) land-cover mapping initiative, produced a land cover map of South America and Mesoamerica using remotely sensed satellite data (Bartholome et al. 2002). Satellite images were processed to produce a land cover map that includes humid tropical forests, flooded tropical forests, dry tropical forests, and other types of land cover following the UN Food and Agriculture Organization's Land Cover Classification System at a minimum mapping unit of 1 kilometer.

Figure 3-1 shows the distribution of tropical dry forests in the Neotropics by including forested area detected by the GLC2000 within the tropical dry forest biome limits in the Americas, as defined by Olson et al. (2001).

In Mesoamerica, tropical dry forests are distributed among all countries, mostly near the Pacific coastline with the exception of important fragments in Honduras and the Yucatán peninsula. The GLC2000 calculates that tropical dry forests in Mesoamerica comprise 188,995 square kilometers. In South America, tropical dry forests are distributed among Venezuela, Colombia, Peru, Bolivia, and Brazil, with larger and more continuous fragments occurring in an area between eastern Bolivia and southwestern Brazil. Colombia and Venezuela also have important extensions of dry forests. The GLC2000 estimates that 1,467,200 square kilometers of tropical dry forests exist in South America. That gives a total of 1,656,195 square kilometers of land covered by tropical dry forests in the Neotropical region.

On the other hand, Miles et al. (2006) produced a distribution map of SDTF by looking at the percentage of tree cover in 500 meter by 500 meter grid cells based on the MODIS Vegetation Continuous Fields product from the University of Maryland Global Land Cover Facility (http://www.landcover.org). Tropical dry forest was defined as those tropical grid cells with at least 40 percent forest cover, based on the MODIS data set. It included cells found within the "tropical and subtropical dry broadleaf

FIGURE 3-1. Geographical distribution of SDTFs in Mesoamerica and South America based on the GLC2000 product (Bartholome et al. 2002). The map includes all forest land covers within the tropical dry forest biome, as defined by Olson et al. (2001).

forest," the "Mediterranean forest, woodland, and scrub," the "desert and xeric shrublands," and the "tropical and subtropical grassland, savanna and shrub" biomes defined by Olson et al. (2001). Their results showed considerably different area estimates. They report that SDTF extent in Latin America covers only 699,482 square kilometers, where 81 percent belongs to South America and the rest to Mesoamerica and the Caribbean.

The work by the GLC2000 is based on processing satellite imagery with coarser spatial resolution (1 kilometer by 1 kilometer grid cells). Large proportion errors can arise as landscapes are represented at increasingly coarse scales (Moody and Woodcock 1995). The mixture of agricultural lands, urban landscapes, and other vegetation covers with forested areas in a single class increases as the spatial resolution of the imagery decreases.

Therefore, area estimates improve by decreasing the scale of land cover aggregation (Moody and Woodcock 1995). As a result of the use of MODIS imagery at 500-meter spatial resolution for mapping the forest extent, we could agree that the SDTF extent estimate from Miles et al. (2006) is indeed a better approximation of the actual extent of SDTF in the Americas. Nonetheless, their work made several assumptions and generalizations (grid cells with greater than 40 percent tree cover within four different semiarid biomes) that imply the aggregation of shrublands and woodlands with tropical dry forest in a single class, suggesting the possibility of important overestimation of SDTF extent.

Further improvement of SDTF extent estimates is needed. The problems associated with incorrect estimations of tropical dry forests go beyond the simple estimation of extent and spatial configuration. Kalacska et al. (2008) have documented the significant problems with the use of MODIS-derived data sets when applied to approaches to estimate the cost of payments for environmental services in both the Chamela-Cuixmala Biosphere Reserve in Mexico and the Santa Rosa National Park in Costa Rica. Miles et al.'s (2006) accuracy for the estimation of the extent of forest cover has been estimated by Kalacska et al. (2008) to be 34 percent for Santa Rosa and 66 percent for the Chamela-Cuixmala Biosphere Reserve. When MODIS-derived data sets are used to estimate payments for environmental services in the context of carbon sequestration, biodiversity, and water supply to local communities, sharp differences are observed in the order of US$100,000 to US$3,500,000, depending on the site.

Nevertheless, despite the preliminary nature of these previous assessments, we can derive some salient characteristics of the SDTF extent and spatial distribution in the Americas. For Mesoamerica, tropical dry forest remnants are distributed among all countries, and most of it is distributed along the Pacific coastline, with the exception of important fragments in Honduras and the Yucatán peninsula, Mexico. SDTF along the Pacific coastline is highly fragmented and exposed to higher degrees of human degradation (Galicia et al. 2008). The majority of SDTF occurs in South America, where tropical dry forest remnants are distributed among six countries: Venezuela, Colombia, Ecuador, Peru, Bolivia, and Brazil. The largest and most continuous fragments occur in Bolivia (the Bolivian montane dry forest and the Chiquitano dry forest) and eastern Brazil (Atlantic dry forest). SDTF natural fragmentation is higher in Venezuela and Colombia, where SDTFs expand as a transitional ecosystem located between the eastern Andean slopes and the Venezuelan and Colombian llanos, or grasslands (the Apure-Villavicencio dry forests). Important remnants are also found along the Cauca, Sinú, and Magdalena river valleys in Colombia, the Lake Maracaibo basin in Venezuela, eastern Ecuador, and northeastern Peru in the western side of the Andean Cordillera, as well as in Caribbean islands such as Cuba, Jamaica, Haiti, Dominican Republic, and Puerto Rico.

The study of Miles et al. (2006) suggests that the total extent of SDTF in the Americas is 699,482 square kilometers, much smaller in extent than the 1,650,000 square kilometers suggested by the GLC2000 study at 1-kilometer spatial resolution (Bartholome et al. 2002). Even though this might be a much better approximation to the actual extent, other recent studies suggest that this is still an overestimation (Kalacska et al. 2008). Further research is necessary in order to reduce error sources and to offer a more confident estimate of SDTF extent for regional and national decision-making as a baseline for studies on biodiversity, as well as national or regional policy regarding ecosystem protection and restoration, payments for environmental services, and carbon capture and flux research.

Deforestation Rates

Deforestation analysis in the Neotropical region has focused on the humid tropical forest (Achard et al. 2002; Sánchez-Azofeifa et al. 2005; Hansen et al. 2008). Scientific studies for the determination of global deforestation rates for tropical dry forest are very scarce. It has been suggested, however, that at least 48 percent of its extent has already been converted to other land uses, compared with the 32 percent that has been converted from humid tropical forests (Hoekstra et al. 2005). Fajardo et al. (2005) explain that this ecosystem has been subjected to higher degradation because of several features attractive to human use: (1) many tropical forests are established in relatively flat landscapes, (2) the soils are often relatively fertile because of a low rate of nutrient lixiviation and pedogenetic development, (3) the marked climatic seasonality allows for the development of an agricultural lifestyle based on short-cycle crops, (4) there is lower structural complexity and lower aerial biomass when compared with wet forests, and (5) drier climates limit the transmission of pathogens carried by several vectors.

At the global scale, studies on deforestation are limited or not very recent. For example, Achard et al. (2002) estimated that annual forest loss for Latin America ranges from 0.5 to 4.4 percent, but these results are based on humid forest deforestation analyses. On the other hand, the 1981–1990 FAO report (discussed by Aldhous 1993) concluded that the global SDTF deforestation rate was considerable: 2.2 million hectares, or 0.9 percent per year, greater than the percent annual rate of humid forests (0.6 percent per year). The global assessment of Miles et al. (2006) compared the loss of SDTF area in three major regions (continents), including Africa, Asia, and Latin America, and concluded that the SDTF of Latin America experienced the greatest declines between 1980 and 2000 in percentage area forested (relative to total habitat area), with an estimated figure of 12 percent (220,000 square kilometers decrease). A few scientific assessments of deforestation rates within Latin America are available at the national or site level. However, these studies have been performed using different methodologies, and also for periods that are considerably different among regions, all of which makes comparisons difficult. For example, for Mexico, Masera et al. (1997) present deforestation rates at the national level and show that 2 percent of SDTF has been lost annually since the 1980s. They also report a study from Chamela, Jalisco, on the Pacific coast of Mexico, where deforestation rates of SDTF reach 3.8 percent per year. For Venezuela, Fajardo et al. (2005) report results for different sites. Deforestation rates in northwestern and northeastern Venezuela are 2.6 and 2 percent per year, respectively, mainly driven by cattle grazing and agriculture. In Bolivia, Mertens et al. (2004) report annual deforestation rates ranging from 3.0 to 4.1 percent in the colonization frontiers of the Santa Cruz Department. Steininger et al. (2001) reported an annual deforestation rate of 4.56 percent per year, during the time period 1990—1998, in the landscape of Tierras Bajas, Santa Cruz, Bolivia.

According to the results of the assessments discussed above, annual loss of SDTF ranges from 2 percent to 4.56 percent in different parts of Latin America, with higher rates reported for the Bolivian dry forests. A multisite regional assessment of deforestation rates in SDTF landscapes using a standardized methodology, just like the ones available for humid forests (Skole and Tucker 1993; Achard et al. 2002), would ultimately avoid the effects of using different methodologies and would possibly show comparable results between countries. Nevertheless, given the human preference for SDTF environments for development, the historical contraction of its extent, and the underrepresentation of SDTF by protected areas (Masera et al. 1997; Sánchez-Azofeifa et al. 2005), there is no doubt that deforestation threats in absolute terms are higher for deciduous vegetation than for humid forests in the Americas.

A different but complementary perspective can be derived from a recent, detailed study, based on a supervised classification of MODIS surface reflectance imagery at 500-meter resolution, which included a systematic assessment of the current extent and the magnitude of conversion of SDTF in continental and insular countries of Latin America (Portillo-Quintero and Sánchez-Azofeifa 2009). The results of this study, in terms of the proportional conversion of SDTF into other land uses, are shown in figure 3-2. In continental countries, the magnitude of SDTF conversion ranges from 45 percent in Bolivia to 95 percent in Peru, with the next highest conversion in Guatemala with 86 percent. With the exception of Bolivia, all other continental countries have converted at least 52 percent their SDTF coverage, and the overall conversion for these countries is an impressive 66 percent. The five insular countries for which data are available show a magnitude of conversion that ranges from 58 percent in the Dominican Republic to 68 percent in Haiti, with an overall conversion of 66 percent. In sum, this analysis suggests a dramatic rate of conversion, with only 44 percent of SDTF remaining in the region.

Causes and Drivers of Change

Causes of degradation of tropical dry forests in the Neotropics are different from those of tropical rain forests, based on our current available knowledge derived from tropical rain forests. In general, tropical dry forests are not the last but the first frontier for economic development policies, and the degradation and deforestation rates in this ecosystem have been greater than for other types of tropical forests (Sánchez-Azofeifa et al. 2005). Furthermore, when dealing with causes and drivers of change in tropical dry forest, it is important to separate the effects of continental and insular degradation forces, since they are different.

The main driver of change in tropical dry forests is the expansion of the agricultural frontier, since these forests are located in areas with excellent soils for agricultural industries and cattle expansion (Sánchez-Azofeifa 2000; Kalacska et al. 2005). Arroyo-Mora et al. (2005), examining deforestation processes in Guanacaste, Costa Rica, concluded that the expansion of the cattle frontier was the main force driving tropical deforestation in the late 1950s and 1960s, the former contributing to the creation of a highly fragmented landscape. Trejo and Dirzo (2000), studying tropical dry forests in central Mexico, and later Sánchez-Azofeifa et al. (2009), studying deforestation processes in the coastal region of Jalisco and the areas surrounding the Chamela-Cuixmala Biosphere Reserve, arrived at similar conclusions.

Although the forces driving land use/cover change in tropical dry for-

FIGURE 3-2. Map showing percent converted area of SDTF in Latin America and the Caribbean. Pie graphs show the percent of the potential extent currently converted (in black) and the percent remaining (in white) by country. Derived from data in Portillo-Quintero and Sánchez-Azofeifa 2009.

est environments seem to be, on the surface, similar to those for other sites in the tropics, the WWF's Global 200 ecoregions (Olson and Dinerstein 2002) scientific reports provide important insights into the nature of those forces. Drivers of change reported for 35 Neotropical dry forest ecoregions were identified from the scientific reports and tabulated for comparison. Tropical dry forest sites were selected in both island and continental sites. When drivers of change are analyzed at a global level, expansion of the agricultural frontier, cattle ranching, and selective logging emerge as the main three dominant forces on more than 50 percent of all tropical dry forest sites studied at the continental and insular levels. When island and continental sites are decoupled, it is possible to observe a sharp difference in the forces of change in each region (fig. 3-3). For insular regions (fig. 3-3A) the main driving forces (defined as the forces that are identified on more than 50 percent of all sites) are invasion of exotic species, urban sprawl, selective logging, agriculture, tourism development, and road construction. In the continental sites (fig. 3-3B) the main identified driving forces are agriculture expansion, cattle ranching and grazing, selective logging, urban sprawl, and hunting. Though an indirect driver of change, alien species invasion is an important agent that affects natural restoration processes and increases the impact of fires in deciduous ecosystems (Cabin et al. 2002; Holzmueller et al. 2008), therefore having an important impact on the dynamics of land cover. Hunting, a phenomenon that disrupts plant-animal interactions such as seed dispersal and seed/seedling predation that drive forest dynamics, is likely to affect SDTF, but this is an inference largely based on studies in moist forest (S.J. Wright et al. 2007; Núñez-Iturri et al. 2008), and studies of this sort in SDTF are badly needed.

When sites are compared against each other, only two drivers of change emerge: selective logging and urban sprawl are common to each region, showing a sharp difference not found previously in the literature, which has been driven mostly by the belief that expansion of the agricultural frontier and cattle ranching/grazing were the main forces of land cover change in this ecosystem.

Fires and Tropical Dry Forest Land Cover Change

Although the presence of fires is not considered part of the natural habitat of tropical dry forests, as in the Brazilian Cerrado, it is considered one of the main factors of change in this ecosystem (Otterstrom 2006). Fire in tropical regions generally is considered a landscape disturbance that impoverishes ecosystems while it serves as an important element in the human domination of the landscape (Ehrlich et al. 1997). Eva and Lambin (2000) found strong linear relationships between humid forest cover change and fire frequency in Central Africa and Amazonia. However, they also found that there is not always a positive feedback between fires and humid forest cover change but that the type of land use controls the impact of fires on natural vegetation. Di Bella et al. (2006) found that the Neotropical ecosystem that is most affected by the occurrence of fires is actually the tropical dry forest, a fact that was later confirmed by S.J. Wright et al. (2007). Given that the majority of the population in the Neotropics lives within the dry

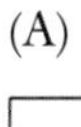

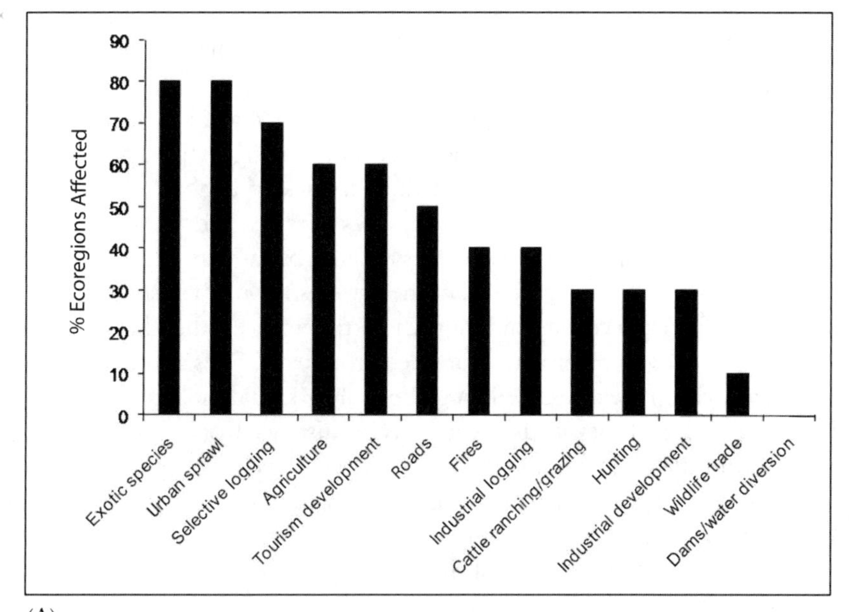

(A)

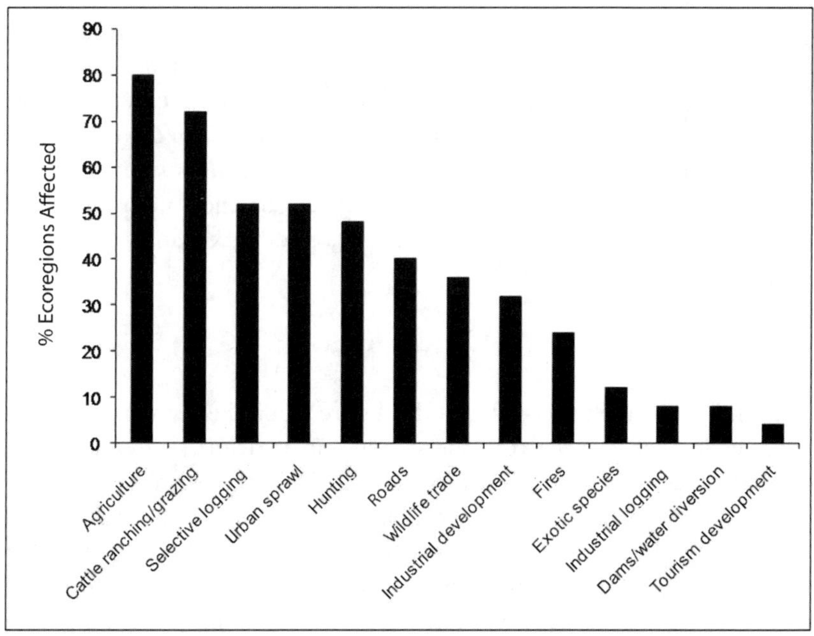

(B)

FIGURE 3-3. Drivers of change in tropical dry forests in (A) insular America and (B) continental America. Derived from Olson and Dinerstein 2002.

forest ecosystem in a matrix of high-intensity agriculture (Fajardo et al. 2005), it is expected that fire occurrence will be higher here than in other tropical ecosystems.

Fire in tropical dry forests is, then, an anthropogenic process rather than a naturally induced phenomenon distributed along all tropical dry forests in the Neotropics. Nonetheless, the relationship between fire occurrence and tropical dry forest cover change has not been addressed so far. In general, fires in tropical dry forests are initiated by farmers, either aiming to control pasture lands at the end of the dry season or as a slash-and-burn technique after clear-cutting, although the latter practice rather than the former is more used in tropical dry forest environments. Fires also tend to be more frequent in the areas bordering national parks and biological reserves in the tropics, and they decline with greater distance from parks (Wright et al. 2006). Fire control, in fact, is considered to be the most important achievement in the restoration of tropical dry forests at the Santa Rosa National Park after other significant initiatives for forest restoration failed.

Fire is also a clear example of the way humans develop control mechanisms in tropical environments. Tropical dry forests, because of their unique characteristics, are more prone to this disturbance, which probably is one of their more important land-cover change drivers. Unfortunately, our current knowledge of fire's impact on SDTFs and their long-term resilience is limited (Otterstrom et al. 2006), and additional work is necessary to better understand the long-term effects of this land-cover change driver. In general, fire is destructive and considered one of the most important threats to the survival of tropical dry forests. Fighting fires around and inside SDTF protected areas is one the most important tasks for park managers.

Conclusions

Although not widely recognized, SDTFs are tropical ecosystems of great biodiversity, particularly from the standpoint of richness of endemisms, diversity of life-forms and functional groups, and their considerable species turnover (beta diversity) (Trejo and Dirzo 2000). In light of this biodiversity, the dramatic transformation described in this synthesis, the magnitude of deforestation and fragmentation, and the threats from the current drivers of global environmental change (both in isolation and in synergism; see Miles et al. 2006), conservation and sustainable use of SDTFs in Latin America should clearly be a priority in national, regional, and global agendas. The deployment of conservation agendas, however, rests on more and better scientific information regarding the spatiotemporal patterns and processes of SDTF alteration. It is our hope that this study highlights some of the critical lacunae of tropical forest conservation that need urgent attention.

Acknowledgments

This work was carried out with the aid of a grant from the Inter-American Institute for Global Change Research (IAI) CRN II # 021, which is supported by the U.S. National Science Foundation (Grant GEO 0452325). Logistical support by the University of Alberta is also acknowledged.

PART II

Animal Biodiversity of Seasonally Dry Tropical Forests

Chapter 4

Seasonally Dry Tropical Forest Soil Diversity and Functioning

DIANA H. WALL, GRIZELLE GONZÁLEZ, AND BREANA L. SIMMONS

Knowledge of biodiversity is key to sustainable management. For tropical dry forests, information on soil animal diversity is sparse compared with tropical wet forests, and considerably less than for grasslands and deserts. Nevertheless, given the rapid transformation of dry tropical forests, primarily due to land use change (Janzen 1988a), it is important to examine the uniqueness of its faunal diversity and functioning as well as its vulnerability to increasing rates of global changes (Sala et al. 2000).

Why are there so few studies on soil faunal diversity and ecosystem functioning in dry tropical forests compared with other ecosystems? Reasons include (1) an emphasis on wet tropical forests due to the high biodiversity and rapid deforestation (Lawton et al. 1996; Lawton et al. 1998; Ineson et al. 2004; Bignell et al. 2005); (2) the small area of dry forests relative to wet tropical forests (Murphy and Lugo 1986a); and (3) poor recognition of ecosystem services that soil biota provide to humans (see table 4-1; van der Putten et al. 2004; Wall 2004; Wardle et al. 2004), which include pollination and wild food when the edaphic phase of the life cycle of aboveground organisms is considered (Daily et al. 1997; van der Putten et al. 2004; Kremen 2005; Barrios 2007). Further, the assumption that belowground faunal diversity will follow a latitudinal gradient seen aboveground (sensu Swift et al. 1979) of highest biodiversity and abundance in the tropics may be unfounded (Boag and Yeates 1998; Bardgett et al. 2005; Maraun et al. 2007). There is evidence that local factors such as microclimate, resource quality, and habitat complexity have a greater importance to diversity than regional and latitudinal factors (Hansen and Coleman 1998; Wardle 2002; Bardgett et al. 2005). Additionally, evidence is accumulating that several groups of soil taxa (earthworms, collembolans, nematodes, oribatid mites) do not follow this gradient (Seastedt 1984; Judas 1988; Foissner 1997a; Boag and Yeates 1998; Maraun et al. 2007), and a standardized assessment across latitudes is needed (Culik and Zeppelini-Filho 2003; Maraun et al. 2003). Oribatid mites, for example, have highest species richness in warm temperate systems (Maraun et al. 2007), and abundance of microarthropods is higher in boreal forests (greater than 300,000 per square meter) than in tropical forests (less than 50,000 per square meter) (Seastedt 1984).

Taxonomic impediments should also be considered for determining soil animal biodiversity (Wall et al. 2005; chap. 5). Worldwide, most soils have a high (greater than 95 percent) proportion of unknown or improperly described species (Lawton et al. 1998; Wall and Virginia 2000). Assessing soil animal biodiversity at the species level requires taxonomic

TABLE 4-1. Ecosystem goods or services, processes involved, and estimated contribution of faunal diversity in dry tropical forests

Good or service provided for humans	*Ecosystem process*	*Relative contribution of faunal biodiversity*
Food production	Bioturbation, wood decomposition	Small
Water quality	N-retention in biomass, physical stabilization, interception of runoff	Large
Watershed flow	Moisture retention by organic matter, evapotranspiration	Medium to small
Fiber production	Decomposition, organic matter/nutrient cycling, nutrient availability, N-fixation	Large
C sequestration	Organic matter formation, inorganic C deposition	Large
Trace gas regulation	Maintenance of C and N balances	Large

expertise and several types of extraction methods for the many different taxonomic groups (Adis 1988; Coleman et al. 2004; Barberena-Arias 2008). Use of different methods and identification categories contributes to an inadequate database for soil biodiversity and ecosystem analysis. Although summaries are available for a few taxa such as collembolans and oribatid mites in Brazilian ecosystems (Culik and Zeppelini-Filho 2003; Oliveira et al. 2005) and earthworms of Neotropical systems (Fragoso et al. 1995), Adis (1988) noted that soil animal biomass and population densities in the Neotropics were based on partial inventories, collected by various methods. This is in contrast to standardized techniques used in wet tropical forests by global projects (e.g., Giller et al. 1997; Bignell et al. 2005; TSBF 2007). Despite these inconsistencies, there are several studies that provide examples of drivers of diversity and the roles of soil biota in biogeochemical cycling and ecosystem service. The examples highlighted throughout this chapter are a basis for more-comprehensive and standardized studies of soil biodiversity and functioning in dry tropical forests.

Species Diversity and Distribution in Seasonally Dry Neotropical Forests

Light (1933) and Thorne et al. (1994) published termite species lists for western Mexico, which includes some dry tropical forest habitats. Within the tropics, termites are known to be numerically and ecologically important (Thorne et al. 1996). Fragoso et al. (1995) synthesized earthworm diversity and biogeography in northern Neotropical countries and proposed a framework for regional and global distribution of earthworm functional groups based on multiple interacting factors of phylogenetic constraints, moisture, temperature, and soil fertility, which helps explain the presence of earthworms in dry and seasonally dry forests. Biogeographical distribution of larger soil macrofauna appears to vary at local and regional scales of dry tropical forests, as has been noted in other ecosystems (Fragoso et al. 1995). Hanson (chap. 5) notes that while some groups of insects with soil-inhabiting larvae are common in dry tropical forests, others are rare or absent, which may be linked to inherent characteristics of this particular ecosystem, such as soil type. However, factors affecting the distribution of smaller biota (microbes and protozoa) are just being examined. Foissner (1995) found 80 ciliate species in a single soil sample from a seasonally dry tropical forest (SDTF) in Puerto Rico and from this described four new genera and seven new species. Soils vary on the microhabitat scale, which suggests many more species of protozoa to be discovered.

Most studies of soil fauna have concentrated on microarthropods (Acari and collembolans). Neotropical collembolan diversity is about 1200 species (Mari Mutt et al. 1996—2001), which is considered an underestimate (Culik and Zeppelini-Filho 2003), compared with 7500 species globally and 812 species for North America. Culik and Zeppelini-Fihlo's (2003) synthesis of Brazil's collembolan diversity (199 species, of which 127 are endemic) across ecosystems indicates about 66 percent are from forests, and the remainder are found in other or undescribed habitats. The Brazilian dry tropical and seasonally dry forests have about 46 collembolan species, of which 13 are considered endemic to Brazil (Culik and Zeppelini-Filho 2003). Underestimates of species diversity occur not only with Collembola but also with other groups. A yearlong study of prostigmatid mites living in litter of two dry forest watersheds in the Chamela Biological Station in Mexico noted 31 of 43 species in the family Cunaxidae as new to science (Mejía-Recamier and Palacios-Vargas 2007; Palacios-Vargas et al. 2007). These are important findings as the Cunaxidae are predators of smaller microarthropods in the food chain (Walter and Kaplan 1991) found in soils, litter, and decomposing bark.

Many factors, such as seasonality, soil heterogeneity, and plants, can affect faunal species diversity and abundance (Swift et al. 1979; Coleman et al. 2004). Adis et al. (1989) examined wet and dry season soil arthropod densities (58,000 per square meter) in a moist campinarana forest near Manaus, Brazil, and found no differences and no evidence of animal migration to deeper depths during the dry season. Sixty-three percent of all arthropods measured at 0 to 14 centimeters depth were in the top 3.5 centimeters of soil. This was in contrast to studies during the dry season in seasonal tropical soils, showing lower arthropod abundance and vertical migration (Wallwork 1976; Lieberman and Dock 1982). In a dry forest of Mexico, Palacios-Vargas et al. (2007) examined arthropod abundance in soil and litter on a monthly basis for a year and noted different responses of microarthropod taxa to wet and dry seasons. Mite abundance was greater in the dry season, while collembolans were more abundant during the wet season. As with other arthropod studies (Adis 1988; Barberena-Arias 2008), Acari and collembolans dominated in soil and litter, comprising 90 percent of total arthropods, with total abundance greater in soil than litter (Palacios-Vargas et al. 2007). Seasonality of plant root growth and death as a source of carbon to soils (Kummerow et al. 1990) could be a factor regulating soil abundance in this dry forest but has yet to be quantified.

In many ecosystems, plant litter quality and quantity appear to govern diversity and density of soil and litter animals. For example, within a dry subtropical forest site, taxonomic composition of soil fauna varied in wood and litter of the same tree species (González and Seastedt 2000; Torres and González 2005). Whether diversity and composition of plants and litter quality across sites govern belowground (soil and plant litter) diversity and abundance is inconclusive for soil fauna in many ecosystems (see Maraun et al. 2007). Southwood et al. (1979) proposed that higher tree diversity could explain higher aboveground arthropod diversity. However, evidence varies as to whether plant diversity, identity, or composition drives soil diversity and abundance in ecosystems. Maraun et al. (2007) and others (Hansen 2000; St. John, Wall, and Behan-Pelletier 2006; St. John, Wall, and Hunt 2006) note that microarthropods have limited possibilities to adapt to a particular tropical or temperate site, since food resources (fungi, plant litter, physicochemical composition) are similar. Salamon et al. (2004) tested the relationship of collembolan diversity with plant species diversity and found little correlation and suggested that coevolutionary processes with plant species probably were not important.

In dry semitropical systems, Barberena-Arias (2000) examined litter arthropods in three forest ecosystems of differing tree diversity (and management) across Mona Island, Puerto Rico, and found faunal diversity was positively related to a gradient of higher tree diversity. The managed mahogany plantation and a native plateau were dominated by a single tree species and exhibited lower litter microarthropod diversity, compared with the more diverse native coastal forest. Arthropod abundance was not related to litter quantity across sites and was highest at the native plateau forest where arthropod predators were considerably lower (Barberena-Arias 2000). In a study of two secondary dry forests (natural and planted) of Guadeloupe, Loranger-Merciris et al. (2007) reported that tree litter identity was related to greater abundance of macrofauna and microarthropods in a plantation of *Tabebuia heterophyla* than in other trees at the site. They attributed this to soil types and chemistry and lower levels of leaf tannins in the *T. heterophyla* plantation. Two anecic earthworm species occurred in soils only at the planted sites. Soil microarthropod (Collembola and Acari) composition varied in the soil profile at the two sites, with collembolans dominating the planted forest (Loranger-Merciris et al. 2007). These limited examples are insufficient to explain drivers of animal diversity in dry tropical forest soils but do illustrate that plant species and composition are a major factor for litter animal diversity, though soil fauna may be more influenced by physicochemical characteristics. Heterogeneity of soil habitats within meters can govern faunal distribution, particularly for smaller biota such as protozoa and nematodes (Foissner 1995, 1997b; Anderson 2002). These few examples also indicate that seasonal responses to precipitation differ with animal group and may be characteristic of life history and physiological and behavioral response.

While the above examples describe the species richness of some groups of soil taxa, they were not designed to quantify or compare the effect of soil animal abundance and diversity on ecosystem processes among different forests; only a few studies compare soil fauna and ecosystem processes in wet and dry forests. Here we highlight some of these.

Soil Fauna and Ecosystem Processes

Wood debris is a major component of inputs to soils in dry tropical forests, and its decomposition rate varies with ecosystem (Eaton and Lawrence 2006). Wallwork (1976) postulated that soil fauna effects on wood decomposition were minimal in dry forests because of moisture constraints and in wet forests because soil fauna avoid the massive extent of fungal hyphal networks. Torres and González (2005) compared soil taxa in decomposing wood logs in wet and dry subtropical forests in Puerto Rico to test this hypothesis. They identified organisms, many to species level, in decomposing *Cyrilla racemiflora* logs after 13 years in the field at the Guánica dry tropical and the Luquillo wet tropical forest. They found higher species diversity and different taxonomic and functional groups (table 4-2) in the Guánica site (25 species) compared with the wet forest (12 species). For example, termites and ants occurred in greater abundance in wood in the dry forest (table 4-3), but termites were of different species in the wet site. Earthworms and fungi were more frequently associated with logs in the wet than dry forest. Reptiles occurred only in the dry tropical forest: the coastal blind snake, *Thyphlops hypomethes*, feeds on termites and ants and inhabits soils and logs, and the gecko, *Sphaerodactylus nicholsi*, lives in shaded but thick litter of dry forests (López-Ortiz and Lewis 2004). The presence of reptiles illustrates that soils are also habitats and food sources for vertebrates, a dependence that could be affected by land use change. The decomposition of logs also differed: the dry tropical forest logs had lower moisture content but higher decomposition rates (an average of 61 percent and 54 percent mass loss, respectively) than the Luquillo wet forest. Nutrient composition of the decaying bark of logs in the dry tropical forest appeared to be higher in calcium, phosphorus, and nitrogen than in the Luquillo wet forest. González and Seastedt (2000, 2001) and González et al. (2001) hypothesized that the presence of certain functional groups, rather than taxonomic identity (order level), whose activities were not constrained by climatic conditions such as high temperatures and low moisture, were important determinants of litter decay in SDTF compared with other ecosystems they examined (e.g., alpine forest).

While the Torres and González (2005) study provides evidence that faunal biodiversity is higher in decaying wood in a dry compared with a wet forest, evidence varies on whether diversity is also higher in soil and litter of dry tropical forests. Barberena-Arias (2008) found no difference in taxonomic richness in soils and litter in wet and dry forests analyzed at a coarse level—class and order—but noted differences with extraction methods. Overall abundance, however, was considerably greater in the dry forest, with Acari and collembolans dominating the soil community (tables 4-2, 4-3). For certain taxonomic groups abundance was low or similar in dry and wet forests. These examples illustrate the need for comparative studies in dry and wet forests.

TABLE 4-2. Comparison of soil fauna diversity in wood, litter, and soil in wet and dry Neotropical forests

Location	*Wet forest*	*Dry forest*	*Taxonomic level*	*Habitat type*	*Reference*
Puerto Rico	12	29	Species	Wood	Torres and González 2005
	13 m^{-2}	13 m^{-2}	Order	Soil	Barberena-Arias 2008
	5 g^{-1}	8 g^{-1}		Litter	
	4 g^{-1}	1.5 g^{-1}	Order	Litter	González and Seastedt 2000
Mexico		32 m^{-2}	Order	Litter	Palacios-Vargas et al. 2007
		28 m^{-2}	Order	Soil	
Guadeloupe		71	Species	Soil (native forest)	Loranger-Merciris et al. 2007
		61		Soil (planted forest)	

A variety of collection methods were used. Diversity is represented per unit measure.

TABLE 4-3. Soil fauna abundances in litter and soil from wet and dry Neotropical forests

Location	*Wet forest*	*Dry forest*	*Taxonomic level*	*Habitat type*	*Reference*
Puerto Rico	10,189 m^{-2}	23,210 m^{-2}	Order	Soil	Barberena-Arias 2008
	3.5 g^{-1}	4.5 g^{-1}	Order	Litter	
	80 g^{-1}	2 g^{-1}	Order	Litter	González and Seastedt 2000
Mexico		26,497 m^{-2}	Order	Soil	Palacios-Vargas et al. 2007
		15,756 m^{-2}	Order	Litter	
Guadeloupe		100 m^{-2}	Species	Soil (native forest)	Loranger-Meciris et al. 2007
		183 m^{-2}	Species	Soil (planted forest)	
		51,000 m^{-2}	Order	Soil (native forest)	
		61,000 m^{-2}	Order	Soil (planted forest)	

A variety of collection methods were used. Abundance expressed as number per unit.

Global Change and Belowground Biota in Seasonally Dry Tropical Forests

Few studies examine how global changes will affect the linkages of soil biota in SDTFs to aboveground biota, and little evidence exists for assessing the vulnerability of soil biota to climate change in either wet or dry tropical forests (but see Wall et al. 1999). However, in both wet and dry tropical forests many studies show that soil biodiversity and abundance change with disturbance (Fragoso et al. 1997; Giller et al. 1997; Johnson and Wedin 1997; Allen et al. 1998; Lawton et al. 1998; Chauvel et al. 1999; Bignell et al. 2005). In dry tropical forests, community composition can be altered (Barberena-Arias 2000) and single species may dominate abundance with varying effects on ecosystem processes (Barberena-Arias 2000; Decaëns et al. 2004; Palacios-Vargas et al. 2007). For example, Yamada et al. (2006) studied nitrogen as an input to decomposition in two tropical forests, a dry deciduous forest and a dry evergreen forest in Thailand, and found that of a diverse group of termites, only wood- and litter-feeding termites fixed nitrogen. Two species, *Microcerotermes crassus* and *Globitermes sulphreus*, had nitrogen fixation rates of 0.21 and 0.28 kilogram per hectare per year, respectively (see also Genet et al. 2001; Decaëns et al. 2004). Disruptions of soil habitats and inhabitants may alter nutrient cycling such as nitrogen fluxes (Yamada et al. 2006) and carbon sequestration (Singh et al. 1991) as well as porosity, aeration, water capacity, and other physical and chemical properties of the soil habitat (Höfer et al. 2001; Decaëns et al. 2004; Barrios 2007). In fact, it has been hypothesized that the loss of termites, and the general decline in the abundance of macroinvertebrate decomposers along broad latitudinal gradients, reduces the ability of the fauna to uniformly affect decomposition rates in colder regions as well (González 2002). Alterations of rates of decomposition have rippling effects on other biota beyond soils (Wall and Moore 1999; Ineson et al. 2004).

Land use change is a major driver affecting soils. For example, human population pressures and fertile soil conditions are resulting in the conversion of dry tropical forest to agriculture (Murphy and Lugo 1986a). Agricultural lands increased in Mexico by 64 percent between 1977 and 1992; while forested areas decreased by 26 percent (Cairns et al. 2000). As more land is converted from dry tropical forest to agriculture, the benefits of managing for complex and complete soil food webs that provide many ecosystem services will become more important to growers, especially in areas where access to pesticides is limited by availability, economics, or environmental concerns.

Conclusions

Dry tropical soil biodiversity and abundance may be greater than in wet forests. The ecosystem services provided to humans by soil animals and microbes are a key reason for considering animals in the context of soil sustainability. While clearing of forests for agriculture may be an immediate need, examples presented here illustrate that changes occurring in the wealth of undescribed biodiversity belowground can be much longer lasting. We suggest the following are urgently needed: (1) a network of taxonomists and ecosystem scientists studying a critical ecological process in several dry tropical forest sites, (2) establishment of comprehensive and long-term studies of management effects on the vulnerability of macrofauna and selected groups (e.g., nitrogen fixers) as linked to soil carbon and nutrient fluxes and management, (3) research efforts comparing nearby tropical wet and dry forests, (4) assessment of conservation priorities at local and regional scales that include consideration of belowground biodiversity, and (5) assessment of ecosystem services for animals in soil habitats of dry tropical forests.

Chapter 5

Insect Diversity in Seasonally Dry Tropical Forests

PAUL E. HANSON

The primary objective of this chapter is to provide a preliminary overview of insect diversity of Central American seasonally dry tropical forests by comparing the proportions of species that either are restricted to dry forests versus wet forests or occur in both. One of the major impediments to evaluating insect diversity, especially in the tropics, is that the majority of species are still undescribed. However, over the past 20 years an intensive inventory has been carried out in Costa Rica, and several taxonomists have dedicated considerable effort towards describing this fauna (Hanson 2004). Although the proportion of the country occupied by dry forests is relatively small (about 15 percent), Costa Rica provides some of the best data presently available; additionally, the diversity of its entomofauna is considerable.

The dry forests of Costa Rica are located in the northwestern part of the country, and their proximity to both high- and low-altitude wet forests permits migration out of the dry forest during the dry season. Several species of insects migrate to lowland wet forest to continue breeding, while others migrate up into the mountains, either to continue breeding or to become dormant. In the Central American dry forests the presence of adjacent volcanoes is especially conducive to altitudinal migration followed by dormancy. Finally, the presence of rivers running down from the volcanoes through the dry forest allows some species to move shorter distances during the dry season to the moister habitats adjacent to creeks and springs.

Many species of insects are restricted to seasonally dry tropical forests, but at higher taxonomic levels the list is much shorter. Examples of genera that appear to be dry forest specialists in Central America include three soil inhabitants: *Tenuirostritermes* (Termitidae), *Spiroberotha* (Berothidae), and *Quemaya* (Tiphiidae). Several higher-level taxa, while not restricted to dry forests, are generally considered to be more species rich in dry forests. Examples include ant lions (Mymeleontidae), hide beetles (Trogidae), certain click beetles (Elateridae: Cebrioninae), spider beetles (Anobiidae: Ptininae), darkling beetles (Tenebrionidae), blister beetles (Meloidae), certain longhorn beetles (Trachyderini and Acanthocini; Noguera et al. 2002), and bee flies (Bombyliidae).

Ant lion larvae are predators that live in the soil, especially in sandy soil. Hide beetles and spider beetles colonize late successional stages of cadavers (when they are dry), and the latter are also associated with detritus in vertebrate and ant nests. The larvae of cebrionine click beetles and many darkling beetles live in the soil, the former feeding on roots, the latter primarily on decomposing plant material. Blister beetle larvae feed in bee nests or on grasshopper eggs in soil, while longhorn beetles feed in dead branches. The larvae of bee flies are parasitoids of beetle larvae in soil or wood, caterpillars, nest-building wasps and bees, or grasshopper eggs. One of the few patterns to emerge from this brief compilation is that many of these taxa have larval stages that live in the soil.

A few of the groups mentioned above are known to possess physiological and/or behavioral adaptations for survival under hot, dry conditions. For example, diurnal darkling beetles in arid regions possess mechanisms for reducing respiratory water loss (Duncan et al. 2002). Blister beetles apparently lack such water conservation mechanisms but instead feed on plants to obtain water; it is thought that this, combined with a tolerance to high temperatures, allows them to inhabit dry habitats (Cohen and Pinto 1977).

The majority of insect families are probably more species rich in wet forests than in dry forests. Given that species richness of plants is greater in wet forests than in dry forests (Gentry 1995) and that a large proportion of all insects are phytophagous or parasitoids of phytophagous insects, a similar trend is expected in insects. For example, 75 percent of the grasshopper species of Costa Rica are confined to wet forests, and the latter also harbor a disproportionately large proportion of grasshopper species showing restricted distributions. The grasshoppers of dry forests include relatively fewer species and are more widely distributed (Rowell 1998). Among the few grasshoppers that are restricted to dry forests is the Central American migratory locust (*Schistocera piceifrons*).

Insights Derived from Insect Surveys in Costa Rican SDTF

For these surveys, dry forests were defined as those sites on the north Pacific coast below about 600 meters altitude. The choice of taxa was necessarily limited to those that have been well collected and studied taxonomically. For this reason, recent taxonomic monographs were the preferred source of information, since all known species of a particular taxon are included, the identifications are the best available, and the distributions within the country are summarized. The monographs utilized provided data for Dynastinae (Ratcliffe 2003), Ichneumonidae (Gauld 1997, 2000; Gauld, Ugalde et al. 1998; Gauld, Godoy et al. 2002), Encyrtidae (Noyes 2000), Papilionidae and Pieridae (DeVries 1987), and Riodinidae (DeVries 1997). For ants, the Costa Rica Ants Web site (AntWeb; Longino 2006) was used. For the remaining groups, the database of the Instituto Nacional de Biodiversidad (INBio; ATTA 2004) was used. For each of the species examined, the collecting sites were noted to determine whether the species is restricted to dry forests, restricted to wet forests, or is present in both forest types. Although the same methodology was not possible for bees, results from extensive collecting at a wet forest site (Bijagua) and a dry forest site (Cañas) were used to compare species richness between sites (Griswold et al. 1995).

Locality data from 3563 species (14 families) of holometabolous insects were examined. The number of specimens per species ranged from 1 to over 4000. Obviously the data from species represented by very few specimens yield less-reliable results than those with large numbers of specimens. It should be noted that there is no reason to suspect that wet forests have been better sampled than dry forests or vice versa, other than the fact that wet forests occupy a much larger portion of the country.

Of the 3563 species included in this study, approximately 8 percent are restricted to dry forest, 72 percent are restricted to wet forest, and 20 percent are present in both. Beetles show a greater proportion (14.7 percent) of species restricted to dry forests than do the other three orders, but very few families from each order were included in the analysis, and thus comparisons between insect orders will only be meaningful when many more families are included.

Beetles (Coleoptera)

Larvae of most tiger beetles (Carabidae: Cicindellinae) dig burrows in the soil where they await their prey. Only three species are restricted to dry forests, and all belong to the very large cosmopolitan genus *Cicindela*. Eight other species in this genus occur in both wet and dry forests, and thus 11 of the 13 tiger beetle species that are recorded from the dry forest belong to this genus. The other large genera in Costa Rica, *Ctenosoma* and *Odontocheila*, are poorly represented in dry forests. Adults of the former inhabit the canopy, and larvae burrow in rotting logs, while adults of the latter rest on foliage of the forest understory between hunting forays on the ground (Pearson and Vogler 2001).

Within the family Scarabaeidae, the subfamily Melolonthinae shows the greatest relative species richness in dry forests (table 5-1). The only data available were for the largest genus, *Phyllophaga*, whose larvae feed on plant roots. Moreover, the only major group of Rutelinae that feeds on roots is the genus *Anomala*, which has a greater proportion of species restricted to dry forests than does the subfamily as a whole, most of which feed in rotting logs. Dung beetles (subfamily Scarabaeinae), whose larvae also live in soil, also show a relatively high proportion of species restricted to dry forests. At the other extreme, the group showing the lowest relative number of species in dry forests is Dynastinae, whose larvae inhabit rotting tree trunks and other rotting plant material.

Leaf beetles (Chrysomelidae) are the second largest group of phytophagous beetles after the weevils (Curculionoidea). Unfortunately, the data are currently insufficient to analyze the largest group of leaf beetles, the subfamily Galerucinae (including Alticini). However, among the subfamilies for which sufficient data are available (table 5-1), the very high proportion of cryptocephaline species restricted to dry forests is one of the most striking results of this study: 72 percent is much higher than the proportion found for any other taxon analyzed in this study. *Cryptocephalus* and the tribe Clytrini, which account for a substantial portion of the species in the subfamily, are known to be especially diverse in the southwestern United States and northern Mexico (W. Flowers, pers. comm.). Cryptocephaline larvae are unusual among leaf beetles in that the larvae construct protective cases and many species feed in the leaf litter or in ant nests.

Seed beetles (Bruchinae) also show an extraordinarily high proportion of species restricted to dry forests (table 5-1). The larvae of all species feed in seeds, especially in seeds of Fabaceae (Janzen 1980). There are far fewer species in lowland rain forests, where seed-feeding weevils are proportionately more species rich and abundant. At upper elevations, both groups are almost nonexistent except for weevils in the seeds of Sapotaceae and Laura-

TABLE 5-1. Relative percentage of beetle (Coleoptera) taxa in dry versus wet forests

	Restricted to dry forests (%)	*Restricted to wet forests (%)*	*Present in both (%)*
Carabidae			
Cicindellinae (*n* = 35)	8.6	62.9	28.5
Scarabaeidae			
Melolonthinae (*n* = 44)	16.0	61.3	22.7
Scarabaeinae (*n* = 120)	15.0	73.3	11.7
Cetoniinae (*n* = 35)	11.4	65.7	22.9
Rutelinae (*n* = 80)	10.0	76.25	13.75
Dynastinae (*n* = 128)	1.0	71.0	28.0
Chrysomelidae			
Cryptocephalinae (*n* = 25)	72.0	8.0	20.0
Bruchinae (*n* = 61)	49.2	21.3	29.5
Eumolpinae (*n* = 52)	15.4	65.4	19.2
Cassidinae (*n* = 259)	11.2	70.3	18.5
Chrysomelinae (*n* = 45)	8.9	73.3	107.8
Total (*n* = 884)	**14.7**	**65.2**	**20.1**

Cassidinae includes the hispine beetles.

ceae, but seed-feeding wasps are more abundant, especially in the seeds of Rubiaceae (D.H. Janzen, pers. comm.).

The other three subfamilies of leaf beetles included in the analysis (table 5-1) do not show any unusual trends when compared with the overall results, except perhaps Eumolpinae, which have a slightly higher than average number of species restricted to dry forests.

Parasitic Wasps

There are approximately 60 families of parasitic wasps, but very few of these have received species-level taxonomic study in the Neotropics. In Costa Rica three large families—Ichneumonidae, Encyrtidae, and Eulophidae—

have been the subject of major taxonomic monographs, and the first two were included in the analysis. For Encyrtidae, only the subfamily Tetracneminae was included, since taxonomic work on the other subfamily (Encyrtinae) is only partially complete. Similarly in Ichneumonidae, not all subfamilies have received taxonomic treatment.

In an examination of the results (table 5-2), four groups stand out as having a relatively greater proportion of species restricted to dry forests: Tetracneminae, Labeninae, Anomaloninae, and Ctenopelmatinae. Tetracneminae are parasitoids of mealybugs; in Labeninae, *Grotea* species are cleptoparasites of bees that nest in twigs, while *Apechoneura* and *Labena* parasitize beetle larvae in wood; Anomaloninae attack lepidopteran caterpillars, while Ctenopelmatinae parasitize sawfly caterpillars. Most of the groups having relatively fewer species restricted to dry forests also parasitize lepidopteran larvae (although many Pimplinae parasitize spiders).

The observed trends are thus difficult to explain on the basis of hosts or type of parasitism (Labeninae and Pimplinae paralyze their hosts, while the other groups generally do not). However, there does appear to be a

TABLE 5-2. Relative percentage of parasitic wasp (Hymenoptera) taxa in dry versus wet forests

	Restricted to dry forests (%)	*Restricted to wet forests (%)*	*Present in both (%)*
Encyrtidae			
Tetracneminae (n = 146)	15.1	55.5	29.4
Ichneumonidae			
Labeninae (n = 44)	15.9	59.1	25.0
Anomaloninae (n = 69)	14.5	58.0	27.5
Ctenopelmatinae (n = 42)	14.3	81.0	4.7
Cremastinae (n = 138)	8.0	65.9	26.1
Tryphoninae (n = 41)	7.3	90.2	2.5
Metopiinae (n = 130)	6.2	83.1	10.7
Metopiinae (n = 185)	2.2	77.3	20.5
Banchinae (n = 247)	0.8	89.9	9.3
Total (n > = 1042)	**7.0**	**75.0**	**18.0**

general correlation with altitudinal patterns in species richness. Groups having relatively more species restricted to dry forests also tend to reach their maximal species richness at low altitudes. Tetracneminae, Labeninae, Anomaloninae, and Cremastinae reach their peak species richness at altitudes of 500 meters or lower, whereas Tryphoninae, Metopiinae, Pimplinae, and Banchinae all reach their maximum species richness at 1000–1400 meters (Gauld 1997, 2000; Noyes 2000). Ctenopelmatinae show a peak at 1600 meters but a secondary peak at 400 meters (Gauld 1997). The correlation with altitudinal patterns in species richness is hardly surprising, since the dry forests occur at lower altitudes. Nonetheless, certain questions remain; for example, why Anomaloninae is better represented in dry forests than is Cremastinae, despite the fact that both are predominantly low-altitude groups.

Ants

Pseudomyrmecinae and Dolichoderinae were excluded because of insufficient data. Among the subfamilies included in the analysis, nine introduced species were excluded because they occur primarily in highly disturbed habitats. Ants as a whole are very poorly represented in dry forests: only 4 percent of the analyzed species are restricted to dry forest versus 82 percent restricted to wet forest (table 5-3). In a comparison of the number of species per area (number species per 1000 square kilometers, $s = x/a^{0.25}$) between North America (north of Mexico) and Costa Rica, ants show a much more pronounced increase in species per unit area with decreasing latitude than do any other group of Hymenoptera (P. Hanson, unpublished data). Thus, it could be that ants are not depauperate in seasonally dry tropical forests, but they do not show the same latitudinal trend that they do in wet forests. It should also be noted that, unlike the other taxa, ants have been less intensively sampled in dry forests than in wet forests (J. Longino, pers. comm.).

One of the few groups of ants that has an expected proportion of species restricted to each forest type are the fungus-growing ants (Attini; table 5-3). Within this tribe, the genus *Mycetosoritis* appears to be largely absent from most of the wet tropical forests, represented by just one or two species (*M. hartmanni/vinsoni*) from northwestern Costa Rica to Texas and by three species in southern Brazil and Argentina. In the subfamily Formicinae, the only genus having species restricted to dry forests is *Camponotus* (7 of 65, or 11 percent), including *Camponotus conspicuus zonatus*, which

TABLE 5-3. Relative percentage of ant (Hymenoptera: Formicidae) taxa in dry versus wet forests

	Restricted to dry forests (%)	*Restricted to wet forests (%)*	*Present in both (%)*
Attini (*n* = 44)	9.1	75.0	15.9
Formicinae (*n* = 100)	6.0	77.0	17.0
Various tribes (*n* = 285)	4.6	84.2	11.2
Cephalotini (*n* = 28)	3.6	67.9	28.5
Dacetini (*n* = 73)	2.7	83.6	13.7
Poneromorphs (*n* = 104)	1.9	79.8	18.3
Ecitoninae (*n* = 40)	0.0	85.0	15.0
Basicerotini (*n* = 26)	0.0	92.3	7.7
Total (*n* = 700)	**4.0**	**81.6**	**14.4**

Myrmicinae is divided into Attini, Cephalotini, Dacetini, Basicerotini, and "various tribes" (other tribes).

appears to have spread, within historical times, to the Costa Rican Central Valley where it often nests in electrical appliances and is a serious pest. Among the groups that are poorly represented in dry forests are Cephalotini (arboreal ants that feed on nectar), Dacetini (nest in soil and rotting wood; predators with specialized mandibles), poneromorphs (soil nesting, predatory ants), Ecitoninae (army ants, which prey primarily on other ants), and Basicerotini (slow moving, generalist predators of the leaf litter). The last two groups do not have a single species restricted to dry forests.

I have been unable to locate studies comparing the biomass of ants between dry and wet lowland forests. In the dry forests of the Argentinean Chaco, an abundance of leaf litter and terrestrial bromeliads promotes a high density and high species richness of ants (Theunis et al. 2005). Seed-harvesting ants in the genus *Pogonomyrmex* are an ecologically significant group of ants in arid regions of North America and Argentina, but they are absent from the area between southern Guatemala and Colombia (Hanson and Gauld 2006). Among the genera excluded from the analysis, there are at least a couple of dry forest specialists, *Pseudomyrmex cretus* and *Azteca coeruleipennis*, the latter being a (mutualistic) *Cecropia* ant (J. Longino, pers. comm.).

Bees

The data available for bees are not sufficient to make comparisons of the type made in the other taxa included in this study. However, Frank Parker carried out intensive collecting of bees at one dry forest site (Cañas) and a nearby wet forest site (Bijagua). These samples were identified by Terry Griswold, and a brief summary of the results was presented in Griswold et al. (1995). The wet site had 247 species and the dry site 201 species, which is due primarily to the greater species richness of sweat bees (Halicitidae), orchid bees (Euglossini), and stingless bees (Meliponini) at the wet site. Orchid bees appear to make regular flights between the wet forest and dry forest, since there are very few orchid species available for male bees in the dry forest (Janzen et al. 1982). Analyzing the collection data of 49 species of stingless bees reveals that none are restricted to the dry forest; 30 are restricted to wet forest, and 19 occur in both types of forest. These results agree with previous findings that there is a shift from dominance of solitary species in dry forests to eusocial species (i.e., stingless bees) in lowland wet forests.

Bee genera that are more diverse in the dry forest site than in the wet forest site include *Coelioxys* (Megachilidae; 21 versus 8 species) and *Centris* (Apidae; 11 versus 3). *Coelioxys* are cleptoparasites of other bees, while *Centris* harvest floral oils. The latter have been the subject of numerous studies (e.g., Frankie et al. 1998), and oil-collecting bees in general appear to be an important component of nonseasonally dry tropical forests. In one well-studied Costa Rican dry forest site (Bagaces), large bees (12 millimeters or more in length) such as *Centris*, *Epicharis*, *Mesocheira*, and *Mesoplia* (all Apinae) appear to be the predominant pollinators of about one-quarter of the tree species, with small bees the predominant pollinators of another quarter. In plants other than trees, large bees are the predominant pollinators of only 12.5 percent of the species, and small bees 64 percent. Large-bee flowers bloom mostly during the dry season, which is when these bees breed. No such seasonal trends are evident among the small-bee flowers (Frankie et al. 2004).

Flies

Three families of flies—Asilidae ($n = 40$), Tabanidae ($n = 128$), and Syrphidae ($n = 194$)—were included in the analysis, and the last family was separated by subfamily. The larvae of both robber flies (Asilidae) and deer flies (Tabanidae) are generally soil-inhabiting predators. The latter tend to be more associated with wet soil, and thus it is not surprising that more species of Asilidae are restricted to dry forests (17.5 percent) than are Tabanidae (10.9 percent). Among hover flies (Syrphidae), 20.0 percent of Microdoninae, 10.7 percent of Eristalinae, and 1.3 percent of Syrphinae are restricted to dry forests. Microdoninae are associated with ant nests in the larval stage, and Eristalinae tend to be saprophagous, whereas many Syrphinae are predators of aphids, scale insects, and so on. These differences between the three subfamilies of Syrphidae merit further research.

Moths and Butterflies

The two families of nocturnal moths, Sphingidae and Saturniidae, have been especially well studied in the dry forests of Costa Rica (e.g., Janzen 1993, 2003; Janzen and Hallwachs 2005). These studies have shown that while both families include species that spend the dry season as dormant pupae in the leaf litter, many hawk moths (Sphingidae) migrate to wet forests during the dry season. The latter finding is reflected in the results of the present study, where an unusually high proportion of hawk moth species is present in both forest types (table 5-4). Nonetheless, this family also has a relatively high proportion of species that occur only in dry forests, and this is especially notable in *Manduca* (24 percent restricted to dry forests). Hawk moths are not only prominent but also an ecologically important component of the dry forests, pollinating approximately 10 percent of the tree species (Haber and Frankie 1989).

TABLE 5-4. Relative percentage of butterfly and moth (Lepidoptera) taxa in dry versus wet forests

	Restricted to dry forests (%)	*Restricted to wet forests (%)*	*Present in both (%)*
Papilionidae (n = 39)	15.4	59.0	25.6
Sphingidae (n = 121)	13.2	35.5	51.3
Pieridae (n = 66)	9.1	65.1	25.8
Saturniidae (n = 114)	4.4	66.7	28.9
Riodinidae (n = 235)	2.5	84.7	12.8
Total (n = 575)	**6.8**	**66.8**	**26.4**

Thirty-one of the 112 species (28 percent) of Saturniidae occurring in Costa Rica breed in the dry forest (Janzen 2003). In the present study five species were found to be restricted to the dry forest. Two of these (*Automeris io*, *Citheronia lobesis*) are quite polyphagous, which suggests that their absence in wet forests is probably not due to a lack of host plants. The other three (*Adeloneivaia isara*, *Schausiella santarosensis*, *Syssphinx mexicana*) are monophagous, and their host plants are restricted to the dry forest.

Among butterflies, Papilionidae harbors the greatest proportion of species restricted to dry forests (table 5-4). Of the three subfamilies of Pieridae occurring in Costa Rica, Coliadinae has the largest proportion of species (5 of 25) restricted to dry forests: three species of *Eurema* (table 5-5), *Kricogonia lyside* (the only species in the genus), and *Phoebis agarithe*. No species of Dismorphinae occur in dry forests; five species of Pierinae are present in dry forests, but only *Itaballia demophile* is restricted to this ecosystem.

The six species of Riodinidae restricted to dry forests include the only two species of *Apodemia* present in the country (table 5-5) and the only two *Calydna*. Among the 62 species of myrmecophilous riodinids (Eurybiini and Nymphidiini) recorded from Costa Rica, only *Behemothia godmanii* is restricted to dry forests; 11 occur in both dry and wet forests, and 50 occur only in wet forests. Thus, the dry forests (at least in Costa Rica) have relatively few myrmecophilous riodinids.

Nymphalidae was not included in the analysis, but it is clear that the family as a whole is not very diverse in dry forests (DeVries 1987). No species of Danaini, Ithomiini, Morphini, or Heliconiini are restricted to dry forests. Among Brassolini only *Caligo memnon* could be called a dry forest species, although it also occurs in other parts of the country. Three species of Satyrinae (*Cissia simmilis*, *Taygetis kerea*, *T. mermeria*) are restricted to dry forests, as are several other nymphalid species (table 5-5). Notable among the latter is the tribe Melitaeini, with three of eight *Chlosyne* and the only species of *Microtia* being restricted to dry forests. The dry forests have relatively few mimetic butterfly species, compared with other lowland areas in Costa Rica, which is probably due to the paucity of Ithomiini (DeVries 1987).

Many dry forest butterflies undergo population explosions at the beginning of the rainy season and then disappear during the dry season. Over half of the species move out of the dry forest, and these migrations are of three types (Haber and Stevenson 2004): migration across the mountains to lowland wet forest, migration up the mountains where some undergo reproductive diapause while others breed, and southward migration along

TABLE 5-5. Butterfly species restricted to dry forests reaching the southern limit of their distribution in northwestern Costa Rica

Family	*Species*	*Larval host plant*
Nymphalidae	*Adelpha fessonia**	*Randia* (Rubiaceae)
	Anaea aidea	*Acalypha macrostachya* (Euphorbiaceae)
	Chlosyne erodyle	Unknown
	C. melanarge	*Aphelandra deppiana* (Acanthaceae)
	Hamadryas glauconome	*Dalechampia scandens* (Euphorbiaceae)
	Memphis forreri	*Ocotea veraguensis* (Lauraceae)
	Microtia elva	Unknown
	Myscelia pattenia	Unknown
Papilionidae	*Eurytides branchus*	*Annona reticulata* (Annonaceae)
	E. epidaus	*Annona reticulata*
	E. philolaus	*Sapranthus* (Annonaceae)
	Parides montezuma	*Aristolochia* (Aristolochiaceae)
	P. photinus	*Aristolochia*
Pieridae	*Eurema boisduvaliana*	Unknown
	*E. dina**	*Picramnia* (Simaroubaceae)
Riodinidae	*Behemothia godmanii*	Unknown
	Apodemia multiplaga	Unknown
	A. walkeri	Unknown
	Melanis cephise	Unknown (probably Fabaceae)

*species that actually extend into Panama
Source: DeVries 1987

the coast (sometimes to breed in wet forests). Many do not migrate but rather spend the dry season as sexually dormant adults in local moist sites, as is the case in Pyrrhopyginae (Hesperiidae) (Burns and Janzen 2001). It is interesting to note that this group of skipper butterflies, which comprises 25 species in Costa Rica, is one of the few groups of insects that does not have species occurring in both dry and wet forests (three are restricted to the dry forest; Burns and Janzen 2001).

Conclusions

In addition to an expansion of the taxonomic and geographic coverage, it would be helpful to have phylogenies in order to better understand the origins of the dry forest fauna. For example, 4 of the 21 Costa Rican species of *Thyreodon* (Ichneumonidae: Ophioninae) are restricted to dry forests, and the phylogeny suggests that these dry forest species originated three times from wet forest species (Gauld and Janzen 2004). Beyond detecting patterns among the individual taxa lies the more difficult task of explaining the observed patterns. In a few cases there appears to be a simple explanation. For example, taxa that occur primarily at higher altitudes are poorly represented in dry forests. Host-specific insects will be restricted to dry forests or to wet forests if their hosts are found in only one of these forest types. However, the presence of hosts in both forest types does not mean that all insects associated with these hosts are present in both wet and dry forests. For example, among pyrrhopygine skipper butterflies the host plants of some strictly wet-forest species are also present in the dry forest, yet the skippers are not (Burns and Janzen 2001). Among parasitic wasps, *Barylypa broweri* (Anomaloninae) is restricted to dry forests, where it parasitizes *Aresenura armida* (Saturniidae) (Gauld 1997), but the latter also occurs in wet forests.

Certain characteristics of the dry forest itself undoubtedly affect insect diversity. The results of the present study suggest that certain insect taxa with soil-inhabiting larvae (e.g., Melalonthinae, Scarabaeinae, Asilidae) are well represented in dry forests whereas others are not (Cicindelinae, poneromorph ants, Tabanidae). The quality of dead wood, which is an important resource for many groups of insects, must certainly differ greatly between dry and wet forests. For example, Dynastinae, whose larvae inhabit rotting logs, are poorly represented in dry forests. One of the most notable differences between dry and wet forests is the reduced number of epiphytes in the former (Gentry 1995), which could affect ant diversity.

The eusocial insects analyzed here, ants and stingless bees, have very few species restricted to dry forests (stingless bees have none). The same is true for paper wasps (Vespidae: Polistinae), in which only 2 species (2.6 percent) (*Polistes canadensis* and *P. dorsalis*), from a total of 78 for which data are available, are restricted to dry forests. During the dry season some *Polistes* and *Mischocyttarus* migrate from the dry forest to higher elevations and aggregate during the nonnesting phase of their colony cycle (Hunt et al. 1999).

Some species that are more or less restricted to dry forests appear to have expanded their distributions in recent times into altered habitats in areas that were originally wet forest. This is probably true of both potential crop pests and their parasitoids, which has implications for biological control. There are certainly parasitoids that have not expanded their distributions but are nonetheless potential biological control agents in agroecosystems. If this is indeed the case, it provides an additional incentive for conserving seasonally dry tropical forests.

Acknowledgments

I would like to thank Rodolfo Dirzo for inviting me to participate in the Seasonally Dry Tropical Forest Symposium at Stanford University in December 2006. I also thank Wills Flowers, Daniel Janzen, and John Longino for allowing me to cite unpublished results.

Chapter 6

Seasonally Dry Tropical Forest Mammals: Adaptations and Seasonal Patterns

Kathryn E. Stoner and Robert M. Timm

Seasonally dry tropical forests (SDTFs)—with a mean annual temperature greater than 17 degrees Celsius, rainfall ranging from 250 to 2000 millimeters annually and highly seasonal, and an annual ratio of potential evapotranspiration to precipitation of less than 1 (Holdridge 1967)—represent a unique combination of challenges for the living biota contained within them. The harsh abiotic factors create an environment that is hot, with little water, and that generally has an extremely variable but often sparse resource base. It is likely that both intra- and interspecific competition are greater in the severe environments encountered in SDTFs. As a consequence of these abiotic and biotic factors, many species that we think of as "tropical" cannot survive in the dry forest. Despite the biological, cultural, and long-standing economic interest in dry forests, these habitats and the mammals inhabiting them remain poorly known.

SDTFs represent some 42 percent of tropical ecosystems worldwide (Murphy and Lugo 1986a), yet only 14 percent of articles published on tropical environments in the last 60 years focus on dry forests (Sánchez-Azofeifa et al. 2005). This lack of research is changing rapidly, however. Of the more than 10,000 papers published on tropical ecosystems between 1998 and 2008, some 26 percent focus on dry forests (ISI Web of Knowledge, 1998–2008), but only 4 percent of these involve studies of mammals

(n = 103). Recent research documents the great diversity and high number of endemic mammal species in SDTF (Ceballos and Navarro 1991; Ceballos and Garcia 1995; Stoner and Timm 2004; Ceballos and Oliva 2005; Timm and McClearn 2007; Timm et al. 2009).

A number of studies focus on adaptations of particular mammal species to seasonality in different environments (Fleming 1971b, 1977, 1988; Fleming et al. 1993); however, only a few papers attempt to review this theme at a broader level. Fautin (1946), in his classic study of the western Utah deserts, described the mammal community in this harsh environment, detailing adaptations that allow different species to persist in this strongly seasonal habitat. In reviewing the mammalian fauna in the SDTF along the Jalisco coast in western Mexico, Ceballos (1990) and Ceballos and Miranda (1986, 2000) present detailed ecological data, including adaptations for coping with seasonality. Similarly, the mammalian fauna and the ecology and adaptations for dealing with seasonality for dry forests in northwestern Costa Rica are summarized by Stoner and Timm (2004). In recent years, a wealth of studies published in widely scattered sources shed considerable light on the ecology of mammals in seasonally dry forests.

Herein, we review and synthesize the information that exists about the mechanisms of how mammals cope with and survive in SDTFs. Our emphasis is the Neotropics, but we also provide examples from other regions, focusing on unique adaptations to seasonally dry environments. We begin with a summary of the diverse array of physiological adaptations exhibited by mammals living in SDTFs. We then review behavioral adaptations utilized by mammals to cope with this harsh environment. We first describe the adaptation and then provide specific examples. We close each section by pointing out gaps in our knowledge and avenues for further research. We conclude with a discussion of the most serious threats to mammals found in SDTFs—forest destruction and fragmentation and the associated consequences of this threat to SDTF ecosystems.

Adaptations for Dealing with SDTF Environments

The mechanisms by which animals deal with extreme environmental conditions were perhaps best summarized by Fautin (1946, 295) in his classic study of the western Utah deserts:

> The survival value of any organism is dependent on its physiological and morphological adaptation to the environmental conditions to

which it is subjected or to its ability to escape those conditions not favorable to its existence.

The primary challenges that mammals face in SDTF may be broadly divided into two basic groups: (1) physiological adaptations that allow animals not just to endure but to thrive in the environment and (2) behavioral adaptations that allow animals to seasonally escape the environment. Nevertheless, it is important to note that some behavioral adaptations (e.g., changing diet seasonally) require physiological adjustments, and thus behavioral and physiological adaptations are often intimately linked.

Physiological Adaptations

SDTF environments generally are characterized by high ambient temperatures and seasonal restrictions on the availability of water and food resources. Mammals living in these ecosystems must have adaptations not just to survive but also to reproduce. Several physiological adaptations to enhance survival and reproduction have been identified in mammals inhabiting dry forests that are similar to those found in desert environments, including daily changes in body temperature, seasonal torpor or hibernation, water conservation, and delayed reproduction.

Changes in Body Temperature

Body temperature (T_b) is the balance between heat production and heat dissipation and is relatively constant for an individual in most environments (Aschoff 1982). Nevertheless, T_b may undergo adjustments during the daily rhythm and be maintained at a lower average in hot environments, providing reduced energy expenditure and reduced water loss (Dawson 1955).

Lower body temperatures have been reported for several mammals inhabiting hot, dry climates, including marsupials (Lovegrove et al. 1999; Geiser et al. 2003; Warnecke et al. 2008), bats (Audet and Thomas 1997; Cruz-Neto and Abe 1997), primates (Fietz and Ganzhorn 1999; Schmid 2000), and rodents (McNab and Morrison 1963), and it is to be expected in armadillos. Day length may function as an anticipatory cue allowing the gray mouse lemur (*Microcebus murinus*; Cheirogaleidae) a mechanism to perceive upcoming environmental changes and adapt to them (Perret et al. 1998). More detailed studies with other mammals are needed to determine which and to what extent environmental changes (e.g., day length, precipitation, temperature) may function as cues for adjusting T_b during the most critical times of year.

Torpor and Hibernation

Torpor is perhaps best defined as a state of reduced physiological activity usually accompanied by a reduction in body temperature, metabolic rate, respiration, and heart rate. Torpor that is short-term, for either part of a day or a few to several days, is referred to as daily torpor. Long-term winter torpor is referred to as hibernation and long-term summer torpor as estivation. Torpor and hibernation/estivation are best viewed as a continuum rather than discrete categories. Although torpor is most often described as an adaptive strategy to deal with cold weather (Lyman et al. 1982), whether of short duration (e.g., nighttime torpor) or long duration (hibernation), it is used by some mammals to deal with food shortage or large daily temperature fluctuations in seasonal environments, especially seasonally dry environments (Coburn and Geiser 1998; Kelm and Helversen 2007).

Diurnal torpor in tropical regions has been reported in blossom bats (Pteropodidae), leaf-nosed bats (Phyllostomidae), mouse lemurs (*Microcebus*), and dwarf lemurs (*Cheirogaleus*; Cheirogaleidae) (Audet and Thomas 1997; Bonaccorso and McNab 1997; Coburn and Geiser 1998; Schmid 2000; Dausmann et al. 2004). In the SDTF in northwestern Costa Rica, Audet and Thomas (1997) demonstrated that the short-tailed fruit bat (*Carollia perspicillata*; Phyllostomidae) and the little yellow-shouldered bat (*Sturnira lilium*; Phyllostomidae) frequently enter torpor while resting, depending upon their foraging success and body mass. Similarly, Kelm and Helversen (2007) document, under experimental conditions, that the long-tongued bat (*Glossophaga soricina*; Phyllostomidae) enters torpor under conditions of food restriction.

The most extreme cases of torpor in mammals are found in the species that hibernate. The best example of hibernation in SDTF is the Madagascar fat-tailed dwarf lemur (*Cheirogaleus medius*). Dwarf lemurs undergo true hibernation for nearly 6 months each year during the hot, dry season (Dausmann et al. 2004, 2005). Prior to hibernation, dwarf lemurs reduce their activity level and consume fruits high in sugars, resulting in an almost doubling of their body mass before entering hibernation (Fietz and Ganzhorn 1999), and as the common name suggests, fat is stored in the tail (incrassation). It is not clear why this phenomenon is unique to this one primate species in Madagascar, given that a few to several species of primates occur in most dry forests, but we suspect that the small size of the dwarf lemur (130 grams), combined with the extreme seasonality of forests in which it is found (8-month dry season; average annual precipitation 800 millimeters), drove the evolution of this unique adaptation. This pattern of hibernation/estivation has not been described elsewhere for any Neotropical mammal in seasonally dry forests, but it might be a strategy used. The small South American arid-adapted, fat-tailed mouse opossums of the genus *Thylamys* (Didelphidae) may undergo a daily (diurnal) torpor, and they clearly store fat in the tail (Creighton and Gardner 2007). This species should be further investigated to determine whether hibernation/estivation occurs.

Most studies documenting torpor in SDTF mammals are recent. It is important to note, however, that many small-bodied mammals, including mouse opossums (*Marmosa*, Didelphidae) and pygmy skunks (*Spilogale*, Mephitidae), have not yet been adequately studied to determine whether they undergo torpor or fat storage. The possibility that torpor or even more-extended hibernation exists in other Neotropical mammals should not be ruled out until additional studies are undertaken.

Water Conservation

Given the high temperatures and extreme seasonality of rainfall found in tropical seasonal forests, water conservation is crucial to survival. Water conservation in hot, dry environments may be obtained through a variety of physiological mechanisms, including more-efficient kidney function, production of dry feces, concentrated urine, low evaporative water loss, and behavior. Although these adaptations exist in many mammal groups, they have been best studied in rodents. Some combination of these mechanisms is used by most heteromyid rodents that inhabit dry environments in the southwestern United States and western Mexico (Schmidt-Nielsen et al. 1948; Schmidt-Nielsen and Schmidt-Nielsen 1950). The extent to which they are developed is quite variable among species, however. Tracy and Walsberg (2001) suggested that evaporative water loss is the single most important trait associated with conserving water in Merriam's kangaroo rat (*Dipodomys merriami*; Heteromyidae). Fleming (1977) documented that the dry forest Salvin's spiny pocket mouse (*Liomys salvini*; Heteromyidae) tolerates greater weight and water reduction than the closely related wet forest spiny pocket mouse (*Heteromys desmarestianus*; Heteromyidae) and attributed these abilities to the fact that *Liomys* is found in a strongly seasonal environment. Hudson and Rummel (1966) reported that the ability to occupy a burrow is the most critical adaptation for maintaining total water balance in *Liomys*. The importance of these mechanisms for water conservation in dry forest mammals has been inadequately explored; these and additional mechanisms almost surely will be found when they are more thoroughly studied.

Reproductive Delay

Physiological adaptations have been reported in SDTF mammals that enhance reproductive success in these harsh environments. Reproductive delays allowing parturition to occur when food resources are greatest have been suggested as an important physiological adaptation to deal with seasonality (Sandell 1990). Delayed reproduction may be divided in general into three main mechanisms: (1) delayed fertilization combined with sperm storage, (2) obligate delayed implantation, and (3) delayed development (Racey and Entwistle 2000). Delayed reproduction has been documented for more than 50 species of mammals representing 10 families (Sandell 1990); of these, 5 families—leaf-nosed bats (Phyllostomidae), flying foxes (Pteropodidae), armadillos (Dasypodidae), weasels (Mustelidae), and skunks (Mephitidae)—contain species that regularly inhabit SDTFs. With the exception of temperate bats, in which experimental work indicates that sperm storage may be for nearly 200 days (Racey 1982), sperm storage is not common in mammals and rarely exceeds a few days in length (Birkhead and Møller 1993; van der Merwe and Stirnemann 2007). Delayed implantation and delayed development, on the other hand, are better documented and may be widespread in SDTF mammals.

Reproductive delays in bats were first documented in north-temperate species that cope with short summers and long, harsh winters requiring the natal and lactation periods to coincide with the flush of insects appearing in the late spring and early summer. Having the period of mate selection and copulation, potentially expensive behaviors, coincide in late summer when food resources (and fat levels) are the highest may be critical to the reproductive cycle. Reproductive strategies in bats include sperm storage by the female after copulation, accompanied by the female's providing nourishment to the sperm, delayed implantation of the embryo, and delayed embryonic development (embryonic diapause). Reproductive delays are now documented in tropical bats inhabiting regions with long dry seasons (Bernard and Cumming 1997). Delayed embryonic development has been found in two of the most common Neotropical dry forest bats, the short-tailed fruit bat and the common fruit bat (*Artibeus jamaicensis*; Phyllostomidae). Gestation in short-tailed fruit bats is approximately 4 months if delayed development is not employed, whereas gestation is 9 months when delayed embryonic development occurs (Fleming 1971a). Similarly, in the common fruit bat gestation is 113 to 119 days in nondelayed development, whereas it is 160 to 229 days when delayed development occurs (Wilson et al. 1991; Rasweiler and Badwaik 1997). In the seasonal forest of Barro Colorado Island, Panama, common fruit bats alternate episodes of normal embryonic development with delayed development, a pattern seen in no other bat (Wilson et al. 1991). In addition to delayed embryonic development, delayed implantation has been documented in the Paleotropic fruit bat *Eidolon helvum* (Pteropodidae) (Mutere 1967; Fayenuwo and Halstead 1974). All of the above-mentioned bat species are common in SDTFs, but they are not exclusive to this habitat. Very few species of bats in either the temperate or tropical regions breed throughout the year. The common vampire (*Desmodus rotundus*; Phyllostomidae) is one of the few species that is known to breed year-round, and this includes populations that occur in seasonally dry forests. Vampires that are located close to domestic livestock have essentially an unlimited food resource in the ready availability of blood and are thus able to breed irrespective of the climate.

The existence of reproductive delays in most mammals inhabiting SDTF has not been evaluated, and this phenomenon is undoubtedly more common than it appears. The proximate mechanisms responsible for the pattern of delayed reproduction observed are unknown; nevertheless, it is likely that seasonality of food and water resources may be important pressures affecting the evolution of these traits (Sandell 1990). The physiological mechanisms that control the processes of delayed development in mammals are very poorly known (Mead 1993) and represent an important avenue for future research.

Behavioral Adaptations

In addition to physiological adaptations, mammals that inhabit SDTF display a variety of behavioral adaptations that allow them to survive in these harsh environments. These adaptations allow animals to obtain sufficient resources, cope with the hot, dry environment, and reduce competition. Examples of these include changing diets, long-distance migration, short-distance migration, local movements, timing of activity and/or foraging, and seasonality of reproduction (Stoner and Timm 2004). As competition is frequently more intense when fewer resources are available, animals in SDTF must cope with this as well.

Dietary Flexibility

One of the most important environmental factors affecting the distribution and abundance of mammals is the availability of food (Peres 1997; Mendes Pontes 1999). The seasonality in rainfall, which is characteristic of SDTFs, results in considerable variation of resources at all trophic levels (vegetative and reproductive plant material, insects/invertebrates, and vertebrates) throughout the year (van Schaik et al. 1993). As a consequence, most mammals in SDTF are either dietary generalists that adapt and change diet as necessary or dietary specialists that move in and out of seasonally dry forests when resources are abundant or scarce (Stoner and Timm 2004).

The majority of seasonally dry forest mammals appear to be resident dietary generalists. Carnivores are one of the best examples of generalists in these habitats, and several studies document considerable variation in diet between seasons. The pygmy skunk (*Spilogale pygmaea*), for example—an endemic, threatened species found in seasonally dry forests in western Mexico—shows little overlap in diet between rainy and dry seasons. Pygmy skunks, which principally forage on invertebrates, incorporate a wider array of food items in their diet during the dry season (Cantú-Salazar et al. 2005). Similarly, the coyote (*Canis latrans*; Canidae), a generalist carnivore, shows considerable variation in diet between seasons both within and between sites (Vaughan and Rodríguez 1986; Hidalgo-Mihart et al. 2001). Other carnivorous/omnivorous species showing great dietary flexibility, as well as differences between seasons, include coatis (*Nasua*; Procyonidae: Valenzuela 1998; Alves-Costa et al. 2004), raccoons (*Procyon lotor*; Procyonidae: McFadden et al. 2006), and maned wolves (*Chrysocyon brachyurus*; Canidae: Bueno and Motta-Junior 2006).

Many species of frugivorous and nectarivorous bats exhibit considerable dietary flexibility in SDTFs. For example, Heithaus et al. (1975) show that several omnivorous and frugivorous species in the seasonally dry forest of Costa Rica adopt a diet almost exclusively of nectar and pollen during the peak of the dry season, explaining this phenomenon as an adaptation, not only to resource availability, but also as a way to increase water intake (i.e., from liquid nectar) during the driest time of year. Tschapka et al. (2008) document that the diet of the SDTF endemic long-nosed bat (*Musonycteris harrisoni*; Phyllostomidae) varies significantly between the dry and wet seasons along the Pacific coast of Colima, depending upon resource availability. Dietary flexibility of bats is determined not only by resource availability but also by physiological requirements and limitations. The nectarivorous bat *Glossophaga soricina*, for example, meets its protein requirements by ingesting insects during most of the year in the seasonally dry forest of Chamela, western Mexico (Herrera et al. 2001). When *G. soricina* concentrates on plant parts, it obtains nitrogen from pollen (Herrera and Martínez del Río 1998). Other bats, such as the frugivorous *Carollia perspicillata*, have low metabolic requirements allowing them to survive on a low-nitrogen diet (Delorme and Thomas 1996), and they also consume insects at a rate that is more prevalent than was previously understood for frugivorous species based on studies of nitrogen isotope analyses of fur (York 2007).

Similar to bats, several rodents show considerable dietary flexibility in seasonally dry forests related to phenological changes and resource availability, with some showing significantly different gut morphology related to adaptations to particular diets. Seasonal differences in the size of digestive organs in caviomorph rodents in arid and semiarid habitats in Argentina occur as a consequence of variation in diet quality over seasons (Sassi et al. 2007). Similarly, an African savanna rodent community shows a gradient in gut structure allowing more granivory in some species than others and thus more dietary flexibility in seasonal environments (Kinahan and Pillay 2008). In sum, there is a complex interaction between behavioral and physiological adaptations that allows mammals in seasonal environments to cope with a changing resource base.

Several primate species found in SDTF also have flexible generalist diets. Capuchin monkeys (*Cebus*; Cebidae), for example, switch their diet to include more seeds and/or invertebrates during the dry season (Chapman 1987; Brown and Zunino 1990; Galetti and Pedroni 1994). Similarly, squirrel monkeys (*Saimiri*; Cebidae) consume more insects, flowers, and exudates during the dry season when less fruit is available (Stone 2007). Although howler monkeys (*Alouatta*; Atelidae) frequently are described as having one of the most flexible diets among Neotropical primates (Glander 1978, 1983), they are absent (Miranda 2002) or only found in low densities in the most extreme SDTF (Moura 2007). In spite of their apparent dietary flexibility, howler monkeys are quite selective in foraging on specific plants and ages of leaves, both daily and seasonally, in order to maximize total protein and essential amino acids and to minimize secondary plant compounds and fiber (Glander 1981, 1983).

Although much information exists about dietary flexibility of some mammals in seasonally dry forest, there is still a paucity of information for most species. For example, published dietary information exists for the pygmy skunk from only one locality within its distribution in west central Mexico. Similarly, little quantitative information documents seasonal variation in diets of many phyllostomid bats, and the majority of studies focus on only a few of the most abundant frugivores and nectarivores. The extent of dietary flexibility in many small mammals in seasonally dry forest is still unknown, as basic diet information is just being documented for many species (Ceballos 1990). Finally, even in species that have dietary changes over seasons, how these may be affecting the physiology of the organisms has only been addressed for a few species of bats and rodents.

Local Movements and Long- and Short-Distance Migrations

Dietary specialists that consume a narrow range of items in their diet often cope with seasonally dry forests by shifting habitats within the forest or by migrating out of SDTFs. Spider monkeys (*Ateles*; Atelidae) appear to move seasonally, tracking availability of resources within seasonally dry forests as ripe fruits become less abundant (Stoner and Timm 2004). This highly frugivorous species does not switch its diet, as many SDTF mammals do, but rather invests more time in procuring ripe fruit resources (Wallace 2005). The abundance of many frugivorous and nectarivorous bats also varies seasonally within SDTF, suggesting that they move to different habitats within or adjacent to SDTF that provide more resources (Herrera 1997; Timm and LaVal 2000; Stoner 2001, 2002; Stoner et al. 2002; LaVal 2004a; Stoner and Timm 2004; Timm and McClearn 2007; Tschapka et al. 2008).

Local movements frequently involve the exploitation of riparian habitats within seasonally dry forests or adjacent higher altitudinal areas that contain more resources. For example, in the SDTF of Brazil's Caatinga region, howler monkeys are more common in canyons than cliff habitats during the dry season, probably because more tree species retain their leaves in this more humid environment (Moura 2007). In seasonal environments, both collared peccaries (*Pecari tajacu*; Tayassuidae) and white-lipped peccaries (*Tayassu pecari*; Tayassuidae) travel great distances, tracking resources and exploiting specific microhabitats that contain more food (Mandujano 1999; Stoner and Timm 2004; Mendes Pontes and Chivers 2007).

Long-distance migrations have been well documented for populations of the nectarivorous lesser long-nosed bat (*Leptonycteris curasoae*; Phyllostomidae). Each spring, long-nosed bats migrate from Mexico to the southwestern United States where they congregate in extremely large maternity colonies. The young are born in caves that have high humidity and a stable thermal environment; after the young are fully volant, the entire colony returns to Mexico, usually in August (Hayward and Cockrum 1971). This migration follows closely the availability of several species of cactus and agave flowers that constitute critical food resources (Howell 1979; Cockrum 1991; Fleming et al. 1993). In seasonally dry forests along the western coast of Mexico, the lowest density of *L. curasoae* coincides with the greatest scarcity of chiropterophilic resources (Stoner et al. 2003). Rojas-Martínez et al. (1999) show that only populations north of 30 degrees follow this migration pattern, whereas the more southern populations consist of year-round residents. They attribute this pattern to the greater availability of resources year-round for the more southern resident populations. A similar pattern is observed in the insectivorous Mexican free-tailed bat (*Tadarida brasiliensis*; Molossidae). Free-tailed bats migrate in the spring from Mexico to the southwestern United States, form maternity colonies, and return southward in October (Cockrum 1969). The migrations are thought to be closely linked to the seasonal availability of preferred food items (Lee and McCracken 2005).

Seasonal migrations of nectarivorous bats of the arid United States and western Mexico are seen only in the larger species, whereas the smallest nectarivorous bats are permanent residents. Thus, body size, and presumably the ability to store energy during migration movements, may be a critical factor in allowing some species to migrate out of an area (von Helversen and Winter 2003). Conversely, or perhaps concomitantly, the evolution of smaller body size in nectarivorous bats might be an adaptation allowing these species to occupy areas that experience seasonal food shortages by reducing caloric requirements.

Several conclusions about mammal movements within seasonally dry forests are made by indirect evidence (e.g., absence or lower density during part of year). The true extent to which local movements to adjacent habitats occur for many mammalian species inhabiting SDTF is unknown. More detailed studies involving both telemetry data and indices of resource availability for particular species would be useful in evaluating the extent and frequency of mammal movements to surrounding habitats.

Activity Patterns

Adjusting activity patterns, in terms of both distance covered during the day and timing of activity, also represent an important behavioral adaptation allowing mammals to adjust to SDTF conditions. Some mammals increase their home range during the dry season when food is scarce, to increase the search area for food, whereas others reduce their ranging patterns, perhaps to conserve energy. During the dry season, home ranges in white-nosed coatis are twice as large as during the rainy season, and their daily path is longer (Valenzuela and Ceballos 2000), presumably because they need to forage farther to obtain sufficient resources. Similarly, white-lipped peccaries expand their home range and foraging area when resources are scarce in tropical humid areas (Fragoso 1998), and they likely follow a similar pattern in areas of seasonally dry forests (Stoner and Timm 2004).

Large carnivores also show seasonal variation in their ranging patterns; however, they show a different pattern than the midsized mammals. Pumas (*Puma concolor*; Felidae) and jaguars (*Panthera onca*; Felidae) have significantly larger home ranges during the rainy season in SDTF along the coast of Jalisco than during the dry season (Núñez Pérez 2006). Presumably this occurs because prey is concentrated in higher densities around water sources during the dry season, whereas prey is more dispersed and found in lower densities during the rainy season. This same pattern has been described for leopard cats (*Prionailurus bengalensis*; Felidae) in dry forests in Thailand (Rabinowitz 1990).

In contrast, some primates utilize the strategy of energy conservation rather than foraging greater distances during periods of scarce resources in the dry season. For example, Verreaux's sifakas (*Propithecus verreauxi*; Indriidae) in SDTF in Madagascar significantly reduce home range, core area, and daily path length during the dry season. Additionally, they display a daily period of inactivity during the dry season as a strategy to conserve energy (Norscia et al. 2006). Rodents also employ a strategy of foraging in smaller areas during periods of lower resources. Tomes' spiny rats (*Proechimys semispinosus*; Echimyidae), for example, in seasonally dry forests in Panama have larger and greater overlap of home ranges in the rainy season compared with the dry season (Gliwicz 1984).

Several mammals in seasonally dry forests are crepuscular or nocturnal, allowing them to escape high ambient temperatures during the day. Both jaguars and pumas are more active at night or early morning in the SDTF of Chamela (Núñez Pérez 2006). Other carnivores that are primarily nocturnal or crepuscular include civets (*Viverra*; Viverridae: Rabinowitz 1991), crab-eating foxes (*Cerdocyon*; Canidae: Vieira and Port 2007), and leopard cats (*Prionailurus bengalensis*; Felidae: Grassman et al. 2005). Several dry forest ungulates are principally nocturnal or most active in the early morning, including brocket deer (*Mazama*; Cervidae: Rivero et al. 2005) and Baird's tapir (*Tapirus bairdii*; Tapiridae: Foerster and Vaughan 2002).

Another technique for adjusting activity to cope with seasonally dry environments is to remain underground during the day, as found in many fossorial rodents, or remain underground continuously, as seen in many of the mole-rats (Bathyergidae), especially the naked mole-rat (*Heterocephalus glaber*: Burda et al. 2000) and silvery mole-rat (*Heliophobius argenteocinereus*: Galliard Šklíba et al. 2007). In spite of being encompassed in a relatively constant environment, the subterranean silvery mole-rat adjusts daily and seasonal activity based on small temperature fluctuations in the burrow, being less active in the dry season when resources are less abundant (Šklíba et al. 2007).

Seasonality of Reproduction

Most mammals, whether in tropical, temperate, or arctic regions, synchronize parturition to coincide with periods of greater resources (Clutton-Brock et al. 1989). Mammals found within seasonal tropical forests are especially sensitive to timing birth with food availability, since lactation is the most energetically costly activity of the reproductive cycle (Dall and Boyd 2004; Korine et al. 2004). In particular, many mammals that inhabit dry regions show clear patterns of birth related to seasonality and thus resource abundance. The physiological mechanisms by which mammals achieve reproductive delays are discussed above (see Reproductive Delay). Here, we present information about the seasonality of reproduction.

Most rodents that inhabit seasonal forests show strongly seasonal reproduction. In strongly seasonal forests in Panama, Tome's spiny rats give birth to young at the end of the dry season, presumably allowing lactating females and developing young to obtain more resources at the beginning of the rainy season (Gliwicz 1984). A similar pattern is observed in northwestern Costa Rica for Salvin's spiny pocket mouse, whose reproduction is seasonal, with young being born at the beginning of the rainy season (Fleming 1974; Janzen and Hallwachs 1996). Similar seasonality of reproduction has been described for rodents and small marsupials in the dry region of the Cerrado (Mares and Ernest 1995) and in the SDTF of Chamela (Ceballos 1995). Reproduction also has been linked to seasonal availability of food resources in several species of African rodents in the family Muridae that inhabit extremely seasonal forests, including Stella hylomyscus (*Hylomyscus stella*), Peters's hybomys (*Hybomys univittatus*), smokey heimyscus (*Heimyscus fumosus*), praomys (*Praomys* cf. *misonnei*), and Percival's spiny mouse (*Acomys percivali*) (Neal 1984; Nicolas and Colyn 2003).

Seasonality of reproduction in tropical bats was first documented more than 70 years ago (Baker and Baker 1936; Baker 1938) and subsequently has been confirmed by many other studies (see Wilson 1979, Racey 1982, and Racey and Entwistle 2000 for reviews). In the Paleotropics, bimodal polyestry is the most common pattern of reproduction in several families, including Emballonuridae, Hipposideridae, Megadermatidae, Nycteridae, Pteropodidae, and Rhinolophidae (Bernard and Cumming 1997; Heideman and Utzurrum 2003). Bimodal polyestry also is common in many Neotropical leaf-nosed bats (family Phyllostomidae) that inhabit seasonal tropical forests, especially frugivores and nectarivores (Fleming et al. 1972; Wilson 1979; Stoner et al. 2003). The timing of parturition is such that lactation coincides with greatest resource availability in seasonal forests (Timm and Lewis 1991; Bumrungsri et al. 2007). Although specific timing may vary between sites, usually parturition occurs at the beginning and middle of the rainy season, coinciding with greatest resource abundance for both frugivorous and nectarivorous bats (Bernard and Cumming 1997). Wilson (1979) associated the weaning of young in polyestrous species with the beginning of the rainy season, a period of maximal food abundance. Seasonal births during the beginning of the rainy season, however, have been seen in insectivorous bats in the Neotropics, including both of the families Vespertilionidae (van der Merwe and Stirnemann 2007) and Mormoopidae (Stoner et al. 2003). Bernard and Cumming (1997) suggested that weaning of young is a more important factor than the energetic costs of lactation in affecting the seasonality of reproduction in both frugivorous flying foxes and insectivorous bats in Africa, documenting that weaning is closely correlated with peaks in both fruit abundance and insect abundance.

In a comprehensive review of Neotropical primates, Di Bitetti and Janson (2000) documented that most New World monkeys studied so far show some seasonality in parturition. Nevertheless, several ecological and life history characteristics appear to affect the strength of the birth peak within species. In particular, they suggest that diet, body size, and latitude are all important factors that influence seasonality of births in Neotropical primates. Frugivores, for example, have more seasonality in births than folivores, larger primates are more seasonal than smaller ones, and primates farther from the equator show more seasonality in births than those closer to the equator. Additionally, parturition is timed to allow peak lactation (small-sized species) or weaning (large-sized species) to occur at a time of greatest resources. A recent study on captive brown capuchin monkeys (*Cebus apella*) comparing births in zoos over a latitudinal gradient concluded that brown capuchins breed seasonally, regardless of latitude and constant food provisioning (Bicca-Marques and Gomes 2005). They suggest that latitude and the associated change of photoperiod, as well as food availability, are not important factors affecting seasonality of birth for this species. The black and gold howler monkey (*Alouatta caraya*) in northern Argentina, however, shows facultative behavioral response in timing births to coincide with greatest food availability (Kowalewski and Zunino 2004). We can conclude that significant variation exists in reproductive timing both between and within species of Neotropical primates and that further research is needed to identify both the proximate and ultimate causes of this variation.

Seasonal births also have been well documented for several species of lemurs in Madagascar in seasonal tropical forests. For example, Verreaux's sifakas give birth when resources are scarce at the end of the dry season but wean when resources are greatest at the end of the rainy season (Lewis and Kappeler 2005; Randrianambinina et al. 2007). Similar patterns have been recognized in several other lemurs in seasonal environments, including the black lemur (*Eulemur macaco macaco*; Lemuridae: Bayart and Simmen 2005), red-fronted brown lemur (*Eulemur fulvus rufus*; Lemuridae: Ostner et al. 2002), Milne-Edwards's sportive lemur (*Lepilemur edwardsi*; Lemuridae: Randrianambinina et al. 2007), and gray mouse lemur (*Microcebus murinus*; Perret 1992).

Although seasonal births have been documented for many of the species described above, few studies have documented the proximate mechanisms that allow the seasonality of reproduction to occur. Future investigations should evaluate variation in timing of reproduction for species that are found in more- and less-seasonal habitats.

Forest Destruction and Fragmentation

The most serious problem confronting tropical mammals today is habitat destruction from human activities and the resulting fragmentation that occurs with human disturbance (Stoner and Timm 2004; Timm et al. 2009). Largely as a result of being the most accessible tropical habitat for human colonization (see chap. 14), SDTF has been and is continuing to be destroyed at alarming rates worldwide—the last 50 years has seen the loss of most of the world's dry forests (Stott 1990; Steininger et al. 2001; Quesada and Stoner 2004; Sánchez-Azofeifa et al. 2005 and chap. 3; Miles et al. 2006).

In addition to lowering species diversity for most mammal groups, forest destruction and fragmentation directly impact particular mammal species in several ways. First, forest destruction often results in range reductions, especially for large-bodied animals that require large expanses of continuous forest and for dietary specialists that require large foraging areas to obtain food resources. Consequently, range reductions may ultimately cause local or regional extirpation/extinction simply by eliminating essential habitat. Forest destruction also may affect mammal communities by providing the opportunity for a few generalist mammals to expand their populations in disturbed areas, thus resulting in greater population densities of particular species and sometimes increased competition both within and among species. Additionally, fragmentation and the consequent human colonization that commonly follows increase exposure of wildlife to hunting pressure (Cullen et al. 2000; Peres and Lake 2003; Reyna-Hurtado and Tanner 2005). Population dynamics and management of wildlife are critical issues in seasonal dry forests, and hunting, often illegal, continues to have a negative impact on both the mammals and forest structure (Dirzo and Miranda 1991). Examples of forest fragmentation and its impact on mammals in SDTF are discussed below.

Range Reduction and Species Extirpation/Extinction

Range reduction occurs as a consequence of habitat fragmentation for those species that have large home range requirements. Large and some midsized mammals are particularly susceptible to the negative effects of habitat fragmentation because of their habitat requirements (Redford and Robinson 1991; Dalecky et al. 2002; Ibáñez et al. 2002; Mendes Pontes 2004; da Silva and Mendes Pontes 2008). Different foraging guilds of large and midsized mammals may suffer the negative affects of range reduction in fragmented landscapes, including carnivores, herbivores, and frugivores.

Several studies have shown that large and midsized carnivores suffer from range reductions as a consequence of forest fragmentation (Chiarello 1999; Laidlaw 2000; Asquith and Mejía-Chang 2005). Home ranges of jaguars in the SDTF of Jalisco's Chamela-Cuixmala Biosphere Reserve are considerably smaller than those found in the dry regions of Parque Nacional Kaa-Iya del Gran Chaco in Bolivia (Maffeí et al. 2004; Núñez Pérez 2006). These differences are likely attributable to the greater size of the protected area in Bolivia (34,400 square kilometers) versus Chamela (13,142 square kilometers) and the greater habitat fragmentation found within the immediate area surrounding Mexico's biosphere reserve. Large carnivores, such as jaguars and pumas, are always among the first species to disappear from highly fragmented landscapes (Chiarello 1999; Laidlaw 2000; Asquith and Mejía-Chang 2005; Timm et al. 2009).

Forest fragmentation generally has a negative impact on populations of large and medium herbivores (but see discussion about elephants below) by reducing the availability of habitat and by increasing their susceptibility to hunting as fragmentation brings animals closer to human settlements (Cullen et al. 2000). Large grazing herbivores that have been particularly affected by fragmentation in dry regions in the Paleotropics include mountain gazelles (*Gazella gazella gazella*; Bovidae: Manor and Saltz 2004), Nubian ibex (*Capra nubiana*; Bovidae: Attum 2007), blue duiker (*Philantomba monticola*; Bovidae: Lawes et al. 2000), and African elephant (*Loxodonta africana*; Elephantidae: van Aarde et al. 2006). In Neotropical dry regions, several medium and large herbivores are suffering the consequences of fragmentation and the subsequent reduction in range, including tapirs (*Tapirus*), collared and white-lipped peccaries, and brocket deer (Chiarello 1999; Galetti et al. 2001; Ibáñez et al. 2002). In the seasonal forest of southeastern Brazil there is a direct relationship between fragment size and mammal species diversity—the largest species drop out of the smallest fragments first, explaining much of the lower diversity observed in small fragments (Chiarello 1999).

Large and medium terrestrial and arboreal frugivores are extremely susceptible to forest fragmentation and show smaller populations in smaller-sized fragments; these include agoutis (*Dasyprocta*; Dasyproctidae), pacas (*Cuniculus paca*; Cuniculidae), and spider monkeys (*Ateles*) (Chiarello 1999). In addition, the negative effect of forest fragmentation is exacerbated by hunting, which is frequently more common in forest fragments with adjacent human populations (Cullen et al. 2000). In a meta-analysis using an array of tropical habitats, Harcourt and Doherty (2005) show a universal negative effect of forest fragmentation on primates. Specifically, in Neotropical seasonal forests, several studies document that frugivorous primates, in particular, are not found in smaller forest fragments (Chiarello 1999, 2003). As their ranges become restricted by forest fragmentation, density is reduced, and then populations disappear within the smaller fragments.

In addition to large and medium-sized mammals, one group of smaller-sized mammals that frequently is affected by habitat fragmentation in tropical dry regions is bats. Since bats are volant and often have considerably larger home ranges compared with similarly sized terrestrial mammals, they may suffer the consequences of reduced range caused by habitat fragmentation. For example, several studies in SDTF show that bat species diversity is lower in fragmented or disturbed areas and that some species are completely absent (Stoner et al. 2002; Quesada et al. 2003, 2004; Gorresen and Willig 2004; Timm and McClearn 2007; Ávila Cabadilla et al. 2009).

Increase in Species in Disturbed Habitats

Some mammals thrive in altered environments and actually increase in abundance in fragmented areas. For example, several generalist species such as common opossums (*Didelphis*; Didelphidae), coyotes, raccoons, and coatis frequently are found in higher densities in areas of human disturbance (Hidalgo-Mihart et al. 2001; Stoner and Timm 2004). In South Africa, elephants frequently increase in density as a result of fragmentation because regions are cut off that formerly represented routes of movement between different areas. This results in unusually high densities in fragments, and elephants can cause considerable property and habitat damage, often resulting in their removal by humans (van Aarde et al. 2006).

Competition has been identified as an important element of mammal communities living in SDTFs, since seasonality characterizes limited resources during part of the year. Increased densities in fragmented habitats will result in greater competition between similar species that are already competing in this harsh environment. For example, the low density and small group size of howlers in the dry region of the Caatinga is attributed to their competition with capuchins (Zunino et al. 2001). Gray and red brocket deer (*Mazama*) achieve niche separation by segregating through space and time along Bolivia's Chaco-Cerrado border. In the dry region of Brazil's Cerrado, crab-eating foxes (*Cerdocyon thous*) and the pampas fox (*Lycalopex gymnocercus*; Canidae) forage in different habitats and have different peak activity times (Vieira and Port 2007). Niche partitioning of two bovids (*Boselaphus tragocamelus* and *Gazella bennettii*; Bovidae) and two cervids (*Axis axis* and *Cervus unicolor*; Cervidae) in dry regions of western India is achieved by selection of different habitats; the cervids prefer forest habitats, whereas the bovids are found primarily in scrub lands and are more tolerant to disturbance (Bagchi et al. 2003).

The one "dietary specialist" that has increased its abundance manyfold with the increase in disturbed habitats is the common vampire bat. *Des-*

modus rotundus, an obligate blood feeder. It is the dietary generalist of the vampires and uses an extremely wide array of mammals, birds, and even reptiles as food sources. Clearly part of the increase in abundance can be attributable to the introduction by humans of domestic livestock, a ready source of easily obtained food for vampires.

In sum, there is no doubt that habitat fragmentation and forest destruction are having severe consequences on both the structure and the composition of mammal communities in dry forest regions throughout the tropics. The implications of changes in mammal communities for regeneration and forest structure and composition are discussed below.

Consequences of Lower Mammal Diversity in SDTFs

Range reductions of particular species, lower abundance, extirpation, extinction, and increased competition in altered human environments are having severe negative consequences on both the structure and the composition of mammal communities in dry regions throughout the tropics. Given the importance of mammals in biotic interactions such as seed dispersal (Howe and Smallwood 1982; Lobova et al. 2009) and pollination (Ghazoul 2005) in tropical environments, changes in their distribution and abundance may have severe consequences for SDTF ecosystems.

Mammals are one of the most important vertebrate groups that disperse seeds in tropical forests (Jordano 2001); in particular, primates (Lambert and Chapman 2005) and bats (Lobova et al. 2009) are recognized as the most important mammalian dispersers. The Central American spider monkey (*Ateles geoffroyi*) consumes and potentially disperses as many as 40 species in some SDTFs (González-Zamora et al. 2009). LaVal (2004b) reported that bats disperse more seeds than do birds, that pioneer species of plants are especially well represented in bat-dispersed seeds, and that these seeds are dispersed primarily to disturbed areas. Bats, because of their high mobility, can disperse large numbers of seeds for considerable distances. Altered mammal communities and the subsequent change in seed dispersal may have cascading effects and ultimately change SDTF plant species structure and composition and affect the natural process of regeneration within these ecosystems (Stoner, Riba-Hernández et al. 2007; Stoner, Vulinec et al. 2007).

In addition to their seed dispersal, mammals are important in tropical ecosystems also as external vectors for moving pollen (Ghazoul 2005). Bats are undoubtedly the most important mammal group responsible for pollination in tropical forests. Bats visit and presumably pollinate more than 590 plants in the Neotropics and 160 in the Paleotropics (Winter and Helversen 2001). Pollination of chiropterophilic plants in seasonally dry forests differs in fragmented environments, which generally receive fewer bat visits and have consequent lower reproductive success (Stoner et al. 2002; Quesada, Stoner, Rosas-Guerro et al. 2003; Quesada, Stoners, Lobo et al. 2004; Arias-Cóyotl et al. 2006; Quesada, Sánchez, Azofeifa et al. 2009a). The ultimate long-term effects of this lowered reproductive success for plant community structure and composition are unknown.

Conclusions and Future Research

Dry forests originally were believed to be less diverse than wet forests, but recent research is documenting that some dry forests are as diverse as wet forests (see chap. 1 for a case of plant diversity). A wealth of excellent research has been conducted on mammals in the SDTFs, and this is especially true for the last decade. However, we must emphasize how much remains to be learned about dry forest mammals, how they function in their environment, and how critical it is to continue and expand studies as dry forests continue to disappear and the surrounding buffer habitat matrix is increasingly degraded. The dry forests and their associated floras and faunas are critically important to conserve, as their associated species have ecosystem impacts far beyond the dry forest habitats.

Mammals generally are believed to be much better known than other groups of animals, yet even the most basic question—How many species are there?—remains poorly answered. The discovery of mammal species new to science continues at an amazing rate, and in fact the rate of discovery has increased in recent years with the advent of modern systematic techniques. Some 25 to 30 species are now recognized as valid new species each year. A surprising number of these newly recognized species are primates and larger mammals (greater than 5 kilograms), groups that one might think are the best known. An analysis of recently described species suggests that the majority of newly recognized species are restricted to threatened areas of high endemism (Reeder et al. 2007).

The first level of research needed for SDTF mammals is modern systematic revisionary work to better understand species-level taxonomy and relationships of dry forest mammals. There are few preserved specimens of mammals from seasonally dry forests in the world's systematic collections, especially specimens that have preparations adequate for use with modern DNA analyses; therefore, collections need to be augmented. Additionally, older specimens that are available need to be critically reevaluated. What is a species and which species occur in seasonally dry forests are basic questions that will provide the foundation for all future research and conservation efforts.

Compared with our knowledge of rain forest mammals, our knowledge of SDTF mammals is at a rudimentary state in terms of species diversity, ecology, and adaptations. Information about physiological adaptations of dry forest mammals comes predominantly from studies in desert environments that are extrapolated to closely related species occurring in dry forests or from studies of individual species. It will be critically important in future research to conduct comparative studies of mammalian clades among a gradient of habitats such as desert environments, SDTF, tropical moist forest, and tropical rain forest. Broadscale comparative-type studies will help document the variation that exists in particular physiological and behavioral adaptations, and they will allow the distinction between those adaptations that are due to phylogenetic relationships and those that are true physiological and anatomical adaptations to particular environments. In particular, studies that focus on comparing one or only a few closely related species groups in different environments will be most useful in elucidating how important specific adaptations are for any one particular environment.

The study of the world's mammals is in an exciting phase of discovery, with new species still being described each year. Nevertheless, because of human development and land use throughout the tropics, and in seasonal forests in particular (Sánchez-Azofeifa et al. 2005, chap. 3), conservation issues faced by tropical mammals have never been more critical than the issues those species and environments face today. Although far more studies involving mammals in seasonally dry forests have been conducted in the Neotropics than in the Paleotropics, almost 80 percent of this research originated in only four countries—Mexico, Costa Rica, Brazil, and Panama. Although this may partially be explained by the distribution of seasonally dry forests (Miles et al. 2006), the mammalian faunas of several countries that contain dry forest, especially in Central America, have received little study. Future research should concentrate on including some of the lesser-studied regions that contain seasonally dry forests. Additionally, first-rate descriptive studies providing new information and critically evaluating previous studies will help clarify the role of mammals inhabiting seasonally dry forests. Well-documented, basic research is the foundation for science and the basis of intelligent conservation efforts.

Acknowledgments

We thank H. Ferreira, J.M. Lobato-García, G. Sánchez-Montoya, and A. Valencia-García for technical assistance and Gerardo Ceballos, Rodolfo Dirzo, and Hillary Young for a critical review of the manuscript. This project was supported by grants from the Consejo Nacional de Ciencia y Tecnología (grants 2005-51043 and 2006-56799), the Centro de Investigaciones de Ecosistemas, and UNAM and a UC MEXUS Visiting Scholar Fellowship to KES.

PART III

Ecosystem Processes in Seasonally Dry Tropical Forests

Chapter 7

Primary Productivity and Biogeochemistry of Seasonally Dry Tropical Forests

VÍCTOR J. JARAMILLO, ANGELINA MARTÍNEZ-YRÍZAR, AND ROBERT L. SANFORD JR.

Primary productivity in seasonally dry tropical forests (SDTFs) is controlled largely by the amount and timing of rainfall. Because water availability determines leaf production as well as photosynthesis, both interannual and within-rainy-season precipitation constrains and controls ecosystem productivity and nutrient dynamics. Subsequently, soil water availability also regulates organic matter decomposition, fine-root production, and microbial dynamics; hence the timing for and the amount of available nutrients are strongly coupled to the seasonal and annual variations in rainfall. The second important driver for SDTF productivity and nutrient dynamics is land use change (see chap. 10). In many areas secondary succession starts on abandoned agricultural fields; thus, secondary forests with different recovery times are becoming a prevalent feature in the dry tropics (Miles et al. 2006). As a result, research in secondary forests has significantly increased (Lawrence and Foster 2002; Campo and Vázquez-Yanes 2004; Urquiza-Haas et al. 2007).

Here we synthesize results related to primary productivity, biomass, and carbon and nutrient dynamics studies for primary SDTF in the Neotropics. Whenever possible we address how water availability functions as an important driver of these fundamental processes. Our review covers sites with a pronounced seasonal rainfall and 5 to 8 months of continuous drought, during which the majority of the species remain leafless, and where mean annual precipitation varies between 500 and 1500 millimeters. To highlight the extent of local environmental heterogeneity (i.e., effect of topography and soil conditions) and variation at the regional scale, sites where the dominant vegetation is defined as semi-evergreen or semideciduous tropical forest were also included for comparison. These forests are also classified as dry tropical or subtropical forests sensu Holdridge (1947).

Primary Productivity

Primary productivity measurements in SDTF are scarce. As in other temperate and tropical forests, litterfall represents the most common measure of primary productivity, but belowground productivity has been rarely estimated.

Total Net Primary Productivity

Despite the importance and growing number of primary productivity studies in terrestrial ecosystems worldwide, its direct measurement in tropical forests is limited, and only a few sites have been comprehensively investigated. Clark et al. (2001) assembled a set of current estimates of total net primary productivity (NPP) for 39 tropical forest sites throughout the world (from dry to wet, lowland to montane, nutrient-poor to nutrient-rich soils). From these sites, 23 were located in the Neotropics, but only 2 were for dry forests (Chamela, Mexico, and Guánica, Puerto Rico). Total NPP at Chamela was estimated from the sum of above- and belowground components within a small watershed at three landscape positions (70, 130, and 150 meters above sea level). The values were 13.5, 11.5, and 11.2 megagrams per hectare per year, respectively, of which about 44 percent on average was allocated belowground (Martínez-Yrízar et al. 1996). At Guánica, aboveground NPP was 9.0 megagrams per hectare per year (Clark et al. 2001), based on productivity components as reported by Murphy and Lugo (1986b). No new data on total NPP of Neotropical SDTF have been published to date.

Above- and Belowground Litter Production

One of the most important and frequently measured components of total NPP in tropical forests is aboveground fine-litter production (i.e., litterfall: leaves, flowers, fruits, seeds, and woody debris less than 1.0 centimeter in diameter; Clark et al. 2001). Since our previous review (Martínez-Yrízar 1995), only seven new studies have been published for Latin America (table 7-1). The majority of these are from Chamela, where three landscape positions along an elevation gradient within a small watershed were monitored for more than 20 years (Maass et al. 2002a). Additional litterfall information has been published from Sierra de Huautla Reserve, Morelos (Saynes et al. 2005) and the Yucatán peninsula region, in both deciduous and semi-evergreen SDTF in Mexico (Lawrence and Foster 2002; Campo and Vázquez-Yanes 2004). Annual total litterfall for deciduous SDTF in Latin America is 4.9 plus or minus 2.1 megagrams per hectare per year (mean plus or minus SD, $n = 12$) with a range of 2.9 to 8.5 megagrams per hectare per year, of which 65 to 90 percent is the leaf fraction (table 7-1). Values are higher for both semideciduous SDTF (7.0 plus or minus 1.3 megagrams per hectare per year, $n = 5$) and semi-evergreen SDTF (6.9 plus or minus 0.8 megagrams per hectare per year $n = 3$; table 7-1).

There is still little information to compare year-to-year fine-litterfall variation for Neotropical SDTF. Only two multiyear studies have been published to date. Martínez-Yrízar and Sarukhán (1990) found low and high annual leaf-fall values in successive years, independent of the variations in annual rainfall during a 5-year period at Chamela. For the semi-evergreen forests of the Yucatán peninsula, Whigham et al. (1990) reported yearly leaf-fall variation for a 4-year period but found a negative correlation between annual leaf fall and current-year precipitation.

Compared with the number of studies on aboveground litter production, belowground litter production (roots) has been a neglected aspect of productivity studies in Neotropical SDTF. The single recent study (Castellanos et al. 2001) found considerable dry season fine-root mortality at Chamela. Furthermore, the fine-root turnover rate was 3.1 per year, which is three times higher than the leaf litter turnover rate of 0.72 per year (estimated from Martínez-Yrízar 1995). Based on these values, Castellanos et al. (2001) suggest that the relative importance of belowground processes for carbon and nutrient supply to the soil in Chamela could be greater than the aboveground returns.

Seasonal and Spatial Variation in Litterfall

Pronounced seasonality in fine litterfall is a general phenological feature for all SDTF. The largest monthly leaf fall occurs at the onset of the dry season, although maximum leaf drop may be delayed 3 to 4 months

TABLE 7-1. Leaf litter and total litter production in Latin American seasonally dry tropical forests

Country/Site	*Forest type*	*Latitude*	*Longitude*	*MAT (°C)*
Mexico				
Sierra Huautla, Morelos	D	18°28'N	99°01'W	24.5
Chamela, Jalisco				
Plano	SD	19°30'N	105°03'W	24.9
Inclinado	D	"	"	"
Arroyo	SD	"	"	"
Upland	D	"	"	"
Arroyo	SD	"	"	"
Hills	D	"	"	"
W1-up	D	"	"	"
W1-middle	D	"	"	"
W1-bottom	D	"	"	"
Yucatán peninsula				
Dzibilchaltun	D	21°06'N	89°17'W	25.8
Puerto Morelos	SE	20°49'N	87°07'W	25.0
El Refugio	SE	18°49'N	89°23'W	25
Nicolás Bravo	SE	18°27'N	88°56'W	"
Arroyo Negro	SE	17°53'N	89°17'W	"
Puerto Rico				
Guánica	D	18°66'N	66°92'W	25.1
Guánica	D	"	"	"
Costa Rica				
Palo Verde	D	10°26'N	85°23'W	26.0
Venezuela				
Calabozo	D	8°48'N	–	27.0
Brazil				
Cuiabá	SD	15°42'S	56°06'W	26.8
Serra do Japi	SD	23°12'S	46°53'W	20.8

*Studies in previous review by Martínez-Yrízar 1995
Forest type: D = deciduous, SD = semideciduous, SE = semi-evergreen. MAT = mean annual temperature (°C), MAP = mean annual precipitation (mm). Litterfall is in Mg ha^{-1} yr^{-1}. Duration = period of study in years.

MAP (mm)	*Litterfall*		*Duration*	*Ref.*
	Leaf	*Total*		
851	—	4.2	1	1
735	3.6	5.1	2	2*
"	2.4	3.4	2	2*
"	—	6.9	1	3*
"	—	4.2	1	3*
"	4.8	6.6	5	4*
"	2.7	4.0	5	4*
"	2.8	3.3	10	5
"	2.6	3.2	10	5
"	3.1	4.2	10	5
760	—	8.5	4	6
1181	5.3	—	4	7
892	—	6.3	1	8
1144	—	6.6	1	8
1418	—	7.8	1	8
860	4.3	4.8	1	9*
"	2.5	2.9	1	10*
1500	—	8.2	Several	11*
1330	—	8.2	—	12*
1380	5.3	7.7	1	13
1340	5.5	8.5	1	14

Sources are 1: Saynes et al. 2005; 2: Vizcaíno 1983; 3: Ceballos 1989; 4: Martínez-Yrízar and Sarukhán 1990; 5: Martínez-Yrízar et al. 1996; 6: Campo and Vázquez-Yanes 2004; 7: Whigham et al. 1990; 8: Lawrence and Foster 2002; 9: Lugo and Murphy 1986; 10: Cintrón and Lugo 1990; 11: Gessel et al. 1980; 12: Medina and Zewler 1972; 13: Haase and Hirooka 1998; 14: Morellato 1992.

depending on soil moisture availability, landscape position, rainfall distribution, and natural disturbances (Martínez-Yrízar and Sarukhán 1990; Whigham et al. 1991; Campo and Vázquez-Yanes 2004). Leaf drop may start earlier during years with low total precipitation (Whigham et al. 1991). Litterfall seasonality also varies from site to site. For example, in semideciduous forests from Brazil and Mexico, 76 percent and 62 percent of the annual total litterfall, respectively, was collected during the dry season (Martínez-Yrízar and Sarukhán 1990; Haase and Hirooka 1998). Litterfall seasonality has been associated with the relative abundance of deciduous and semideciduous tree species (Bullock and Solís-Magallanes 1990; Morellato 1992).

Other aspects of environmental heterogeneity are important for litterfall dynamics at a local scale. For example, at Chamela, watershed topography modifies the potential annual interception of photosynthetically active radiation such that soil organic matter and soil moisture content are greater at lower- than at upper-elevation areas within the watershed (Galicia et al. 1999). Following the trend in soil moisture, total annual litterfall is 30 percent higher at the lowest elevation (Martínez-Yrízar et al. 1996), suggesting a direct relationship between forest productivity and topography, mediated by soil water availability.

Coarse Woody Debris Net Accumulation

Net annual accumulation of aboveground coarse woody debris (twigs and branches 2 or more centimeters in diameter) is a poorly quantified component of NPP for tropical forests in general. Working within a watershed at Chamela SDTF, Maass et al. (2002b) showed that annual net accumulation of coarse woody debris (2 to 20 centimeters circumference) was significantly lower (1.30 megagrams per hectare per year) at the bottom than at upper and middle elevations (2.26 and 3.26 megagrams per hectare per year, respectively). These results suggest that conditions at the lower-elevation plot may reduce the rate of transfer from live to dead biomass, probably because higher soil moisture and phosphorus availability (Rentería et al. 2005) result in greater branch longevity. Measurements should be extended beyond a single year, since causes of wood mortality (i.e., droughts, storms, and hurricanes) may vary greatly from year to year (Maass, Jaramillo et al. 2002).

Wood Productivity in Stems and Branches (Radial Increment)

Compared with Neotropical humid tropical forest (Malhi et al. 2004), stem and branch productivity are rare in SDTF. In a 10-year study of two deciduous tree species at Chamela, Bullock (1997) showed that annual girth growth was highly correlated with precipitation in the mid wet season for a period of less than 2 months and was not affected by total annual precipitation or by heavy rainfall in the dry season. At the stand level, wood increment at Chamela was 2.4 megagrams per hectare per year, which represented about 24 percent of total NPP (Martínez-Yrízar et al. 1996; A. Pérez-Jiménez, unpublished data).

Fine-Root Productivity

Except for data from Chamela, there are no studies on fine-root productivity in Neotropical SDTF. Kummerow et al. (1990) had reported a fine-root productivity of 4.2 megagrams per hectare per year in the upper 40 centimeters of soil, while Castellanos et al. (2001) measured 1.8 megagrams per hectare per year in the top 10 centimeters of soil. Taken together, these data suggest that 43 percent of fine-root productivity occurs within the uppermost 10 centimeters of the 40-centimeter profile. Interestingly, Castellanos et al. (2001) reported that 70 to 81 percent of root productivity at 0 to 10 centimeters depth was concentrated in the first 5 centimeters of soil. Also, fine-root productivity responded strongly to seasonality of rainfall at this depth, a response not evident at the 5-to-10-centimeter soil depth.

Forest Biomass

Recent biomass estimates for SDTF provide a better representation of the variability in forest structure compared with our previous review (Martínez-Yrízar 1995), although few of the studies are from outside Mexico. Nevertheless, as with belowground productivity, quantification of belowground biomass continues to be very scarce.

Aboveground Live Biomass

Martínez-Yrízar (1995) presented five studies with aboveground biomass data for Neotropical SDTF. New information has recently been published, for a total of 22 SDTF sites in three Latin American countries, from northwestern Mexico at 28 degrees N to Venezuela at 10 degrees N (table 7-2). Most of the work has been conducted in two areas of Mexico (with seven forest sites each), one located at Sonora, which is the northern limit of SDTF distribution, and the other at Chamela, the most intensively studied SDTF in the Neotropics. Aboveground live biomass in the deciduous SDTF varies from 35 to 140 megagrams per hectare, with a mean of 77.4 plus or minus 30.5 megagrams per hectare (plus or minus SD, $n = 16$; table 7-2). In contrast, aboveground biomass for the semi-evergreen SDTF in the Yucatán peninsula is on average twice as high with 159.1 plus or minus 37.5 megagrams per hectare ($n = 7$, range 125 to 225 megagrams per hectare; table 7-2). In this region of Mexico, differences in aboveground biomass among old-growth forest plots (greater than 50 years old) may be partly related not only to past human disturbances but also to successional status, which is unknown for some sites beyond 20 to 30 years of recovery (Urquiza-Haas et al. 2007). At the local level, deciduous upland and the semideciduous floodplain forests in Chamela show pronounced differences in biomass associated with topography and water availability. According to Jaramillo et al. (2003) the mean aboveground live biomass of the deciduous forest is 69.7 plus or minus 10.5 megagrams per hectare ($n = 3$), but the semideciduous forest is more than four times greater at 247 to 390 megagrams per hectare. Floodplain forests at Chamela are at the higher end of the range of biomass values for different tropical forest types within Mexican SDTF (table 7-2).

Belowground Biomass (Roots)

Information regarding total belowground biomass of Neotropical SDTF is still very limited (table 7-2). In addition to the few values included by Martínez-Yrízar (1995), there are three new studies on root biomass in Latin American SDTF. One comes from a site in Venezuela, where total root biomass (66.8 megagrams per hectare) was estimated using a root to shoot ratio value (0.48) calculated for tropical forests reported in the literature (Delaney et al. 1997). The other study is from Chamela, where mean root biomass (17 megagrams per hectare live plus dead) was measured via excavation (Jaramillo et al. 2003). This value is lower than the 31 megagrams per hectare obtained from another forest site in Chamela (Castellanos et al. 1991). This difference could be related to variation in sampling procedures between studies or simply the result of site selection, reflecting the degree of spatial variation in root biomass across the Chamela landscape (Jaramillo et al. 2003).

Not surprisingly, root biomass also varies greatly between forest types within the same region. At Chamela, for example, there was a twofold difference in belowground biomass between the upland deciduous and the floodplain semideciduous forests (17 vs. 32 megagrams per hectare; Jaramillo et al. 2003). Also, root biomass distribution within the soil profile varied between the two land cover types, so on average, 88 percent of the total root biomass was in the top 40 centimeters of the profile in the upland forest with shallower soils, compared with 85 percent in the top 60 centimeters in the deeper arroyo soils (Jaramillo et al. 2003). The high concentration of root biomass in the upper 40 centimeters of soil in Chamela is similar to Guánica, where 90 percent of root biomass was distributed within the uppermost 40 centimeters (Murphy and Lugo 1986b).

The only new data for fine-root biomass (less than 2 millimeters in diameter) in SDTF comes from two studies at Chamela. Castellanos et al. (2001) reported a total fine-root biomass (live and dead, less than 1 millimeter) in the upper 0 to 10 centimeters of soil in upland forest of 108.3 grams per square meter, with 46 percent dead roots. Fine-root density (mean biomass per centimeter of soil) was greatest in the uppermost 2 centimeters (about 50 percent of the total). The other study, by Jaramillo et al. (2003), reported a total fine-root biomass (live and dead, less than 1 millimeter) in the upper 0 to 40 centimeters of soil of 160 grams per square meter in the upland forest and 180 grams per square meter in the floodplain forest, with a significant reduction with increasing depth. The values from Chamela are lower than the global SDTF average for total fine-root biomass, which is 570 grams per square meter to 1-meter soil depth and about 239 grams per square meter in the upper 30 centimeters of soil (Jackson et al. 1997). For mature semi-evergreen forests in southeastern Mexico, fine roots (diameter less than 3 millimeters) varied from 4.1 plus or minus 0.6 to 6.4 plus or minus 0.6 SE megagrams per hectare (Vargas et al. 2008), which is closer to the global average value (Jackson et al. 1997).

Aboveground Dead Phytomass

The quantification of above- and belowground dead phytomass (necromass) has been a neglected aspect of productivity studies in the tropics (Maass, Jaramillo et al. 2002; Baker et al. 2007). Maass, Jaramillo et al. (2002) measured, through direct harvest, a total aboveground necromass of

TABLE 7-2. Phytomass estimates for Latin American seasonally dry tropical forests

Country/Site	*Type*	*Lat. N*	*Long. W*	*MAT*
Mexico				
San Javier, Sonora				
San Juan 2	D	28°36'	109°44'	18
Alamos, Sonora				
Vara Blanca	D	27°01'	108°56'	24.3
Tempisque	D	27°01'	108°56'	"
La Luna	D	27°01'	108°56'	"
Isleta	D	26°54'	108°54'	"
Sahuira	D	26°58'	108°52'	"
Cieneguilla	D	26°55'	108°54'	"
Chamela, Jalisco				
Biol. Reserve	D	19°30'	105°03'	24.9
Biol. Reserve	D	"	"	"
Baja plot	D	"	"	"
Alta plot	D	"	"	"
Ridgeline	D	"	"	"
Middle slope	D	"	"	"
Lower slope	D	"	"	"
Garrapata	SD	"	"	"
Buho	SD	"	"	"
Yucatán peninsula				
El Edén, Q. Roo	SE	21°12'	87°11'	24.2
Q. Roo & Yucatán	SE	18° to 21°	89° to 87°	26.0
La Pantera, Q. Roo	SE	19°07'	—	—
"	"	"	"	"
El Refugio	SE	18°49'	89°23'	25
Nicolás Bravo	SE	18°27'	88°56'	25
Arroyo Negro	SE	17°53'	89°17'	25
Puerto Rico				
Guánica	D	17°57'	62°52'	25.1
Sites 16, 17, 18	SE	18°19'	65°49'	27.5
Venezuela				
Cerro El Coco	D	10°	66°	27

Method for calculation: [a]allometric equations, [b]harvest, [c]root:shoot ratio, [d]forests ≥ 50–60 years old.
Forest type: D = deciduous, SD = semideciduous, SE = semi-evergreen. MAT = mean annual temperature (°C), MAP = mean annual precipitation (mm), BA = basal area (m^2 ha^{-1}), AGB = live aboveground biomass (Mg ha^{-1}, dry weight), Dbh = minimum stem diameter at 1.3 m included in measurements, BGB = total (live + dead) belowground biomass (Mg ha^{-1}, dry weight), n.d. = not determined.

MAP	*BA*	*AGB*	*Dbh*	*BGB*	*Ref.*
760	28.1	81.5[a]	≥3.0	n.d.	1
712	20.7	60.3[a]	≥3.0	n.d.	1
"	15.4	44.8[a]	"	n.d.	1
"	40.4	117.5[a]	"	n.d.	1
"	16	46.6[a]	≥2.0	n.d.	2
"	11.9	34.8[a]	"	n.d.	2
"	25	72.6[a]	"	n.d.	2
735	25.6	85[b]	all	35.7[c]	3*
"	—	73.6[a]	"	30.9[b]	4*
"	—	105.6[a, b]	all	—	5
"	—	121.4[a, b]	"	—	5
"	20.8	59.9[a, b]	all	n.d.	6
"	22.5	68.5[a, b]	"	18.0[b]	6
"	27.1	80.7[a, b]	"	16.0[b]	6
"	34.7	390.2[a, b]	"	32.0[b]	6
"	29.4	247.5[a, b]	"	32.0[b]	6
1650	—	130.9[a, b]	all	26.1[b]	7
≥1000	30.4	191.9[a]	≥10.0	n.d.	8
1200	31.3	225[b]	≥1.0	n.d.	9
"	"	191.5[b]	≥10.0	"	"
892	—	128[a, b]	≥1.0	n.d.	10[d]
1144	—	150[a, b]	≥1.0	n.d.	10[d]
1418	—	125[a, b]	≥1.0	n.d.	10[d]
860	19.8	44.9[b]	≥2.5	45[b]	11[d]*
1262	26	163[a]	≥10.0	n.d.	12[d]
800	—	140[a]	≥10.0	66.8[c]	13

Sources are 1: Martínez-Yrízar et al. 2000; 2: Álvarez-Yépiz et al. 2008; 3: Martínez-Yrízar et al. 1992; 4: Castellanos et al. 1991; 5: Kauffman et al. 2003; 6: Jaramillo et al. 2003; 7: Vargas et al. 2008; 8: Urquiza-Haas et al. 2007; 9: Cairns et al. 2003; 10: Lawrence and Foster 2002; 11: Murphy and Lugo 1986b; 12: Gould et al. 2006; 13: Delaney et al. 1997.
*Studies in previous review by Martínez-Yrízar 1995.

27.2 megagrams per hectare at Chamela; 71 percent of this was deadwood either hanging or attached to live trees, while the rest was surface litter. More recently, Jaramillo et al. (2003), using deadwood diameter at breast height measurements and the planar intercept technique, measured aboveground necromass (downed wood and standing) by land cover type in Chamela and reported 42.5 plus or minus 4.3 megagrams per hectare (or about 38 percent of total aboveground; mean and SE) and 56.0 plus or minus 5.4 megagrams per hectare (or about 14 percent of total aboveground) in upland and floodplain forests, respectively. Necromass in Chamela deciduous forests is higher than the 4.8 megagrams per hectare reported for SDTF in northern Venezuela (Delaney et al. 1998). For semi-evergreen forests of the Yucatán, aboveground necromass is 37.5 megagrams per hectare (Eaton and Lawrence 2006), which is within the range (17.7–42.7 megagrams per hectare) reported by Harmon et al. (1995) for semi-evergreen forests in northeastern Quintana Roo, Mexico. These few studies suggest that variation of woody dead mass across deciduous, semideciduous, and semi-evergreen forests likely reflects differences in floristic composition, tree mortality rates, substrate quality, and productivity (Harmon et al. 1995; Maass, Jaramillo et al. 2002). Other factors affect the size of the deadwood pool, such as successional status, the frequency and intensity of natural disturbance, management and history of land use, or even the methods used in each study (Delaney et al. 1998; Eaton and Lawrence 2006; Baker et al. 2007).

Forest Biogeochemistry

Biogeochemical studies in SDTF in Latin America span a variety of vegetation and soil processes. Similar to productivity and biomass studies, most recent studies are from Mexico. Thus, the information gaps identified in an earlier review (Jaramillo and Sanford 1995) have been partially filled, but with a heavy geographical bias within the Neotropics.

Leaf Nutrients, Resorption, and Litterfall

Our previous assessment of SDTF leaf nutrients revealed that Chamela forests had the highest leaf nitrogen concentration with a mean of 29.6 milligrams per gram, while mean leaf phosphorus was 2.8 milligrams per gram (Jaramillo and Sanford 1995). Only two subsequent publications report SDTF leaf nutrients. In central Mexico, Cárdenas and Campo (2007) measured leaf nitrogen (23.1 milligrams per gram) and leaf phosphorus (2.0 milligrams per gram) in mature leaves of the dominant legume, *Lysiloma microphyllum* Benth. Rentería et al. (2005) found variation in leaf nitrogen and phosphorus at Chamela greater than previously reported; nitrogen concentrations measured for six species during 3 consecutive years varied from 12 to 66 milligrams per gram, phosphorus concentrations from 1.0 to 4.8 milligrams per gram, and leaf nitrogen-to-phosphorus ratios from 7.5 to 30.1 (mean = 18). For comparison, we summarize recent measurements from the Cerrado and Atlantic forests in Brazil, with deciduous, perennial, and semideciduous species (table 7-3). Maximum values for both leaf nitrogen and phosphorus are greatest at Chamela, and although the mean nitrogen-to-phosphorus ratios are not conspicuously different, some species at Chamela show much lower ratios. If foliar nitrogen-to-phosphorus ratios are indicative of nutrient limitation (Güsewell 2004), these data support the view that phosphorus is limiting in these seasonally dry ecosystems. The data from Chamela indicate that high nitrogen-to-phosphorus ratios may covary with local topography and soil phosphorus availability. These are mediated by local variation in water availability and in annual rainfall (Rentería et al. 2005).

Nutrient resorption, the process by which nutrients are translocated from senescing leaves prior to abscission and moved to other plant tissues (Killingbeck 1996), has been well described for many ecosystems but less so for SDTF (Jaramillo and Sanford 1995). Resorption measures include efficiency, in which R_e equals the percentage of a nutrient withdrawn from

TABLE 7-3. Leaf nitrogen and phosphorus concentrations and N:P ratios in seasonally dry deciduous tropical forest, cerrado, and semideciduous Atlantic forest

	Deciduous (1)	*Cerrado (2)*	*Semidec. Atlantic Forest (2)*	*Dec.-Ever. (3)*	*Dec.-Ever. (4)*	*Semidec. (5)*
N (mg g^{-1})	<12–66	14.8	17.3	5–32	7.2–17.8	14–21
P (mg g^{-1})	1.0–4.8	0.67	0.89	0.25–1.75	0.43–0.72	
N:P	7.5–30.2	22.6	20.5		15–28	
No. spp.	6	14	14	11	10	3

No. spp. = number of species, Dec. = deciduous, Ever. = evergreen, Semidec. = semideciduous.
Data sources are (1) Rentería et al. 2005, (2) Hoffman et al. 2005, (3) Franco et al. 2005, (4) Nardoto et al. 2006, (5) Geßler et al. 2005.

mature leaves before abscission, and proficiency, in which R_p equals the minimum level to which a nutrient is reduced during senescence (Killingbeck 1996).

Recent studies on nitrogen and phosphorus resorption efficiencies in seasonally dry Neotropical ecosystems report high interspecific variability in Mexican SDTF (0-62 percent; Rentería et al. 2005) and in Brazilian Cerrado (38-77 percent; Nardoto et al. 2006). The study by Rentería et al. (2005), in which resorption was measured during one wet and two dry years, showed that nitrogen and phosphorus R_e were controlled more by water availability than by soil nutrients. Moreover, phosphorus R_p was controlled interactively by water and soil nutrient availability. Senesced-leaf phosphorus concentrations decreased (= higher proficiency) with diminishing rainfall and were lowest at sites with lower soil phosphorus and water availabilities. A relationship between phosphorus R_p and soil phosphorus availability has also been documented for *Lysiloma microphyllum* in central Mexico (Cárdenas and Campo 2007).

Litterfall nutrient fluxes in tropical forests and nutrient-use-efficiency estimates have been derived to assess potential nutrient limitation to forest productivity (Vitousek 1984). With the exceptions of estimates from Chamela (Jaramillo and Sanford 1995; Díaz 1997; Campo et al. 2001) and the Yucatán peninsula (Read and Lawrence 2003), litterfall nutrient fluxes have been quantified rarely in Neotropical SDTF. Litterfall nitrogen and phosphorus fluxes measured at Chamela by Díaz (1997) are similar to those reported previously (Jaramillo and Sanford 1995), but phosphorus fluxes are spatially and temporally more variable than nitrogen. Read and Lawrence (2003) show that semi-evergreen forest in the southern Yucatán peninsula has both lower nitrogen (1.09 to 1.2 percent) and lower phosphorus (0.039 to 0.059 percent) than at Chamela, but levels are similar to those reported for mature SDTF in Belize (Lambert et al. 1980). In Mexico, nitrogen-to-phosphorus ratios in litterfall range between 18 and 31 in diverse SDTF. For both Chamela and the Yucatán, nitrogen concentrations and fluxes in litterfall are rather insensitive to spatial and temporal variations in rainfall, whereas phosphorus concentrations and fluxes show considerable response to such variations. Based on the low phosphorus concentrations in litterfall and high phosphorus use efficiency in areas with less rainfall, Read and Lawrence (2003) suggest water (primarily) and phosphorus (secondarily) limitations control ecosystem nutrient cycling in the Yucatán. Rentería et al. (2005) arrive at similar conclusions, but they are based on variation in senesced-leaf phosphorus concentrations in response to water and nutrient availability. For seasonally dry Brazilian Cerrado,

Nardoto et al. (2006) show low litterfall concentrations for both nitrogen (0.7 percent) and phosphorus (0.02 percent) and very high phosphorus use efficiency (4340; litterfall mass/nutrient mass in litterfall). Interestingly, the studies from Mexico and Brazil have both identified phosphorus limitation coupled to water availability.

Water Mediation of Nutrient Release in Litter and Soil

Pulsed nutrient release and its relation to water availability in terrestrial ecosystems was proposed as an important process in SDTF by Singh et al. (1989) and has subsequently received considerable attention (e.g., van Gestel et al. 1993; Lodge et al. 1994; Austin et al. 2004). Briefly, they propose that rewetting dry soil at the onset of the wet season results in large pulses of nutrient mineralization primarily due to the lysis of microbial biomass. Jaramillo and Sanford (1995) further suggested that mineralization of microbial biomass could represent an important pathway for nutrient cycling throughout the growing season in SDTF, due to episodic dry spells within the wet season. There is new evidence concerning the actual role of this process in SDTF (see Microbial Associations below).

A number of studies in forests at Chamela show that nutrients accumulate during the dry season, including soil ammonium (NH_4^+) and nitrate (NO_3^-) (García-Méndez et al. 1991; Davidson et al. 1993; Montaño et al. 2007), soluble and microbial phosphorus in litter and soil (Campo et al. 1998), litter and soil-soluble carbohydrates and proteins (García-Oliva et al. 2003), dissolved organic carbon and microbial carbon and nitrogen in soil (Montaño et al. 2007), and water-soluble organic carbon and nitrogen and microbial carbon in surface litter (Anaya et al. 2007). Inorganic and organic nutrients become available for plant and microbial uptake after the onset of rains. Microbial biomass nutrients are also mineralized as water availability increases such that most labile carbon, nitrogen, and phosphorus decrease during the rainy season. Furthermore, there is greater potential carbon mineralization in litter and soil from dry-season than from rainy-season samples (García-Oliva et al. 2003; Anaya et al. 2007; Montaño et al. 2007). Soil nitrogen mineralization increases in rainy-season soil samples as well (Montaño et al. 2007). Rainfall seasonality also influences nutrient redistribution between soil aggregate fractions such that soil organic matter associated with macroaggregates (greater than 250 micrometers) represents the main source of energy for microbial activity at the beginning of the wet season (García-Oliva et al. 2003). Experimental evidence shows that a simulated rainfall event of 30 millimeters on dry-season litter and soil may release up to 0.358 gram of phosphorus per square meter (Campo et al. 1998). This flux represents 83 percent of the phosphorus returned via annual litterfall. In contrast, the same 30-millimeter rainfall event on litter and soil samples collected during the rainy season promoted phosphorus immobilization, suggesting that litter and soil moisture conditions prior to a rainfall event may determine the pattern of microbially mediated nutrient fluxes. Other experiments in Chamela have shown that a threshold of 12 percent gravimetric soil moisture has to be reached for microbial nutrients to mineralize (González-Ruiz 1997). These results indicate that rainfall seasonality and quantity regulate nutrient dynamics in litter and soil in SDTF, making nutrients available for plant and microbial growth.

Microbial Associations

In SDTF with abundant legume species, the many N-fixers are potentially important contributors for nitrogen inputs to the ecosystem. Only two studies on nitrogen fixation are available for SDTF, and both indicate that water availability is crucial for nodule formation and activity. Teixeira et al. (2006) excavated roots of *Cratylia mollis* Mart. ex Benth. in irrigated and nonirrigated plots in the Caatinga and determined delta ^{15}N signatures in plant tissues. They found nodules only in the irrigated plots, where ^{15}N values were lower than in nonirrigated plots, and concluded biological nitrogen fixation was limited by water deficit. González-Ruiz et al. (2008) found a close correlation between soil moisture and the presence of root nodules in nine leguminous species at Chamela. Nodules were active during the rainy season, and peak nodulation appeared soon after the onset of rainfall in July and decreased thereafter; no nodules were found before the onset of rains (April) nor after the end of the rainy season (November). Thus, the evidence suggests that nitrogen-fixing legumes in Neotropical SDTF fix atmospheric nitrogen only when soil moisture is adequate.

Watershed-Level Input-Ouput Budgets

Nutrient mass balances shed light on the relative importance of atmospheric input and soil leaching for the nutrient economy, the long-term sustainability of an ecosystem, and the relative importance of the soil as a nutrient source in an ecosystem. Bruijnzeel (1991) reviewed nutrient budgets for tropical forests, but there were no studies on SDTF. Even now, there are only two published accounts of watershed studies in Neotropical SDTF. One study documented base cation (calcium, potassium, and magnesium) input and output balances (Campo et al. 2000), and the other documented phosphorus cycling (Campo et al. 2001), both at Chamela.

Phosphorus and cation balances indicated that aboveground return (litterfall) was large relative to atmospheric inputs and runoff water losses, although more so for cations than for anions (table 7-4). Average mineral inputs (i.e., bulk deposition plus throughfall) represented approximately 13, 2.6, 3.0, and 4.2 percent of the annual phosphorus, calcium, potassium, and magnesium aboveground return, respectively. Mean nutrient losses from runoff were much lower than litterfall fluxes, but they were much greater for cations than for phosphorus. García-Oliva and Maass (1998) measured cation and phosphorus pools in the uppermost 6 centimeters of soil at 507 (calcium), 98.5 (potassium), 165 (magnesium), and 8 (orthophosphate phosphorus) kilograms per hectare, suggesting that soil nutrient supply exceeds atmospheric inputs and losses in stream water. A comparison of stream water cation losses to inputs for a 6-year period shows that inputs are lower than outputs (table 7-4). This indicates that soil weathering, and most of the associated nutrient release, exceeds the

TABLE 7-4. Watershed input-output balances in the Chamela seasonally dry tropical forest (kg/ha/yr)

Inputs-Outputs	*P*	*Ca*	*K*	*Mg*
Bulk deposition	**0.156**	**3.030**	**1.310**	**0.800**
Stream runoff	**0.064**	**5.240**	**2.830**	**1.790**
POM	0.002	0.083	0.003	0.008
Sediments	0.002	0.021	0.001	0.003
Net	**+0.092**	**-2.21**	**-1.52**	**-0.99**
Intrasystem fluxes				
Throughfall	0.35*	1.27	2.07*	0.290
Litterfall	3.88	11.39	2.31	1.59

*enrichment compared to bulk deposition
Values for bulk deposition and runoff are the means of 6 years of measurements. POM = particulate organic matter. Net = bulk deposition – stream runoff.
Cation data from Campo et al. 2000; P data from Campo et al. 2001.

nutrient uptake capacity of the forest. In contrast, the 6-year balance shows that phosphorus is conserved within the watersheds.

The role of rainfall variation in nutrient balances at this scale is evident when cation budgets for a wet (1993) and a dry (1994) year are compared, where net losses occur in the former and nutrient accumulation in the latter (Campo et al. 2000). However, for phosphorus, losses are always lower than inputs regardless of rainfall and runoff (Campo et al. 2001). The difference between phosphorus and cation net balances at the watershed scale is particularly striking for the El Niño year of 1992, when a precipitation equivalent to a full rainy season fell in January (649 millimeters), which is normally a dry month (mean = 35 millimeters), and total rainfall was 1186 millimeters, well above the 746 millimeters long-term mean. Net losses from the watersheds were –11.7, –4.9, and –1.16 kilogram per hectare for calcium, potassium, and magnesium, respectively, compared with a small net phosphorus gain of +0.011 kilogram per hectare. This indicates that the phosphorus cycle is very conservative at the watershed scale in Chamela.

Ecosystem Carbon and Nitrogen Pools

Very few studies in the Neotropics have quantified total ecosystem carbon and nitrogen pools. One such study conducted at Chamela (Jaramillo et al. 2003) quantified nearly three times greater biomass carbon and nitrogen pools in the semideciduous forest along the floodplain than in the shorter, upland deciduous forest (table 7-5). This difference is striking considering that both forest types experience pronounced rainfall seasonality with a 6- to 7-month dry season. The soil pools also reflect the differences between these forests, although more so for nitrogen (2.5 times greater) than carbon (1.5 times greater). Total ecosystem carbon and nitrogen pools in the semideciduous forest are more than double those in deciduous forest. Interestingly, the biomass carbon in semideciduous forest is very similar to the evergreen wet forest at Los Tuxtlas, Veracruz, Mexico (Hughes et al. 2000; table 7-5), despite a sixfold difference in the mean amount of rainfall between the two areas (746 millimeters in Chamela vs. 4700 millimeters in Los Tuxtlas). Moreover, in spite of a lower soil nitrogen pool in the semideciduous forest at Chamela, the biomass nitrogen pool is 1.6 times greater than for the evergreen forest (table 7-5). These studies show that soil ecosystem carbon is 37 percent (semideciduous forest), 54 percent (deciduous forest), and 51 percent (evergreen wet forest) of the total ecosystem carbon, while soil nitrogen pools comprise 85 percent (semideciduous forest),

TABLE 7-5. Biomass carbon and nitrogen pools in three forest types

	Deciduous[1]	*Semideciduous*[1]	*Evergreen*[2]
Carbon (Mg ha⁻¹)			
Aboveground	58	180	195
Roots	7	13	9
Total C in biomass	**65**	**193**	**204**
Soil	76	114	210
Total ecosystem C	**141**	**307**	**414**
Nitrogen (kg ha⁻¹)			
Aboveground	940	2,623	1,705
Roots	106	264	140
Total N in biomass	**1,046**	**2,887**	**1,845**
Soil	6,659	16,730	19,800
Total ecosystem N	**7,705**	**19,617**	**21,645**

[1]Forest types located within the Chamela region
[2]Rain forest from the Los Tuxtlas region

86 percent (deciduous forest), and 92 percent (evergreen wet forest) of the total ecosystem nitrogen. Thus, a high percentage of nitrogen in tropical ecosystems is stored in the mineral soils, whereas the proportion of carbon in soil pools is more variable. A study of mature semi-evergreen SDTF in northeastern Yucatán peninsula (Vargas et al. 2008) estimates an ecosystem carbon pool of 149.5 megagrams per hectare, which is more similar to the ecosystem carbon pool of the deciduous than to the semideciduous forest of Chamela.

Conclusions and Remaining Challenges

Earlier reviews identified information gaps concerning productivity and nutrient cycling in SDTF worldwide (Jaramillo and Sanford 1995; Martínez-Yrízar 1995), which included above- and belowground biomass, organic matter decomposition, the role of microbial biomass, fine-root turnover, and nutrient resorption. Published work since then indicates progress, albeit still limited. For example, above- and belowground biomass data for Neotropical SDTF has significantly increased in the last few years, but most come from a couple of regions or a few sites.

Our review indicates that studies on NPP and biogeochemistry of SDTF in the Neotropics are still very scarce, especially outside Mexico. Litterfall, as an index of NPP, is the most commonly measured component of productivity. Other components are rarely measured, particularly root productivity, stem growth, losses to herbivores, and tree mortality. Multiyear litterfall studies are also limited, and only two sites in Mexico provide information for periods longer than 4 years. Long-term data are relevant to understand the responses of functional processes to climatic anomalies such as the El Niño–Southern Oscillation and to climate change. The evidence suggests there is an interactive control of water and soil phosphorus availability and cycling on primary productivity and nutrient dynamics, at least at some sites.

Important gaps remain where there is little or no information relevant to the understanding of the consequences of global change in SDTF. These include key components of the carbon cycle (e.g., soil respiration and vegetation-atmosphere carbon exchange), the contribution of N-fixation to the nitrogen economy of SDTF, and quantification of the major nitrogen fluxes and nitrogen transformations in more sites. Understanding their interaction with water availability is crucial, not only because water is the main limiting resource in SDTF, but because climate change will likely alter the amount and the distribution of rainfall in SDTF regions.

The growing interest in secondary SDTF productivity and nutrient cycling is relevant because secondary forests cover an increasing area in all seasonally dry tropical regions of the world. Moreover, such information allows examination of the paths taken and time needed for these processes to recover, and the extent to which they can recover, with woody species regrowth or invasion after disturbance of primary SDTF. Biogeochemical information on the consequences of anthropogenic conversion of SDTF (see chap. 10) should provide a context for understanding soil processes during secondary succession in the dry tropics. This may be relevant for restoration schemes in disturbed or managed areas in the region.

Chapter 8

Physiological Mechanisms Underlying the Seasonality of Leaf Senescence and Renewal in Seasonally Dry Tropical Forest Trees

JUAN PABLO GIRALDO AND N. MICHELE HOLBROOK

Seasonality in the presence and production of leaves is the defining characteristic of seasonally dry tropical forest (SDTF) ecosystems and has important implications for their functioning. Temporal patterns of shoot activity influence photosynthetic carbon gain and thus affect competition between tree phenological types (Givnish 2002), the seasonal rhythms of tree-herbivore interactions (Leigh 1999; Dirzo and Boege 2009), and the annual ecosystem carbon uptake and energy fluxes (Kucharik et al. 2006). The impact of leaf phenology on tree carbon return is associated with its effect on the length of leaf exposure to herbivore and pathogen damage, the timing of leaf loss for water balance, and the energy investment in leaf construction (Franco et al. 2005). The timing and the length of leaf presence in tropical forests has also been suggested to play a key role in maintaining tree diversity by regulating the abundance of pest pressures (Leigh et al. 2004).

According to a recent study, a high proportion of SDTFs in the Americas are at risk of severe climate change due to changes in rainfall patterns; in African forests, habitat fragmentation and fire are the most significant threats; while in Eurasian forest, agricultural conversion and human population density will have a greater impact (Miles et al. 2006; see also chap. 16). Understanding how these threatened ecosystems will respond as climate change and deforestation further modify natural patterns of precipitation and light availability (Cox et al. 2000; Sampaio, Nobre et al. 2007) is important for the implementation of sound conservation strategies. In tropical rain forests, predictions of carbon and energy fluxes have been significantly improved by incorporating physiological mechanisms into simulation models (Williams et al. 1998). Better understanding of the physiological mechanisms underlying the phenology of tropical trees will enhance our ability to predict the impacts of climate change on SDTF (Cleland et al. 2007).

The developmental transitions controlling seasonal patterns of leaf renewal and senescence are thought to be influenced by environmental and age-related factors (Noodén and Leopold 1988). Soil moisture, stem water storage, light, vapor pressure deficit, and photoperiod have all been implicated (Wright and Cornejo 1990; Borchert 1994a,b; Borchert and Rivera 2001), but the mechanisms by which changes in environmental conditions trigger phenological transitions in SDTF trees are not well understood. Studies of tropical deciduous forest phenology have largely focused on the availability of resources and external conditions needed for photosynthesis and growth. In contrast, work on crops and model plant species highlights a role for plant growth regulators in integrating meristem activity with environmental factors (Srivastava 2002). This chapter seeks to bridge these two research themes, exploring how the plant vascular system may play a role both in sensing changes in the environment and in controlling the delivery of compounds that affect developmental transitions. We focus first on the role of leaf and root hydraulics in controlling leaf senescence and abscission, before examining the relationship between seasonal dynamics of stem hydraulic conductance and the production of new leaves. We believe that an integrative physiological understanding of the mechanisms regulating leaf phenology is fundamental for studies of tropical tree community ecology, tree-atmosphere interactions, and biodiversity conservation under a changing climate.

Mechanisms Underlying Patterns of Leaf Senescence and Abscission

Field studies linking phenological activity with climate conditions support what seems obvious to even a casual visitor to a seasonally dry tropical forest: an important role for water availability in determining patterns of leaf senescence and abscission in SDTF trees (Daubenmire 1972b; Frankie et al. 1974; Opler et al. 1980; Reich and Borchert 1982, 1984; Bullock and Solís Magallanes 1990; Medina and Cuevas 1990; Whigham et al. 1990). Such observational studies, however, are not sufficient to uncover the mechanisms by which seasonal variation in rainfall, humidity, cloudiness, and wind speed influences the timing and intensity of leaf fall. Over the past two decades, a combination of detailed physiological studies and experimental manipulations has begun to illuminate the ways in which the activity of dry forest trees is influenced by the seasonality of their environment. These studies call into question some earlier hypotheses and, based on new observations on the hydraulic properties of roots and leaves, open the door to a more integrated understanding of how water availability influences leaf senescence.

A hallmark of tropical deciduous forests is the range of phenological patterns exhibited by resident trees (Sánchez-Azofeifa, Kalaczka et al. 2003). In most tropical deciduous forests, evergreen, leaf-exchanging (brevi-deciduous), and even wet-season-deciduous species co-occur with the canonical drought-deciduous species. Furthermore, within each phenological category there is substantial diversity. For example, among drought-deciduous species, some lose their leaves at the very beginning of the dry season, while others hold onto leaves well into the dry period. The coexistence of evergreen and deciduous taxa has led to the expectation of cost-benefit trade-offs associated with differences in leaf longevity (Givnish 2002). Because the woody portions of the plant both represent the largest investment in biomass and have a lasting impact on plant performance, studies of tropical deciduous forest phenology have largely focused on the hydraulic properties of stems, with the expectation that differences in xylem hydraulic conductivity, resistance to cavitation, and rooting depth would be associated with leaf phenology (Holbrook et al. 1995).

Despite efforts to test this idea, a clear pattern has not emerged. Some studies report that deciduous species are more vulnerable to cavitation and have lower hydraulic investments than co-occurring evergreen species (Sobrado 1993; Choat et al. 2005). However, others report that evergreen and deciduous species exhibit a similar range of stem hydraulic conductivities (Brodribb et al. 2002) and the absence of a clear association between stem xylem vulnerability to cavitation and leaf phenology (Brodribb et al. 2003). Studies of rooting depth are also diverse in terms of their findings. For example, a number of studies suggest that dry forest evergreen and brevi-deciduous trees have deeper roots than deciduous trees (Borchert 1994b; Borchert et al. 2002); in contrast, others indicate that evergreen trees tend to acquire water in upper soil layers, while deciduous trees tap deeper water sources (Jackson et al. 1999). Although stem structure and rooting depth remain important attributes of tropical deciduous forest trees, they are not sufficient to explain the diversity of phenological patterns exhibited by these trees.

A second area in which recent work has advanced our understanding of SDTF phenology concerns the role of leaf age. Leaf senescence in herbaceous and crop plants is an age-dependent developmental process (Lim et al. 2007), and leaf age has been hypothesized to play a role in the phenological behavior of SDTF species (Reich and Borchert 1984). A recent experimental study in which leaf age was manipulated by a series of artificial defoliations, however, found that the timing of leaf abscission was independent of leaf age (Gutierrez-Soto et al. 2008). A similar effect was observed in the timing of leaf abscission of plants affected by the 1998 El Niño–Southern Oscillation drought, which resulted, in some species, in the production of new leaves partway through the usual rainy period (Borchert et al. 2002). Although leaf age does not appear to play a pivotal role in determining the timing of leaf abscission in SDTF trees, caution is warranted because the effects of leaf age have been studied directly in only a small number of species.

Finally, we consider the role of seasonal changes in soil water availability in light of a long-term irrigation experiment conducted at Barro Colorado Island, Panama. Despite the maintenance of water potentials, a measure of the free energy of water, close to soil field capacity for three consecutive dry seasons, patterns of leaf fall remained unchanged in the majority of species (Wright and Cornejo 1990). The authors of that study hypothesize that increased vapor pressure deficits and irradiance, instead of soil drying, trigger leaf fall in tropical trees. Barro Colorado Island falls on the moist and evergreen end of the spectrum of SDTF, and thus it is possible that forests exposed to greater seasonal variation in rainfall would show a more pronounced response to soil moisture. However, atmospheric conditions play an important role in determining leaf fall in many tropical deciduous forest trees. For example, species with low wood density are typically among the first to shed their leaves, despite the presence of large amounts of water in their stems (Nilsen et al. 1990; Olivares and Medina 1992; Chapotin et al. 2006).

Where does this leave us in terms of our understanding of the phenology of SDTF trees? We believe that an important step forward is to place the phenological behavior of SDTF trees within a developmental context. Studies of herbaceous and crop plants demonstrate that leaf senescence is not a passive degenerative process (Lim et al. 2007). The marked changes in gene expression that accompany the onset of senescence indicate that this must be viewed as a well-regulated developmental transition (Buchanan-Wollaston et al. 2003; Guo and Gan 2005), rather than simply a response to adverse conditions. An important aspect of leaf senescence is the retranslocation of nutrients, a process that involves the degradation of chlorophyll, cell membranes, and other major changes in cell structure and metabolism. Thus, at a minimum, leaf senescence in deciduous species must anticipate conditions that render the leaf unable to carry out these functions, while for brevi-deciduous and evergreen species, the timing of leaf senescence may reflect selection for the most advantageous period for leaf expansion, as well as optimization of whole-plant photosynthesis (Ono et al. 2001).

One mechanism for how reduced water availability can trigger the developmental changes associated with leaf senescence is by altering the hydraulic properties of roots and leaves. In SDTF of northwestern Costa Rica, leaf shedding is preceded by a large decline in the conductance of water in the liquid phase through leaves (K_{leaf}) (fig. 8-1) (Brodribb and Holbrook 2003). For example, the deciduous species *Calycophyllum candidissimum* and *Rhedera trinervis* experience a five- to tenfold decline in K_{leaf} from leaf maturity to leaf abscission. Evergreen *Quercus oleoides* and leaf exchanger *Simarouba glauca* also undergo considerable declines in K_{leaf} prior to leaf fall (Brodribb and Holbrook 2005). Consistent with the idea that K_{leaf} plays a role in triggering senescence, rather than occurring as a consequence of leaf senescence, K_{leaf} begins to decline before changes in stomatal conductance and photosynthetic activity, measured by chlorophyll fluorescence (Brodribb and Holbrook 2003).

Declines in root hydraulic conductance associated with soil drying may also be associated with leaf shedding. As the dry season progresses, tropical savanna trees of Brazil with diverse leaf phenologies experience an increase in root embolisms (Domec et al. 2006). Four savanna tree species with significant differences in root depth exhibit dry-season percent loss of native conductivity between 25 and 45, with roots of the brevi-deciduous species *Qualea parviflora* reaching up to 80 percent loss of native conductivity before the beginning of the wet season (fig. 8-2).

Roots and leaves are well suited to play a major role in transducing changes in water availability. Together they constitute a major part of the resistance of water movement through trees (Kramer and Boyer 1995; Sack and Holbrook 2006), while their locations at the terminal ends of the transpiration stream mean that they are closely coupled to the environment. In particular, changes in root hydraulic properties may reflect variation in soil water availability, while changes in leaf hydraulic properties are subject to changes in atmospheric demand (Domec et al. 2009).

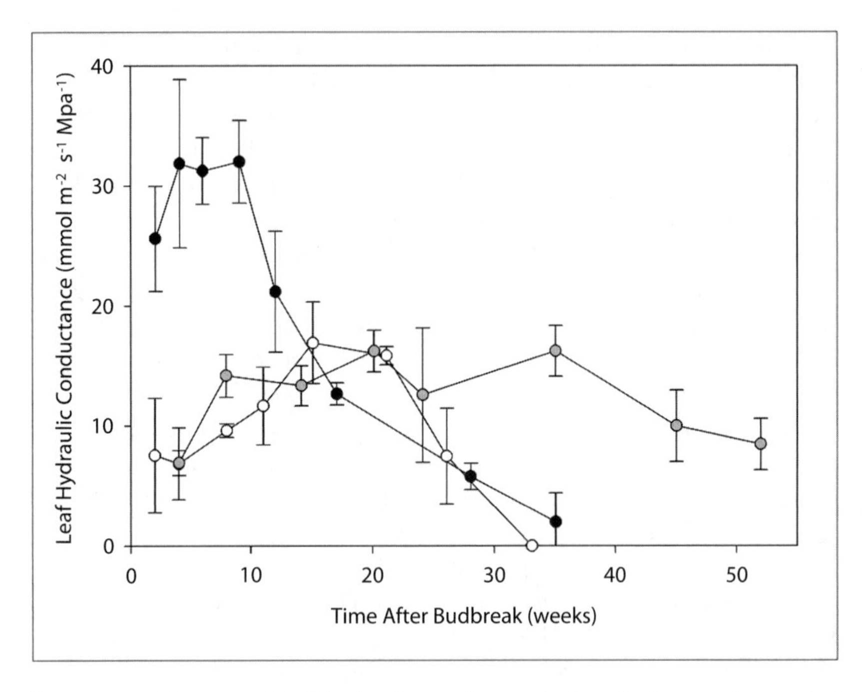

FIGURE 8-1. Patterns of leaf hydraulic conductance, from leaf expansion to leaf senescence in two evergreen species, *Simarouba glauca* (black circles) and *Quercus oleoides* (gray circles), and one deciduous species, *Rhedera trinervis* (open circles). Data represent mean values from five trees plus or minus SD. Redrawn from Brodribb and Holbrook 2005.

A decline in hydraulic conductance, irrespective of where it takes place within the plant, has the potential to amplify seasonal effects of reduced water availability via changes in stomatal aperture. Stomatal closure, in turn, could trigger leaf senescence by affecting carbon availability, the production of toxic forms of oxygen such as reactive oxygen species, or nutrient or hormonal status. Age-related declines in photosynthesis have been proposed as a possible mechanism for leaf senescence in herbaceous plants (Hensel et al. 1993), but the evidence for this remains controversial (Guo and Gan 2005). In *Arabidopsis* mutants and transgenic tobacco plants with reduced photosynthetic activity, age-dependent leaf senescence is delayed, compared with wild type (Miller et al. 2000; Woo et al. 2002). Reactive oxygen species resulting from an imbalance in carbon dioxide and light energy could also play a role. Reactive oxygen species are thought to integrate environmental stresses that induce leaf senescence, such as drought, heat, and intense light (Quirino et al. 2000).

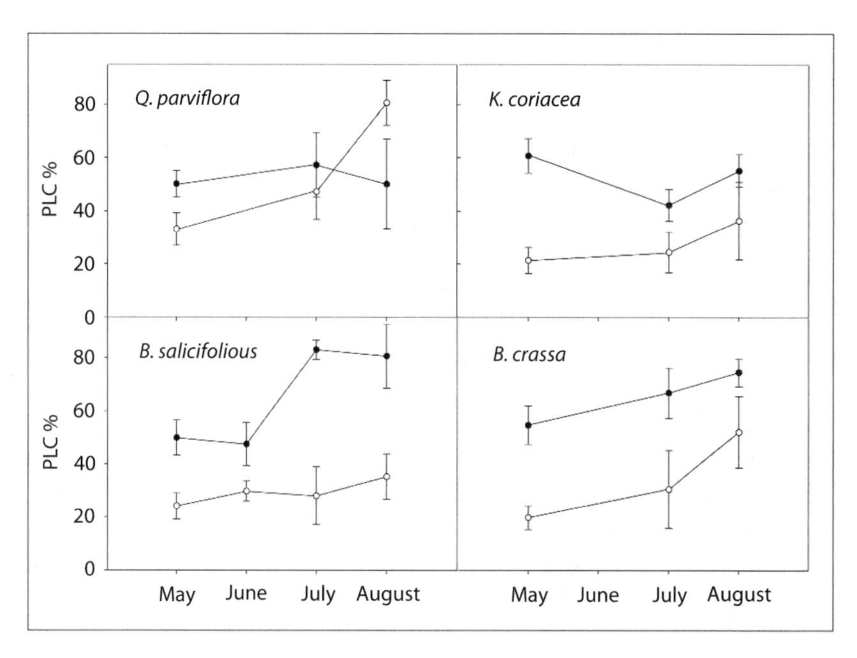

FIGURE 8-2. Diurnal changes in percent loss of native conductivity (PLC) for root xylem of four Cerrado woody species: *Qualea parviflora*, *Kyelmera coriaceae*, *Blepharocalyx salicifolius*, and *Byrsonima crassa* (n = 5–6). Closed and open circles indicate midday and morning predawn measurements, respectively. Redrawn from Domec et al. 2006.

Alternatively, stomatal closure may influence leaf senescence via changes in the delivery of xylem-transported compounds. Nitrogen and hormones produced by the roots, such as cytokinins, are important regulators of leaf senescence (Guo and Gan 2005). In herbaceous species, soil nitrogen deficiency leads to leaf sugar accumulation (Ono and Watanabe 1997), which in turn triggers the expression of senescence-associated genes (SAGs) (Rolland et al. 2006). The highly specific senescence gene *SAG12* encoding a cysteine protease can be induced a hundredfold during leaf senescence by growing *Arabidopsis* plants in 2 percent glucose in combination with low nitrogen availability (Pourtau et al. 2004; Wingler et al. 2004). Cytokinins are also well known for their effects on leaf longevity. The external application of cytokinins to leaves and the regulation by a senescence-specific promoter of a gene encoding an enzyme that catalyses the rate-limiting step of cytokinin synthesis results in delayed leaf senescence (Gan and Amasino 1995, 1996). Moreover, experimental manipulations that impact transpiration but not light levels or stomatal conductance have been shown to induce senescence in *A. thaliana*, with decreased xylem cytokinin delivery being implicated as the causal link (Boonman et al. 2007).

The potential for leaves and roots to act as hydraulic "fuses" that alter the internal flows of both nutrients and growth regulators provides a new perspective on phenological diversity of SDTF trees. This framework points the way towards future studies of how differences between species in the hydraulic sensitivity of their roots and leaves are linked to temporal variation in the timing of leaf senescence. It also suggests a range of experimental manipulations to disentangle how reductions in xylem transport influence the developmental transformations leading to leaf senescence and abscission. Furthermore, the sensitivity of the hydraulic properties of leaves and roots to a variety of environmental conditions, such as water, nutrient, and light availability, makes them good candidates to mediate the effects of climate change on leaf senescence. As the signal of the end of photosynthetic activity, any modification to the timing of leaf senescence will play a key role in determining tree performance under changing environmental conditions.

Mechanisms Associated with the Renewal of Leaves

Two hypotheses have been put forward to explain the timing of leaf renewal among SDTF trees. The first proposes that day length provides an environmental cue that triggers budbreak (Borchert and Rivera 2001). This hypothesis has been advanced for brevi-deciduous and leaf-exchanging species that exhibit low interannual variability in the timing of budbreak (Rivera et al. 2002). Budbreak in these species occurred around the spring equinox within the same 10- to 15-day period for 3 consecutive years despite annual variations in rainfall (Rivera et al. 2002). Day length has also been suggested to regulate vegetative budbreak in stem succulent trees (Borchert and Rivera 2001; Rivera et al. 2002). Highly synchronized budbreak in *Bursera simaruba* and *Pseudobombax septenatum*, stem succulent species from Costa Rica, takes place several weeks before the start of the wet season and occurs without a measurable change in plant water status (Borchert and Rivera 2001). Moreover, the fact that budbreak in stem succulent trees was not induced by significant rainfall during the dry season following a year of an El Niño–Southern Oscillation is consistent with a trigger independent of water availability (Borchert et al. 2002). Nevertheless, the fact that other studies of species that flush prior to the onset of the rainy season do not report such interannual constancy (Rojas-Jiménez et al. 2007; Richer 2008) underscores the need for studies that include the experimental manipulations of mature trees. For example, the timing of leaf flushing in the late dry season of the deciduous tree *Enterolobium cyclocarpum* in Costa Rica can vary from year to year by as much as 3 weeks (Rojas-Jiménez et al. 2007).

An alternative hypothesis focuses on rehydration of stems and buds, caused by either an increase in soil moisture or the abscission of aging leaves (Borchert 1994a). At a minimum, stem rehydration is likely to be a prerequisite for leaf flushing to meet the transpirational demands of new leaves and to restore the flow of nutrients and hormonal signals that initiate budbreak and leaf growth. In herbaceous plants, bud dormancy is known to be influenced by the balance of hormonal signals transported by the xylem, such as growth inhibitors like abscisic acid and jasmonic acid, and growth-inducing hormones such as cytokinins (Horvath et al. 2003). Studies conducted in Costa Rica show that tree species that flush before the first rains of the wet season exhibit significant changes in stem water status prior to the renewal of meristem activity (Borchert 1994b). In Australian tropical savannas the onset of leaf flushing is well correlated with dry-season minimum leaf water potential, with leaf flushing taking place immediately after the initial rise in predawn water potential from the seasonal minimum value (Williams et al. 1997). A decline in leaf area in brevi-deciduous species and leaflessness in deciduous species was associated with this rise in leaf water potential. Budbreak in deciduous hardwood trees can also be induced by soil irrigation. Vegetative buds broke in these trees a few days after rehydration, and leaves were fully expanded after a couple of weeks (Borchert 1994a).

The renewal of leaves followed by stem rehydration reflects seasonal dynamics of stem hydraulic conductance in SDTF trees (fig. 8-3). Evergreens have relatively constant stem-specific hydraulic conductivity (K_{sp}, with only minor increases in K_{sp}) early in the wet season (Brodribb et al. 2002) that may allow them to exchange their leaves in the dry season. K_{sp} remains almost constant in the evergreen *Curatela americana*, while *Simarouba glauca* and *Quercus oleoides* exhibit moderate declines in K_{sp} relative to wet season values. In contrast, the hydraulic conductance of the stems in deciduous species exhibits a variety of responses before leaf shedding (Brodribb et al. 2002). For example, stem hydraulic conductivity in *Rhedera trinervis* drops to 15 percent of the maximum wet season value, while in *C. candidissimum* K_{sp} does not vary significantly between wet and dry seasons. The conservation of hydraulic conductance in some deciduous species may be a mechanism to allow meristem hydration during soil rewetting, as in *C.*

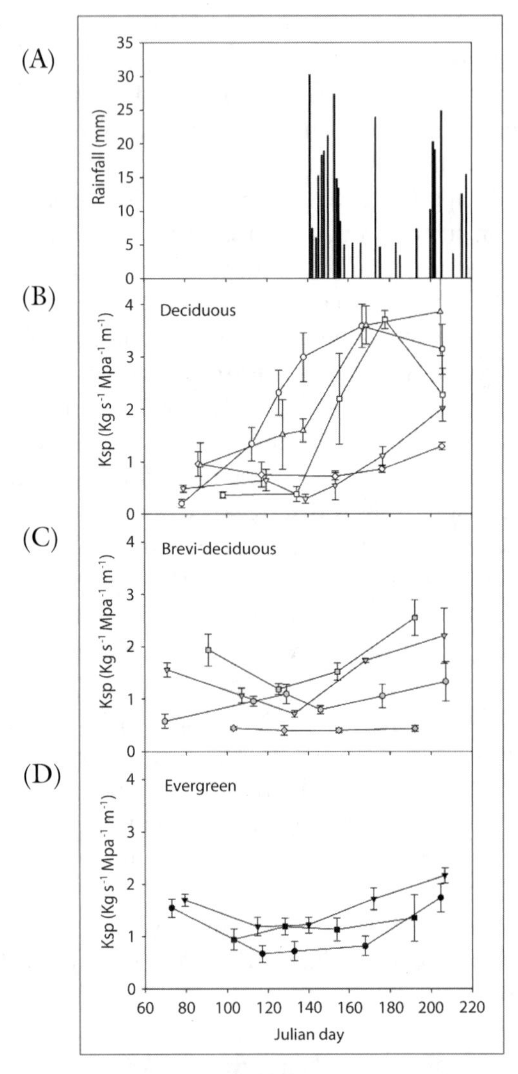

FIGURE 8-3. (A) Rainfall during the study period at Santa Rosa National Park. Lower panels show patterns of mean stem-specific hydraulic conductivity (K_{sp}) plus or minus SE (n = 4) in the 12 species studied. (B) Deciduous species: *Bursera simaruba* (open squares), *Enterolobium cyclocarpum* (open circles), *Calycophyllum candidissimum* (open diamonds), *Rhedera trinervis* (open triangles point down), and *Gliricidia sepium* (open triangles point up). (C) Brevi-deciduous species: *Byrsonima crassifolia* (gray triangles), *Hymenaea courbaril* (gray squares), *Swietenia macrophylla* (gray circles), and *Manilkara chicle* (gray diamonds). (D) Evergreen species: *Curatella americana* (black squares), *Simarouba glauca* (black circles), and *Quercus oleiodes* (black triangles). Redrawn from Brodribb et al. 2002.

candidissimum, a tree that flushes its leaves at the onset of the rainy season. In deciduous stem succulent trees, early leaf flushing during the dry season may rely instead on water stored within their stems and roots (Chapotin et al. 2006; Rojas-Jiménez et al. 2007). The recovery of K_{sp} in dry-season leaf-flushing deciduous trees without large amounts of stem water storage may result from the formation of new wood (Brodribb et al. 2002).

Many SDTF trees recover their hydraulic capacity prior to producing new leaves. The mechanisms by which hydraulic recovery can occur without the formation of new wood are only partially understood. An increase in root hydrostatic pressure caused by the accumulation of solutes in the root stele can refill cavitated vessels by forcing air into solution. A broad range of tropical vines and woody species have been shown to develop root pressures (Fisher et al. 1997), but at present there is only limited evidence for root pressure in SDTF trees (J. Wheeler, unpublished data). The ability to refill cavitated vessels despite tension in the water column in the xylem has been reported in a number of cerrado species (Bucci et al. 2003; Domec et al. 2006), although whether this contributes to seasonal patterns in meristem activity in SDTF trees is not known.

Hydraulic redistribution, which transports water from deep, wetter sources to shallow, drier soil layers, may be another mechanism facilitating leaf flushing near the end of the dry season (Scholz et al. 2002). Water uptake of taproots and reverse flow into the soil in lateral roots have been observed during the dry season in brevi-deciduous tropical savanna trees *Byrsonima crassa* and *Blepharocalyx salicifolius* and the evergreen *Styrax ferrugineus* (Scholz et al. 2002; Moreira et al. 2003). Hydraulic redistribution may contribute to leaf flushing in tropical savanna plants by enhancing nutrient uptake and facilitating symbioses with mycorrhizae and nitrogen-fixing bacteria (Scholz et al. 2002). In nutrient-rich upper soil layers, water supply by hydraulic lift can facilitate soil ion diffusion and uptake by fine roots (Caldwell et al. 1998). Hydraulic lift in herbaceous plants, for example, enhances phosphorus and potassium root uptake in dry and nutrient-poor soils (Rose et al. 2008).

Conclusions

As the nexus of carbon, nutrients, and hormones, the plant vascular system has the potential to act as a master signaling system influencing the activity and longevity of both meristems and leaves. Although complex developmental processes such as leaf flushing or senescence are unlikely to be regulated by a single mechanism, the plant vascular system is well positioned to integrate the wide range of environmental and age-related factors affecting leaf phenology. Furthermore, the environmental sensitivity of plant hydraulic conductance may confer an adaptive advantage to regulating leaf phenology in the diverse microenvironments and climatic variability of SDTF ecosystems.

Conservation of SDTF will rely on models predicting the effects of climate change on ecosystem function and tree community ecology. Leaf phenology is a main factor affecting the interannual variability in carbon uptake of deciduous forests (Goulden et al. 1996), while the incorporation of seasonal dynamics in plant vascular conductance has been shown to have an important contribution to models of carbon and energy exchange with the atmosphere in tropical forests (Williams et al. 1998). A physiological understanding of leaf phenology will enable ecosystem dynamic models to predict changes in the distribution and abundance of SDTF trees with diverse phenological strategies. Elucidating the mechanisms that regulate patterns of leaf production and persistence will also improve our ability to predict seasonality in tree-herbivore interactions and thus shed light on the maintenance of seasonally dry forest diversity in future climate change scenarios (Leigh 1999).

Chapter 9

Water Dynamics at the Ecosystem Level in Seasonally Dry Tropical Forests

MANUEL MAASS AND ANA BURGOS

Water availability is one of the most important factors controlling species distribution in terrestrial ecosystems (Holdridge 1967). Organisms have developed remarkable physiological adaptations to withstand, avoid, or tolerate water limitations. It is crucial to recognize and understand these adaptations, and there is an ample literature addressing this subject. However, very few of these studies have focused their attention on seasonally dry tropical forest (SDTF) species, and fewer still have examined the study of water dynamics at the ecosystem level. This is the aim of this chapter.

The ecosystem approach relates to the study of the movement of energy, water, chemicals, nutrients, and pollutants into, out of, and within ecosystems. With this approach ecosystems are explored as whole, complex systems, and the spatial, temporal, and organizational complexity of the system is included in any attempt to describe or model the ecosystem (Pickett et al. 2005). Specifically, this approach aims to improve our understanding of ecosystem behavior by identifying probable states, thresholds between states, feedback loops, and the overall system capacity to respond to disturbances. Only when all this information is synthesized is it possible to reveal ecosystem dynamic properties (Kay et al. 1999; Milne 1999). Unfortunately, this ecosystem approach is rarely used in the study of SDTFs. Most available literature on hydrology issues in SDTFs refers to water use or water limitations for particular species (ecophysiological approaches) or particular life-history traits of tree species and vegetation structure features, without a clear analysis of how water flows and interacts within the larger ecosystem context. Very few studies deal with hydrological process at the ecosystem level (rainfall, infiltration, evapotranspiration, runoff, leaching, etc.), and almost none look for an integrated analysis of the spatial, temporal, and organizational complexities and the role of water processes on them.

In this chapter we will attempt to organize the current knowledge about water dynamics at the ecosystem level, using two different techniques: a classical forest hydrology description and a more integrated analysis using an ecosystem approach, as described above. In the first part of this review, entrance and movement of water through the SDTF is examined. Particularly, emphasis is placed on reviewing longer-term responses (several years) rather than just single-year, seasonal (wet-dry) dynamics. Major sources of water in the ecosystem are identified, as well as the factors controlling water availability in space and time. A description is given on how water characteristics (i.e., quality, quantity, and regime) change as water flows through the vegetation and soil profile, as well as how topography redistributes water, creating environmental heterogeneity and promoting the development of tree functional types. Finally, a brief discussion is presented on how SDTFs may respond to climate change.

In the second part of the review, we attempt to begin unraveling the hydrological connections in SDTFs, using three different types of analysis that are consistent with an ecosystem approach. First we employ a mechanistic approach, exploring the causal relationships between plant life-history traits and hydrological factors conditioning successional outputs. Second, we explore the available knowledge on hydrological connections underlying the system's behavior based on the analysis of tightly linked variables (e.g., primary productivity and rainfall patterns), the existence of lag effects, system trajectories, alternative states, multicausalities, thresholds, bifurcation points, irreversibility, and the revelation of nonlinear dynamics. Finally, we take a socioecosystem perspective, in which we examine human interactions with and disturbance of natural hydrological processes.

Forest Hydrology in the Tropical Dry Forest Ecosystem

Even though water is the most limiting factor for the SDTF, not much has been written about this key ecosystem component.

Rainfall Variability

Rainfall seasonality is the most conspicuous hydrologic characteristic of SDTF. Frequently, more than 80 percent of annual precipitation occurs within 5 months. During the wet season, rainfall averages greater than 100 millimeters per month, which allows for the development of a dense, if rather short in stature (10 to 15 meters high), forest-type vegetation (Holbrook et al. 1995). In contrast, during the dry season, average monthly precipitation is less than 10 millimeters, creating conditions so dry that most trees drop their leaves as a mechanism to deal with the lack of water. Trees respond fast to the onset of the rainy season, and a leaf area index as high as 4 square meters leaf per square meter can be reached within a few days after the first rains (Moreno 1998).

Interannual rainfall variability is also a common characteristic of SDTF. Figure 9-1 shows examples for three distinct SDTF sites. The differences can be fivefold between the wettest and the driest year, in a 50-year series. Enquist and Leffler (2001) found a mild relationship between the Southern Oscillation Index and annual precipitation in Guanacaste, Costa Rica, with El Niño years being significantly drier than La Niña years. García-Oliva et al. (1991) report a significant influence of the California Current as a source of rainfall variability in the Chamela region.

Small rainfall events (less than 20 millimeters) dominate the frequency

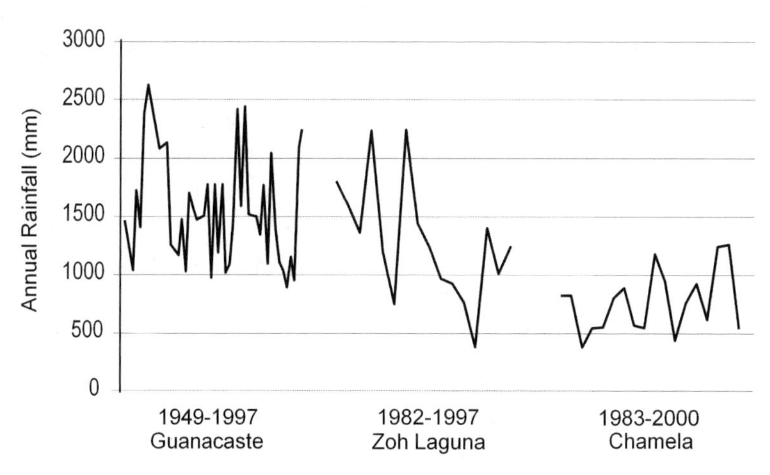

FIGURE 9-1. Interannual variation in precipitation in lowland Guanacaste (Enquist and Leffler 2001), Zoh Laguna, Yucatán, Mexico (Lawrence and Foster 2002), and Chamela, Jalisco, Mexico.

distribution, and few large storms (greater than 50 millimeters) determine total annual rainfall. In Chamela, for example, only 7 percent of the rainfall events are greater than 50 millimeters, but these events, associated with hurricane activity, deliver 42 percent of the total precipitation (García-Oliva et al. 2002).

This high interannual rainfall variability causes high variation in the length of the growing season (Maass et al. 1995), which can be as short as 3 months or as long as 7.5 months. Interestingly, the onset of the rainy season is significantly more regular (plus or minus 7 days) than the end of the growing season (plus or minus 28 days).

Water Entering into the Ecosystem

Lilienfein and Wilcke (2004) reported a significant increase in element concentration as the rainfall water passes through the canopy in the Brazilian cerrado. The source of this element enrichment comes from soil dust carried by the wind and from leaf leaching or dead organic matter (branches and leaves) decomposition. The former source of elements is particularly important, since substantial aboveground dead biomass lies or hangs on the canopy of SDTF. This has been reported by Maass, Jaramillo et al. (2002) for Chamela, where two-thirds of the total aboveground necromass is located (and decomposes) more than 30 centimeters above the soil surface. When all this dead material was experimentally removed from the canopy, there was a significant reduction in nutrient concentration of the throughfall.

A thick leaf litter layer on the forest floor peaks at the end of the dry season (Martínez Yrízar 1995). Therefore, interception can be as high as 100 percent during a small storm (less than 5 millimeters), which describes about one-third of the rainfall events in the Chamela region (Cervantes 1988). Annually, as much as 21 percent of total rainfall is intercepted by the canopy (Burgos 1999), and the leaf litter on the forest floor intercepts an additional 6.5 percent.

In areas of intense hurricane storm events, rainfall erosivity can be as high as 13,000 MJ millimeters per hectare per hour (García-Oliva et al. 1995). Without litter protection, soil erosion rates increase by several orders of magnitude (Maass et al. 1988).

Soil-Water Dynamics

Usually, SDTF soils briefly reach field capacity only during a strong storm event. Plant species in SDTF have developed highly diverse strategies to cope with this low soil water availability. Borchert (1994b) has identified more than 10 functional tree types, based on differences in their wood density, root length, leaf resistance, and access to water during the dry season. One common feature, according to Canadell et al. (1996), is that trees in SDTF have significantly lower average root depth than do those in tropical evergreen forest and even tropical savanna, mainly as a result of negligible groundwater availability. According to Holdridge (1967), total annual potential evapotranspiration in SDTF can be as high as four times its annual precipitation. Therefore, transpiration is the most important output flux of water in the ecosystem, and there is a transpirative water deficit most of the year (fig. 9-2). It is common that as much as 90 percent of the annual rainfall returns to the atmosphere as an evapotranspirative flux, as is the case in Chamela (Burgos 1999).

The combined effect of water availability and the potential evapotranspiration/precipitation ratio ("dryness index") determines the type of dominant tropical biome found in a region. SDTFs occur in sites where the potential evapotranspiration/precipitation ratio is slightly higher than 1 (fig. 9-3). But the particular type of SDTF depends on water availability. On the dry extreme (sites with low storage of soil water and no access to groundwater), a deciduous SDTF dominates the landscape, whereas semideciduous SDTF can develop in areas where some water accumulates in the soil. In areas where trees have access to groundwater, a perennial tree stratum develops in conjunction with a deciduous under-canopy stratum, as is the case in savannas or tropical woodlands.

Runoff and groundwater recharge are minor components of total

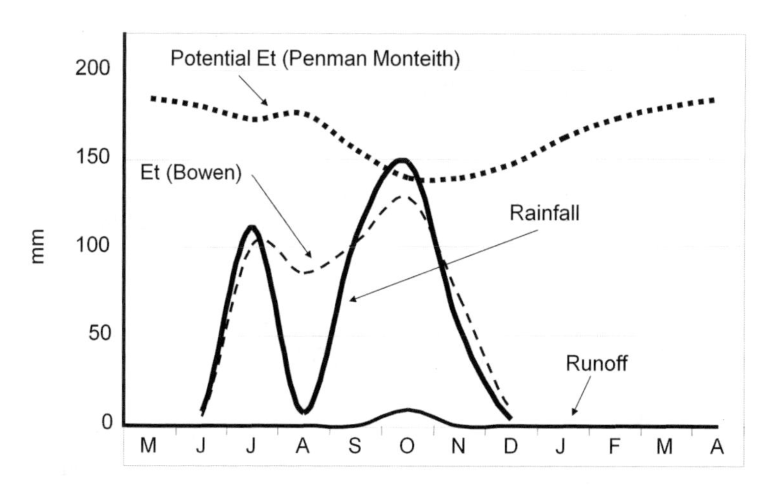

FIGURE 9-2. Water balance over a year in the tropical deciduous forest at Chamela, Jalisco, Mexico (Burgos 1999). Et = evapotranspiration.

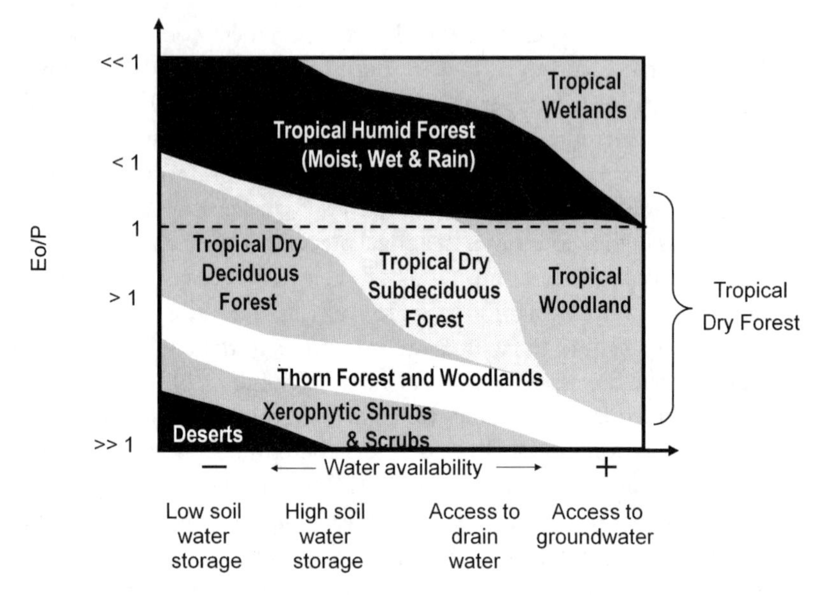

FIGURE 9-3. Abiotic factors controlling tropical ecosystems distribution.

ecosystem water balance (fig. 9-2). In Chamela, a series of 20 years has an average runoff of 45 millimeters, which represents only 7 percent of total annual precipitation (Cervantes et al. 1988; López-Guerrero 1992; Maass, Martínez-Yrízar et al. 2002). However, runoff is the major source of water for human settlers on SDTF environments, making water availability the most limiting factor for socioeconomic development. This water restriction may be exacerbated under climatic change scenarios, as hydrological models forecast an increase in evapotranspiration and a reduction in runoff in higher-temperature scenarios. Even in simulations of plant responses where stomata resistance is doubled, a 70 percent reduction in runoff is forecast for a 20 percent increase in average air temperature (Vose and Maass 1999).

Searching for Hydrological Connections in Tropical Dry Forest Ecosystems

In SDTF, hydrological processes strongly control rates and timing of essential ecosystem processes, such as species establishment, primary production, and nutrient dynamics. Hydrological processes also play a role in the spatial distribution of vegetation and dynamics along environmental gradients (i.e., the spatial dimension) and are important factors that explain vegetation successional pathways after anthropogenic perturbation (i.e., the temporal dimension). In a similar way, humans and their productive activities are closely coupled to rainfall patterns and hydrological processes within the ecosystem. All these connections result in an inseparable "water-regulated" socioecosystem, present all over the dry tropics, with emergent properties at different levels of organization. In this section, three approaches are used as an example of complementary points of view for searching hydrological connections in SDTF.

The Mechanistic Approach: Hydrological Connections and Plant Cycle Stages

Life-history traits of tropical dry forest species have been the focus of many studies on topics as diverse as seed ecology and seedling growth (Khurana and Singh 2001), plant dispersal traits (Gillespie 1999), phenological rhythms (McLaren and McDonald 2005), and species-specific biological effects on forest regeneration (Vieira and Scariot 2006). Water has been recognized as a strong structuring factor in SDTF. However, the identification of connections between hydrological processes and plant cycle stages from a causal point of view has not been explored enough. This kind of mechanistic analysis can be performed using causal-loop diagrams in a kind of systems thinking tool consisting of arrows connecting variables (things that change over time) in a way that shows how one variable affects another (Saysel and Barlas 2001; Spector et al. 2001). In this section, two main plant cycle stages were searched for hydrological connections using causal-loop diagrams: (1) predispersal stage–seed production, seed dispersal/predation (fig. 9-4) and (2) postdispersal stage–seed germination and seedling emergence (fig. 9-5). In this exercise, primary factors that trigger a chain of intermediate conditions are represented by squares. State variables of fast change (i.e., with short-time life) are shown with ovals. Resulting processes are represented by rounded rectangles (rectangles with trimmed corners), and specific supplies (such as seeds and reproductive trees in the first example) are indicated with thick arrows.

Processes of seed production, dispersal, and predation determine propagule availability at the landscape level and play an important role in defining successional pathways (Pickett et al. 1987). In SDTF, the relationship of these processes with ecosystem hydrological dynamics is mainly

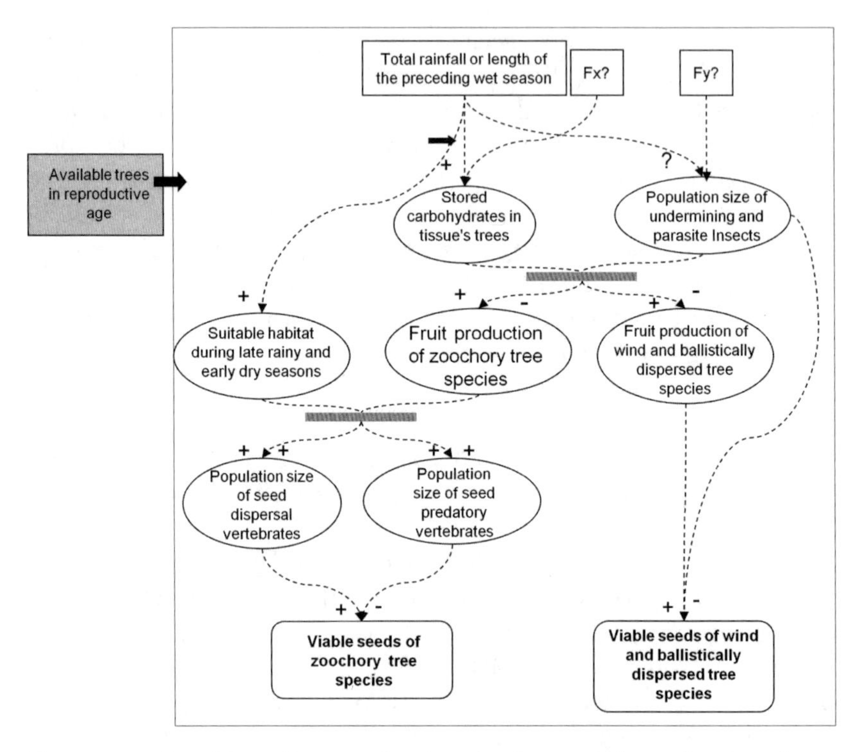

FIGURE 9-4. Causal-loop diagram of seed production, seed dispersal, and predatory effects on seeds.

known in an evolutionary sense (ultimate causes), but ecological connections (proximal causes) are less understood.

Concerning ultimate causes, two general patterns have been described linking seed production with hydrological conditions. First, according to the germination hypothesis (Van Schaik et al. 1993), the peak of seed dispersal occurs mainly between the last 2 months of the dry season and the first month of the rainy season (Gillespie 1999; Justiniano and Fredericksen 2000; Griz and Machado 2001; Grombone-Guarantini and Ribeiro-Rodrigues 2002; McLaren and McDonald 2005), since plants of SDTF adjust the fruiting time to precede the optimal time for germination. The second general evolutionary pattern describes a positive relationship between the percentage of wind-dispersed species and the length of the dry season and recognizes the predominance of animal-dispersed tree species when the rainy season is longer (Ibarra-Manríquez et al. 1991).

In contrast, proximate causes to explain connection of seed produc-

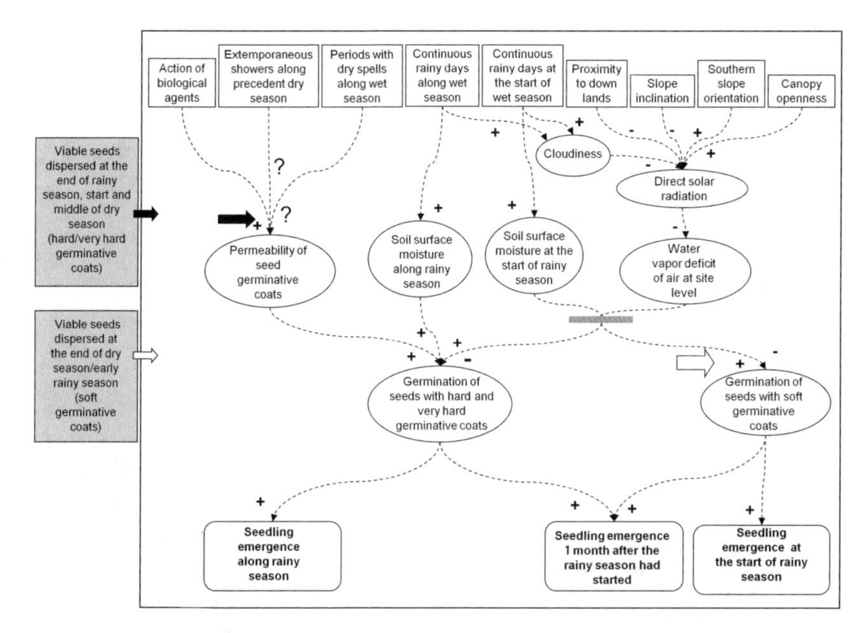

FIGURE 9-5. Causal-loop diagram of seed germination and seedling emergence.

tion with hydrological processes in SDTF have been less evident, and only some elements for building hypotheses can be presented (fig. 9-4). It was observed that the production of viable seeds is highly variable among years but that a regular pattern can be found among species (Fredericksen and Mostacedo 2000; Marod et al. 2002). Most trees species of SDTF require carbohydrates and nutrients for fruit production when water is unavailable in the late dry season; therefore those carbohydrates and nutrients stored in stem and root tissues during the wet season must be translocated to the buds. Available water in the preceding wet season, resulting in tree energy balance and nutrients, could be the most important abiotic factor determining fruit production in these tree species. However, the strength of this relationship can vary among different phenological patterns (e.g., species flowering in dry vs. wet season) or between groups of individuals located at different topographical positions in the landscape (Murali and Sukumar 1994).

In addition to seed production, the stock of available seeds at landscape level is also affected by dispersal and predation. Wind-dispersed tree species are abundant in SDTF. Many of them are attacked by insects that colonize flowers or fruits, promoting, in some cases, the miscarriage of the embryos (see also chap. 6). Some vertebrates, such as rodents and bats, can also impact the stock of viable seeds of particular species by predation when fruits reach the ground (Pizo and Oliveira 2000; chap. 6). In the case of vertebrates, years with longer wet seasons than average have less extreme conditions in the following dry seasons, with more food and water available (Dirzo and Domínguez 1995; Valenzuela 2002). However, these connections between annual rainfall variation and population dynamics processes in predator species are scarcely known. In general terms, the hydrological connections behind seed production, dispersers and predators, and seed availability of trees in SDTF seem to be mostly weak and indirect, and there are several gaps of information to fill before being more conclusive.

Seed germination/seedling emergence is another important plant life stage that can be explored using causal diagrams (fig. 9-5). The microclimate surrounding seeds is affected by both water vapor deficit and soil surface moisture. Water vapor deficit depends on direct solar radiation reaching the soil, a condition affected by primary factors such as canopy openness, slope position, inclination, and exposition. Soil surface moisture, on the other hand, is mainly dependent on the sequence of rainy days along the rainy season, and even small rainfall events are important. For example, rainfall events less than 5 millimeters, common in SDTF, have no impact on soil water recharge. However, small water pulses are capable of moistening a few centimeters of litter or superficial soil, promoting conditions for seed germination.

The rate of seedling emergence in dry forests peaks at the start of the rainy season, probably an adaptation to take advantage of as much of the wet season as possible for growing (Garwood 1983). However, the time and conditions when seeds become permeable to water vary among species, since they are related to factors such as time of dispersal, seed characteristics, length of dormancy, and the previous action of natural agents on their germinative coats (Khurana and Singh 2001; Marod et al. 2002). Therefore, water vapor deficit and soil surface moisture trigger germination only in the cases of seeds with soft seed coats and high permeability. These seed types are frequently dispersed late in the dry season or early in the wet season (Blakesley et al. 2002). In the case of seeds with hard and very hard coats, the permeability must be included as an intermediate condition determining germination (fig. 9-5). Permeability of coats is affected by primary factors acting on seed coats several months prior to the wet season. Physical scarification can be provided by factors such as hydration-dehydration pulses. For example, seeds of *Cordia elaeagnoides* (an abundant wind-dispersed species that fruits at the end of the wet season) show very low or no germination in the case of new seeds (taken directly from trees). However, "older" seeds (taken from the soil under the trees) under similar conditions germinate. It was proposed that rainfall anomalies, such as late rains in the wet season and early rains in the dry season, may control establishment in discrete episodes for many years for *C. elaeagnoides* (Bullock 2002). Such hydration-dehydration pulses can be created by extemporaneous showers during the dry season that moisten the soil surface under intense solar radiation conditions, as well as dry spells during the wet season that create a lapse of dryness within a period of dampness (fig. 9-5). Given these relationships between rainfall pattern and germination types, different associations can be found. The germination of seeds with soft coats is favored by a sequence of small but continuous rainfall events and cloudy days at the start of rainy season. These seeds can also be negatively affected by dry spells (Vieira and Scariot 2006). On the other hand, germination of seeds with hard and very hard coats could be favored by the interruption of rainfall events at the start of the rainy season, and these seeds usually germinate asynchronously along the rainy season (fig. 9-5). Similarly, the interruption of wet conditions by dry spells early in the wet season has been suggested as the major obstacle for seedling recruitment (Ray and Brown 1995), but these conditions promote a gradual reduction and even total loss of dormancy, a phenomenon exclusive to seeds subjected to low-humidity conditions (Khurana and Singh 2001). Each different combination of rainfall conditions during dry and wet seasons provides conditions for different germination types. The high interannual rainfall variability and the rolling landscapes usually associated with SDTF seem to be cofactors creating highly variable hydrological conditions, not only at micro spatial scales (within a few meters), but also at small temporal scales (within months or years). This hydrologic variability is reflected in the high beta diversity observed in these SDTFs (Balvanera et al. 2002) and may explain the lack of a clear regeneration pattern in these forests after disturbances (Kennard 2002).

The hydrological connections to seed germination and seedling emergence are better known and apparently stronger than connections to other plant life cycle stages. As described above, causal diagrams help to reveal the hydrological connections and the existing knowledge about them. Vegetative and reproductive phenology and other plant cycle stages, such as plant establishment, growth, survival, tree coppicing, and mortality, can also be reviewed with this approach. A combination of the analyses of plant cycle stages and hydrological processes can provide a more complete understanding of the role of water in the full dynamics of the ecosystem, and novel hypotheses to understand the functioning of SDTF can be established.

The Ecosystem Approach

From an ecosystem perspective, hydrological connections must be examined using large spatial and temporal scales for capturing spatial heterogeneity and temporal variability in hydrological conditions.

Spatial heterogeneity in dry forest is usually high when rolling topographies dominate the landscape. In this case, site water balance shows important differences according to topographical positions (ridges, slopes, or valleys), slope exposure (north, south), and soil properties (e.g., texture, percent organic matter). The importance of temporal variability in hydrological conditions becomes evident when interannual rainfall patterns are recognized, because contrasting climatic water balance regimes appear between dry and wet years. This interannual rainfall variability has effects on ecosystem function that are not direct and linear. To exemplify these relationships, figure 9-6 shows hypothetical connections between rainfall and ecosystem state variables such as net primary productivity, litter standing crop, recruitment, coppicing, cattle herd size, grazing intensity, and soil erosion, as well as the delayed effects of the preceding year's rainfall. In this analysis, data for each state variable in a particular year are presented under three standard conditions: at the maximum, average, and minimum values of its known range. Using a long series of rainfall data from the locality of Churumuco, Michoacán, Mexico, four common situations were analyzed: two dry years (1941 and 1960) with about 520 millimeters of annual rainfall, which is 18 percent under the average rainfall; and two wet years (1968 and 1992) with about 850 millimeters, which is 34 percent above the average. The first selected dry year (1941) was preceded by two drier years. Under this condition, all state variables show their values at the minimum of the range, meaning that all ecosystem processes were occurring at very low rates, generating no (or little) change of state. However, the second selected dry year (1960) was preceded by two wetter years and thus represents a transitional regressive state where some variables, like overgrazing, cattle herd size, and standing litter levels, remain at high values, as a legacy (inertia) of the preceding wet years. Some other state variables, such as recruitment and coppicing, are in the middle range of values, showing some response to rainfall

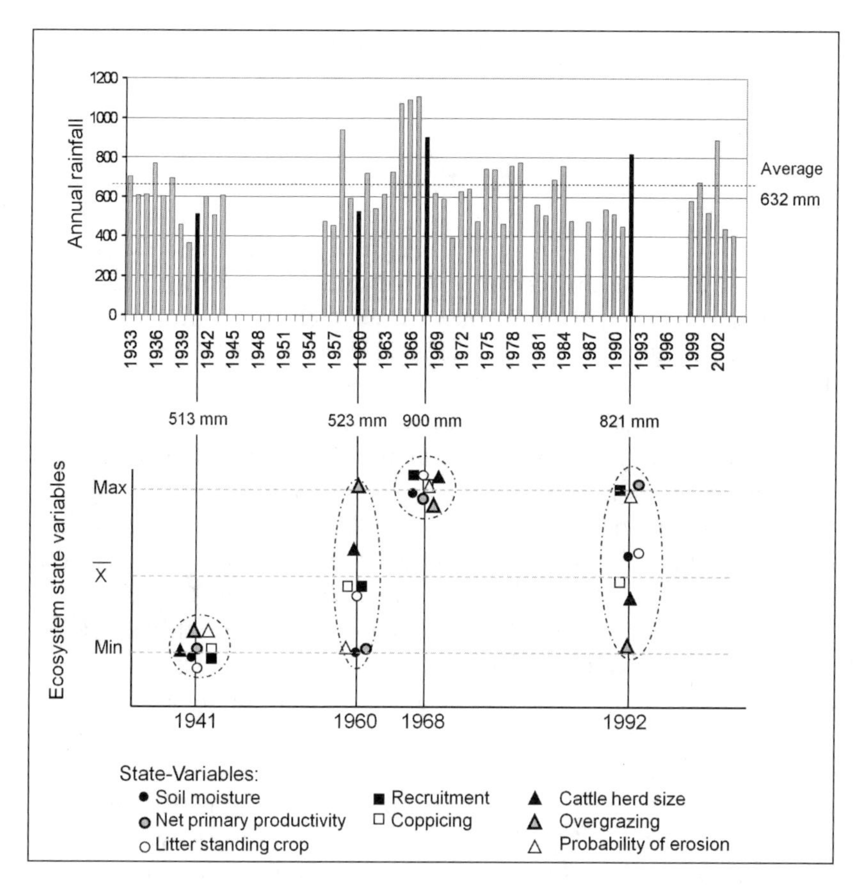

FIGURE 9-6. Hypothetical ecosystem responses to interannual rainfall variability.

conditions. Finally, there are other state variables of immediate response, such as soil moisture and net primary production, which show values at the minimum of the potential range.

The third and fourth cases analyzed correspond to a situation opposite to the previous two. In 1968, a wet year was preceded by two very wet years, and presumably all state variables would show their maximum rates. However, in a similarly high rainfall year (1992), the very dry preceding years should push the ecosystem to another transitional state in which slow-response variables, such as standing litter levels or cattle herd size, would show the legacy (inertia) of previous dry years with their minimum values, while fast-response variables, such as soil moisture and net primary production, would show their maximum (fig. 9-6). The recognition of this interannual dependence on the preceding year's rainfall is a key component to correctly interpreting ecosystem processes occurring at a particular time.

A Socioecological System Approach

Studies on vegetation structure in SDTFs have increased in recent years in an effort to elucidate the successional dynamics of ecosystems affected by anthropogenic land use change (Maass 1995; González-Iturbe et al. 2002; Kalacska et al. 2004; Ruiz et al. 2005). However, the understanding of the structural patterns of these forested landscapes requires a more integrated, multidimensional approach, including the study of social factors driving local management decisions. Studies about resilience, vulnerability, and adaptability of entire ecosystems are important for understanding ecosystems' behavior and evolution and for building scenarios to face the uncertainty of future conditions. Recently, more holistic conceptualizations have emerged in concepts such as socioecological systems (Gallopin 2006; Young et al. 2006), coupled human-environment systems (Turner, Kasperson et al. 2003), or coupled human-natural systems (Walsh 2008). Tropical dry forests around the world are mainly inhabited by poor farmer families with lifestyles highly coupled with environmental dynamics. Thus they can be seen as socioecological systems where people's lives are strongly affected by the intra-annual and interannual rainfall variability. Turner, Matson et al. (2003) used a coupled human-environment system perspective to analyze the vulnerability of the socioecological system of a tropical dry forest in the Yucatán and recognized that water stress and hurricanes are two environmental hazards strongly affecting the farmers' lives. Their productive options are usually restricted to subsistence agriculture and cattle ranching. Annual rainfall variability brings about a high level of uncertainty, making short-term benefits highly unpredictable and making it difficult for farmers to plan and decide about management options. Thus, for example, cattle ranching is a common livelihood, since it offers a more rapid and adaptable way to face climatic uncertainty and because, in case of rainfall failure, it incurs lower costs than agriculture (Burgos and Maass 2004). Agriculture is limited by a very short and variable rainfall period, implying high risks, since crops can be completely lost in years of low rainfall or unusual intra-annual distribution (e.g., dry spells within the wet season), which result in the loss of farmers' investments. The close relationships among annual meteorological conditions, ecological processes, and economic activities create the high vulnerability of these socioecological systems. In Mexico, for example, annual rainfall, forage availability, water availability, and cattle stocks seem to be tightly woven in complex relationships, with lag effects and inertial behavior that are better understood from a long-term perspective. Overgrazing is expected to be greatest during a dry year preceded by two wet years (see 1960 in fig. 9-6), whereas it is expected to be least during wet years preceded by dry years (1992). Other social processes, such as emigration and abandonment of traditional agricultural jobs, also seem to be associated with sequences of dry years, among other factors.

Conclusions

Water is the most limiting factor for the function of dry tropical forest ecosystems. Water inputs occur mainly as rainfall, which is strongly seasonal and highly variable between years. The beginning of the rainy season seems to be more predictable than the end of the growing season. Quantity, quality, timing, and energy of the water change as it passes through the canopy and the soil profile. Soil water is highly dynamic and dependent on rainfall distribution during the rainy season. Topography has a major impact on soil water distribution and availability. Potential evapotranspiration is much higher than precipitation, with transpiration being the most important output flux of water in the ecosystem. Runoff flow values are very low and mostly affect the subsurface. The El Niño/Southern Oscillation has a strong influence on annual rainfall but does not completely determine the water dynamics in these ecosystems. Hydrological models forecast an increase in evapotranspiration and runoff reductions under higher-temperature scenarios. Rainfall history (preceding rainfall) highly influences forest dynamics under natural conditions as well as under management and recovery. Water dynamics in forest succession is a promising field of research.

Beyond the traditional hydrological analysis, this review used a systems approach to search for hydrological connections in the tropical dry forest ecosystem. First, a mechanistic view was used to recognize causal relationships among hydrological factors and determinant processes of plant life cycles. Seed production (including dispersal and predation), as well as seed germination (including early establishment), was used as an example of how to synthesize complex hydrological information and how to formulate new hypotheses. Second, the review explored available knowledge about hydrological connections behind the system behavior. Interactions among rainfall-related variables of different temporal characteristics were recognized. The hypothesis that different ecosystem states exist in years that have the same total annual rainfall but different preceding wet or dry years was analyzed. Finally, the use of a socioecological systems approach was emphasized, in which the human component is analyzed in its interaction with hydrological processes. The close relationships among annual meteorological conditions, ecological processes, and economic activities create the high vulnerability of this socioecological system. The need for new, more integrative and creative approaches was recognized, as they are the keys to fully understanding the connections among hydrological conditions and land use, social variables (such as migration and agriculture abandonment), livestock dynamics, and long-term changes in land cover.

Acknowledgments

The authors are deeply grateful to Hillary Young, Rodolfo Dirzo, and anonymous reviewers of the manuscript. Technical support from Raúl Ahedo, Salvador Araiza, Heberto Ferreira, Alberto Valencia, and Atzimba López is also acknowledged. Part of this review was conducted during a sabbatical visit of Manuel Maass to Rob Vertessy's laboratory at the Land and Water Division, CSIRO Black Mountain, Canberra, Australia. This is a contribution from the Ecosystem Group at the Centro de Investigaciones en Ecosistemas, UNAM, which has received financial support from the Consejo Nacional de Ciencia y Tecnología (Mexico) and the Dirección General de Asuntos del Personal Académico (UNAM).

PART IV

Human Impacts and Conservation in Seasonally Dry Tropical Forests

Chapter 10

Impact of Anthropogenic Transformation of Seasonally Dry Tropical Forests on Ecosystem Biogeochemical Processes

FELIPE GARCÍA-OLIVA AND VÍCTOR J. JARAMILLO

Seasonally dry tropical forests (SDTFs) were recognized as the most endangered major tropical ecosystem in 1988 because of their high deforestation rates (Janzen 1988c). Maass (1995) analyzed the effect of conversion of SDTF to agriculture and pastures on ecosystem processes. He concluded that the main consequences of these conversions were (1) reduction in species diversity, (2) reduction in soil vegetative cover, (3) disruption of the water cycle, (4) changes in nutrient status, and (5) losses of nutrients from ecosystems through different pathways.

Unfortunately, the rates of deforestation of SDTF in Latin America have not decreased since the mid 1990s (chap. 3, 15). We review the state of knowledge on the consequences of SDTF transformations in the Neotropics due to land use change on biogeochemical ecosystem processes, based on the previous review by Maass (1995). Thus, we consider changes in carbon, nitrogen, and phosphorus pools (i.e., biomass and soil) and alteration of nutrient dynamics, including processes controlling soil nutrient availability and mineralization. The effects occur at two temporal scales: in the short term, like the consequences of slash and burn, and in the longer term, after several years of management. The studies covered in the present chapter are based on three SDTF sites: Chamela (Jalisco, Mexico), Caatinga (Pernambuco, Brazil), and the Cerrado (Brasilia, Brazil) because these sites have been studied for several years and their data can be compared. Strictly, Cerrado does not belong in SDTF (see chap. 1) but is used for comparative purposes as a seasonally dry tropical ecosystem.

The disruption of biogeochemical cycles is not only related to depletion of soil fertility and plant production in specific sites, it also affects global cycles by increasing the emission of greenhouse gases or by decreasing the efficiency of greenhouse gas sinks. For example, deforestation in Latin American tropical forests accounts for an important amount of carbon emissions, with an estimated annual rate of 0.75 petagram carbon per year (Houghton 2003).

Aboveground Biomass and Nutrient Pools

The higher storage of nutrient pools in aboveground biomass of tropical forest ecosystems (Trumbore et al. 1995; Kauffman et al. 1998; Hughes et al. 2000) makes them more vulnerable to disturbances (deforestation, fire, etc.) than temperate forests (García-Oliva, Hernández et al. 2006). Slash and burn is commonly used in SDTF conversion in Latin America (Maass 1995) because fire is an inexpensive tool and improves the productivity of cattle pastures (Kauffman et al. 1998). Several authors have reported increases in soil fertility following fire (Raison 1979; Ramakrishnan and Toky 1981; Buschbacher et al. 1988; Bruijnzeel 1990; García-Oliva and Maass 1998), but they are short-lived (Singh 1989; Srivastava and Singh 1989; Tiessen et al. 1992; García-Oliva and Maass 1998). The reduction in available nutrients in the soil is explained by fire disruption of the main soil nutrient sources: aboveground biomass, soil organic matter, and soil microbial biomass (Walker et al. 1986; Giardina et al. 2000b).

Very few studies on the impact of slash and burn on aboveground biomass and nutrients have been conducted in SDTF in the Neotropics (see Kauffman and Kauffman et al. 1993, 2003). Fire in SDTF of Mexico and Brazil decreased aboveground biomass by 62 to 78 percent with low fire severity and up to 80 percent with high fire severity (table 10-1). Consequently, nutrient pools in aboveground biomass were also depleted; losses of carbon were 62 to 96 percent, losses of nitrogen 67 to 96 percent, and losses of phosphorus 4 to 56 percent (table 10-1). But the nutrient pools measured in aboveground biomass after fire included their contents in the ash derived from burned vegetation (Kauffman et al. 1993; Steele 2000), which may not be completely incorporated into the soil. Giardina et al. (2000a) estimated that 74 percent of nitrogen and 55 percent of phos-

TABLE 10-1. Total aboveground biomass and aboveground pools of carbon, nitrogen, and phosphorus before and after fire in two Latin American seasonally dry tropical forest sites

	High fire severity			*Low fire severity*			
	Prefire	*Postfire*	*% loss*	*Prefire*	*Postfire*	*% loss*	*Source*
Aboveground biomass ($Mg\ ha^{-1}$)							
Brazil	73.7 (1.9)	4.0 (1.1)	95	73.8 (2.7)	16.4 (2.9)	78	1
Mexico	134.9 (1.4)	26.8 (4.1)	80	118.2 (2.8)	43.6 (1.4)	62	2
C content ($Mg\ ha^{-1}$)							
Brazil	33.7 (1.9)	2.7 (0.5)	96	33.6 (4.2)	9.0 (1.5)	73	1
Mexico	64.7 (0.6)	13.6 (2.0)	80	56.5 (1.3)	21.6 (8.0)	62	3
N content ($kg\ ha^{-1}$)							
Brazil	547 (33)	21 (4)	96	539 (44)	110 (17)	79	1
Mexico	944 (14)	178 (22)	81	856 (32)	276 (98)	67	3
P content ($kg\ ha^{-1}$)							
Brazil	36.5 (2.0)	15.9 (0.2)	56	35.9 (3.0)	34.7 (1.2)	4	1
Mexico	26.8 (0.5)	14.3 (0.4)	46	25.2 (1.4)	15.0 (2.2)	40	3

Numbers are means and standard errors. Postfire nutrient contents included ash nutrients. Sources are 1: Kauffman et al. 1993; 2: Kauffman et al. 2003; 3: Steele 2000.

phorus contents in ash were lost by wind erosion before the first rainfall events, thus increasing the depletion of nutrient pools due to slash-and-burn management. Additionally, nutrients may also be lost from the upper soil because of root consumption by fire. For example, fine-root biomass (less than 1 millimeter in diameter) was decreased by 51 percent in the top 2 centimeters of soil by fire in the Chamela SDTF region (Castellanos et al. 2001). Although the seeding of pasture grasses increases aboveground biomass after slash and burn, it represents a small proportion of prefire forest biomass. For example, total aboveground biomass in 2-year-old and 13-year-old pastures was 30 percent and 18 percent, respectively, of that in the forest (Jaramillo et al. 2003). Similarly, fine-root production decreased 50 percent the first growing season following the slash fire (Castellanos et al. 2001), suggesting a further reduction in soil organic matter inputs in pasture soil compared with forest.

The decline of aboveground biomass in pasture increased the relative importance of soil concerning the ecosystem nutrient pools. For example, soil comprised 54 percent of ecosystem carbon pools in the forest but 89 percent in pastures (Jaramillo et al. 2003; fig. 10-1). A similar change occurred with nitrogen (fig. 10-1). These results indicate that soils play an increased key role in nutrient dynamics under pasture or agriculture after SDTF conversion, and thus soil processes may be critical for secondary succession in SDTF (chap. 7).

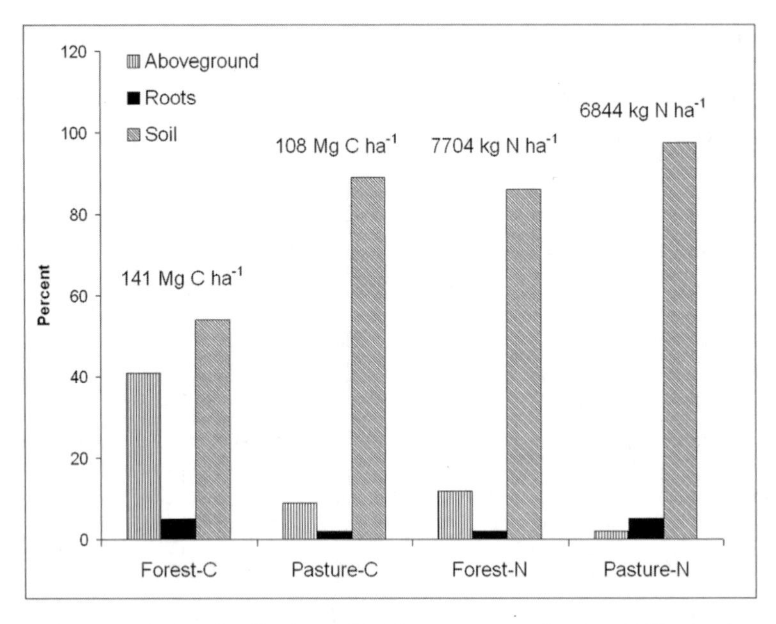

FIGURE 10-1. Relative distribution of total ecosystem carbon and nitrogen among the three main pools in a SDTF and derived pastures at Chamela, Mexico. Numbers over bars represent the total ecosystem pools. Soil values correspond to 60-centimeter depth (Jaramillo et al. 2003).

Soil Nutrient Dynamics

Management strongly affects physical and chemical soil characteristics. However, the depletion of soil fertility cannot be assessed only by the comparison of nutrient concentrations between forest and managed soils. Therefore the study of processes such as disruption of carbon, nitrogen, and phosphorus dynamics may allow a better understanding of soil fertility reduction with management, which in turn can adversely affect both forest regeneration and crop and pasture productivity (Giardina et al. 2000a; Steininger 2000; Davidson et al. 2004). Table 10-2 shows soil properties of four contrasting SDTF sites under forest cover. The site with the highest clay content has the highest concentration of soil organic carbon (SOC) (IBGE, Brazil), while the sandy oxisol has a lower carbon value (IPA-Pernambuco, Brazil; table 10-2). It is well documented that fine-textured soils have a higher capacity for carbon stabilization by organo-mineral complexes than coarse-textured soils (Tiessen and Stewart 1983; Dalal and Mayer 1987; Hassink 1997). Thus, SOC has a lower decomposition rate in clayed soils than in sandy soils, and SOC content is maintained longer in the former soil. Among the sandy soils, the sites with higher pH have higher nutrient concentrations than the site with the acid oxisol (IPA-Pernambuco, Brazil). A lower pH constrains plant productivity by decreasing soil nutrient availability, mainly soil phosphorus (Salcedo et al. 1997; Gallardo and González 2004). Moreover, total phosphorus concentration in the Chamela soil (regosol) is 1.6 times higher than in the acid oxisol (table 10-2). These results suggest that the fine-textured soils favor carbon

TABLE 10-2. Soil properties in four seasonally dry tropical forest sites in Latin America

	IBGE, Brazil (1)	*IPA-Pernambuco, Brazil (2, 3, 4)*	*Pernambuco, Brazil (5)*	*Chamela, Mexico (6, 7)*
Soil type (FAO)	Oxisol	Oxisol	Xerosol	Regosol
% sand; % clay	15%; 74%	78%; 10%	75%; n.d.	60%; 26%
pH	4.5	4.2–5.2	6.7	6.6
C (mg g^{-1})	47.4	9.8–12.2	14.1	29.0–36.0
N (mg g^{-1})	2.6	0.67–0.79	1.1	1.9–3.2
C:N	18	14–16	13	11–17
P (mg g^{-1})	n.d.	0.12	0.17	0.2–0.3

IBGE = Instituto Brasileiro de Geografia e Estatística, IPA = Institute of Agronomic Research, n.d. = no data.
Sources are (1) Varella et al. 2004; (2) Salcedo et al. 1997; (3) Sheng and Tiessen 2000; (4) Tiessen et al. 1992; (5) Wick et al. 2000; (6) García-Oliva et al. 1994; (7) García-Oliva, Gallardo et al. 2006.

accumulation, and soils with low pH could decrease nutrient availability, mainly soil phosphorus. For this reason, in the following sections we consider these soil properties further.

Organic Carbon Dynamics

Soil organic carbon plays an important role in the availability and recycling of soil nutrients and in the nutrient storage capacity of soils by influencing physical soil structure and microbial activity (Singh et al. 1989; Schmidt et al. 1997; Joergensen and Scheu 1999; Six et al. 2002; Montaño et al. 2007). For example, soil carbon availability stimulates the activity of heterotrophic microorganisms, which increases microbial demand for nitrogen and thus promotes nitrogen immobilization, allowing nitrogen protection in microbial biomass (Montaño et al. 2007). In contrast, low soil carbon availability reduces microbial nitrogen immobilization and promotes autotrophic nitrification activity (Hart et al. 1994; Chen and Stark 2000; Booth et al. 2005). This could increase nitrogen loss by nitrate leaching as well as by emissions of nitrogen oxides (Paul and Clark 1989; Davidson and Kingerlee 1997).

Salcedo et al. (1997) found no changes on SOC after fire in the upper 20 centimeters of soil in acid oxisols (IPA-Pernambuco, Brazil). This result may be explained by the low capacity of soil for conducting heat, which may be negligible below the top 5 centimeters of soil (Giardina et al. 2000a). Thus, the heat effect on the soil due to fire may be critical only in the upper 5 centimeters. However, SOC was reduced 46 percent 1 year after slash burning at the same study site (Tiessen and Santos 1989). In contrast, García-Oliva et al. (1999b) reported a decrease of 32 percent by combustion of SOC associated with macroaggregates (greater than 250 micrometers) in the top 2 centimeters of soil at Chamela, Mexico. This reduction in carbon did not affect soil macroaggregates, but it disrupted macroaggregate stabilization processes by reducing energy for microbial activity and destroying hyphae and fine roots. Subsequently, the breakdown of macroaggregates was the main factor affecting soil carbon pools after first rainy season: SOC and macroaggregates decreased 50 percent after this season. Breakdown of macroaggregates results in a lower carbon stabilization capacity of soils and a reduction in microbial activity (Six et al. 2002; Six et al. 2004; García-Oliva et al. 2004). Fire also affected soil microbial groups: there was a decrease in fungi, yeast, and bacterial populations of 80 percent, 64 percent, and 40 percent, respectively, after fire (García-Oliva et al. 1999a). Results from this study showed that ash inputs from the burned biomass modified soil chemical conditions for microbial populations. For example, ash addition to soil samples from the forest increased the pH of soil by one unit and reduced microbial activity compared with that in control forest soil (García-Oliva et al. 1999a). The increment of soil pH after ash input as documented for Pernambuco and Chamela soils (Salcedo et al. 1997; García-Oliva and Maass 1998; Giardina et al. 2000a) results from incorporation of alkaline cations (i.e., calcium, potassium, magnesium) in ash. In conclusion, organic carbon redistribution among physical-size fractions and alteration of microbial communities by fire are the driving factors for changes in SOC dynamics in pasture soils after forest transformation.

Reductions in SOC after several years of management in Neotropical SDTF have been reported in four studies, but two other studies have found no differences in SOC due to changes in land use (table 10-3). It has been proposed that cultivation reduces SOC faster than pasture because pastures have higher soil carbon inputs due to higher root productivity than in cultivated plots (Guo and Gifford 2002; Murty et al. 2002).

TABLE 10-3. Soil carbon and nitrogen concentrations in forest and pasture/crop plots in four SDTF sites in Latin America

Site	*Land use*	*Age of plot (years)*	*Soil depth (cm)*	*C (mg g^{-1})*	*N (mg g^{-1})*
IBGE, Brazil (1)	Forest		0–5	47.4	2.6
	Pasture	18	0–5	24.1*	2.5
IPA, Brazil (2)	Forest		0–20	10.3	0.74
	Cassava	5	0–20	8.9*	0.64*
IPA, Brazil (3)	Forest		0–20	12.2	0.79
	Sorghum	12	0–20	8.8*	0.64*
Pernambuco, Brazil (4)	Forest		0–8	14.1	1.1
	Pasture	14	0–8	12.0	1.0
Chamela, Mexico (5)	Forest		0–5	36.0	3.2
	Pasture	10	0–5	30.0*	1.3*
Chamela, Mexico (6)	Forest		0–10	25.4	1.2
	Pasture	13	0–10	23.6	1.4

*values that are significantly different between land use covers within each site ($P < 0.05$)
Sources are (1) Varella et al. 2004; (2) Lessa et al. 1996; (3) Sheng and Tiessen 2000; (4) Wick et al. 2000; (5) García-Oliva, Gallardo et al. 2006; (6) Jaramillo et al. 2003.

However, none of the studies in SDTF showed an increase in SOC under pasture, as it has been reported for pastures derived from tropical rain forest (Cerri et al. 1991; Neill et al. 1997). Conditions in wet tropical sites promote higher pasture productivity than in SDTF, increasing SOC inputs by roots. The reduction of SOC could be due to the depletion of new soil carbon input. For example, at Chamela, total fine-root productivity at the top 10 centimeters of soil was 42 percent lower in pasture than in forest soils (1.2 and 1.8 megagrams per hectare per year, respectively; Castellanos et al. 2001). It is also expected that aboveground net primary productivity in pastures is lower than in primary forest (9 and 12 megagrams per hectare per year, respectively; Martínez-Yrízar et al. 1996; Jaramillo et al. 2003) and that a fraction is consumed by cattle. This means that only a percentage of aboveground net primary productivity would be incorporated into the soil as dead organic matter (García-Oliva, Hernandez et al. 2006). In general, we would suggest that SOC decreases with pasture age in SDTF, disrupting soil nutrient sources, as mentioned above.

Nitrogen Dynamics

The effect of fire on soil nitrogen occurs as a consequence of high temperatures in soil organic matter and microbial populations. Fire increased by four times the NH_4 concentrations, while NO_3 showed a reduction by 50 percent or more in two SDTF study sites (IPA-Pernambuco, Brazil, and Chamela, Mexico; table 10-4). This high increment of NH_4 is due to thermal decomposition of organic nitrogen, protein hydrolysis, and release of microbial nitrogen due to lysis of microbial biomass (Rusell et al. 1974; Dunn et al. 1979; Raison 1979). In contrast, NO_3 is depleted by volatilization (Raison 1979). Similarly, fire promotes soil organic nitrogen loss as NO_x gases or N_2 at temperatures above 300 degrees Celsius (Raison 1979), resulting in increased nitrogen gas emissions during the burn. Also, biological nitrogen mineralization is reduced by a decrease in microbial activity, as mentioned above. Thus, fire disrupts the soil organic nitrogen cycle by decreasing both microbial immobilization and plant uptake, by substantial increases in soil NH_4, and by greater nitrogen losses due to erosion and leaching and nitrogen gas emissions (fig. 10-2).

The increment of NH_4 is temporary and decreases after the first growing season because of microbial and plant demand and mainly by nitrification, which increases NO_3 concentrations in pasture/crop soils (table 10-4). Increased nitrification has been reported with low soil carbon avail-

TABLE 10-4. Total soil N (N_t), NH_4-N, NO_3-N, and N mineralization for prefire, postfire, and crop/pasture from two SDTF sites in Latin America

	IPA-Pernambuco, Brazil (1)			*Chamela, Mexico*		
	Prefire	*Postfire*	*Crop*	*Prefire (2)*	*Postfire (2)*	*Pasture (2, 3)*
N_t (mg g^{-1})	0.67	0.72	0.60	5.4	4.1	1.1
NH_4 (μg g^{-1})	3.1	13.0	4.6	22.0	84.0	5.0
NO_3 (μg g^{-1})	48.0	25.0	33.3	21.0	3.7	4.0
N mineralization (μg g^{-1} d^{-1})	0.61	0.46	0.45	1.2	0.7	0.1

Sources are (1) Salcedo et al. 1997; (2) Ellingson et al. 2000; (3) García-Oliva et al. 2006b.

ability, which constrains nitrogen immobilization by heterotrophic microbes and promotes nitrifying-bacteria activity (Hart et al. 1994; Chen and Stark 2000; Montaño et al. 2007). Soil nitrogen availability is depleted with management time because soil organic matter inputs are reduced in pasture/crop soils compared with forest. Although nitrogen losses by gas emissions were similar between forest and pasture soils (García-Méndez et al. 1991; Varella et al. 2004), pastures may have net losses due to the reduction of nitrogen inputs by nitrogen fixation, which occurs in legume species of SDTF (González-Ruiz et al. 2008). Additionally, nitrogen losses through leaching could increase, because NO_3 dominates over NH_4. In contrast to the effect of fire after slash burning, NH_4 pools did not increase after fire in a 2-year-old pasture, likely because of the lower fire intensity in pasture compared with the slash fire and of the lower nitrogen pools in pasture soils (Ellingson et al. 2000). In sum, nitrogen pools are reduced under pasture/crop soils because nitrogen inputs decrease, the plant-microbe-soil loop is disrupted, and nitrogen losses increase (fig. 10-2). This trend may be exacerbated through time with pasture management.

Phosphorus Dynamics

Soil phosphorus dynamics are also strongly affected by slash burning. Fire increased by five times the soil-available phosphorus at both Chamela and Pernambuco SDTF sites (fig. 10-3).

Such a response may be due to phosphorus inputs associated with

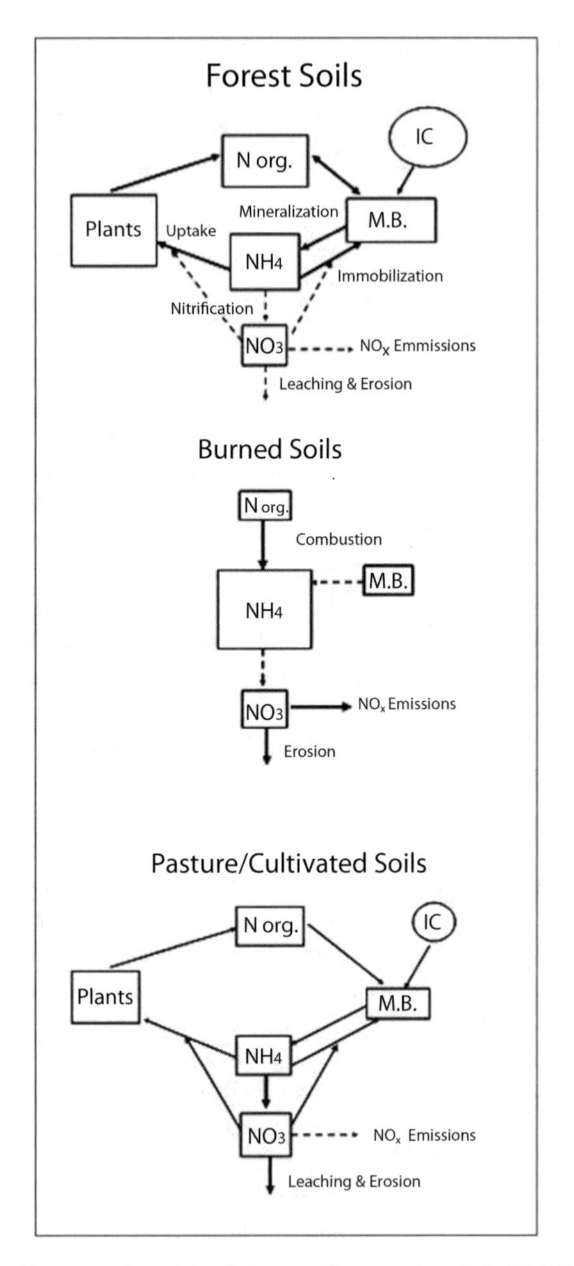

FIGURE 10-2. Conceptual models of nitrogen fluxes and pools in SDTF soil, burned soil, and pasture/crop soil. The box size indicates the relative difference in nutrient concentrations, while arrow type shows the relative importance of fluxes. Solid arrows indicate relatively more important fluxes than dashed arrows. N org. = organic nitrogen, lC = labile carbon, and M.B. = microbial biomass.

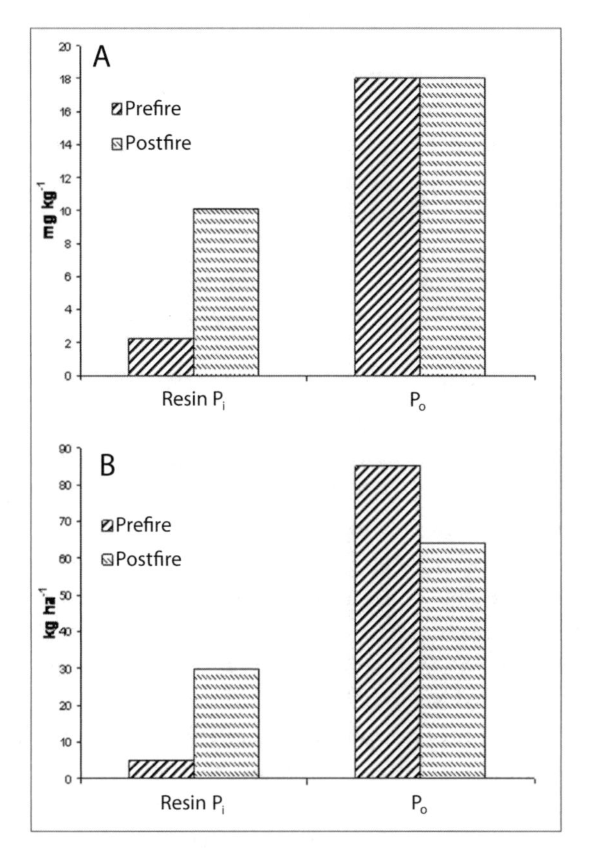

FIGURE 10-3. Effect of fire in two phosphorus fractions: organic (P_o) and inorganic (P_i) at two SDTF sites: (A) IPA-Pernambuco, Brazil (Salcedo et al. 1997), and (B) Chamela, Mexico (Giardina et al. 2000b). Note that Brazil values are concentrations and Chamela values are pools.

ash, thermal transformation of soil organic phosphorus, and desorption of occluded phosphorus by soil chemical changes (Sibanda and Young 1989; Salcedo et al. 1997; Giardina et al. 2000a). The effect of ash inputs could be negligible because a significant amount of ash is lost by erosion and only a small proportion of ash nutrients becomes available for plants (Khanna et al. 1994). For example, Giardina et al. (2000a) estimated that 55 percent of the phosphorus contained in ashes was lost before the first rainfalls at the SDTF in Chamela, Mexico. In contrast, thermal mineralization of organic phosphorus could better explain this increment, because NaOH organic phosphorus was reduced 50 percent after fire (Giardina et al. 2000a). Additionally, the occluded phosphorus pool diminished 20 percent after fire (Giardina et al. 2000a). Chemical changes in soil due to ash inputs and high temperatures promote phosphorus desorption from nonavailable forms (Sibanda and Young 1989). For example, Salcedo et al. (1997) found that pH and exchangeable basic cations were increased after fire, which accounted for increments in resin phosphorus after fire. Results from the studies above suggest that increases of plant-available phosphorus with fire are likely through thermal decomposition of organic phosphorus and by phosphorus desorption from inorganic forms due to soil chemical changes (fig. 10-4).

Similar to soil nitrogen, the high concentration of available phosphorus fractions does not persist with management time. Salcedo et al. (1997) reported that resin phosphorus decreased 71 percent after 5 years of cropping in an acid oxisol at Pernambuco, Brazil, due to organic matter depletion and soil reacidification (Tiessen et al. 1992). Similarly, Alvarez-Santiago (2002) reported a reduction of plant-available phosphorus and an increase of occluded phosphorus in a 10-year-old pasture soil at Chamela, Mexico. This was likely due to depletion of SOC, a lower soil pH buffer capacity, and an increase in soil Ca, which sequesters available phosphorus (García-Oliva and Maass 1998; Nava-Mendoza et al. 2000). Similarly, Lardy et al. (2002) found a positive correlation between soil carbon and phosphorus contents in different Cerrado vegetation types and pasture in Brazil, suggesting that the SOC plays a key role in soil phosphorus cycling in these ecosystems. These results suggest that the reduction of SOC and increments of phosphorus sorption by changing pH buffer capacity are the main factors that decrease soil phosphorus availability (fig. 10-4). Available phosphorus fractions could be reduced with management with a concurrent increase in occluded phosphorus forms.

Conclusions

The main consequences of SDTF conversion to agriculture or pasture for nutrient dynamics are the reduction of ecosystem nutrient pools and the disruption of nutrient protection mechanisms, resulting in a gradual depletion of soil fertility and plant production. The major nutrient pools in SDTF are in the aboveground biomass, thus these forest ecosystems are vulnerable to anthropogenic disturbances such as deforestation. For this reason, the plant-microbe-soil loop is critical for the maintenance of available nutrients and plant productivity.

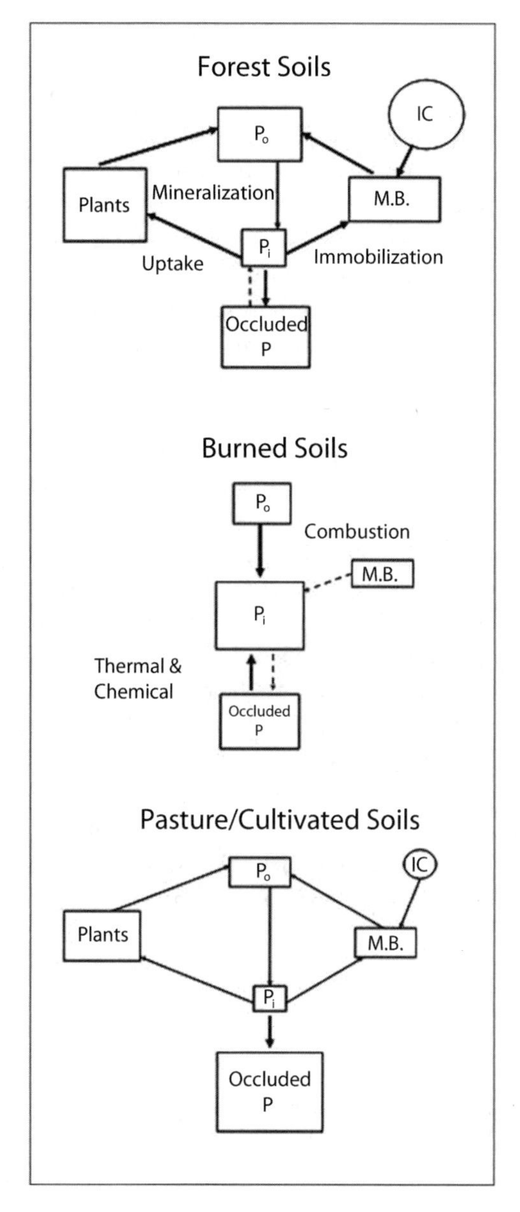

FIGURE 10-4. Conceptual models of phosphorus fluxes and pools in SDTF soil, burned soil, and pasture/crop soil. The box size indicates the relative difference in nutrient concentrations, while arrow type shows the relative importance of fluxes. Solid arrows indicate relatively more important fluxes than dashed arrows. lC = labile carbon, M.B. = microbial biomass, P_o = organic phosphorus, and P_i = available inorganic phosphorus.

Soil type influences the magnitude of biogeochemical processes: deterioration processes are similar, but the rates differ. For example, an acid sandy soil can be degraded faster than a neutral sandy soil. As a consequence, the sites with the first soil type (e.g., IPA-Pernambuco, Brazil) should be used for shorter periods than sites with neutral sandy soils (e.g., Chamela, Mexico). These results suggest that the same driving processes must be improved irrespective of soil type for soil reclamation, but the intensity of actions must be adjusted to the soil type.

Disruption of the plant-microbe-soil loop starts with slash-and-burn practices and continues under pasture/crop management. Fire increases short-term soil nitrogen and phosphorus availability, but it degrades biological and geochemical mechanisms of nutrient protection, such as microbial biomass and soil pH buffer capacity. As a consequence, nutrient losses may dominate over nutrient inputs in pasture/crop soils of dry tropical regions, gradually reducing plant productivity as management continues.

At a global scale, the consequences of SDTF conversion are related to greenhouse gas emissions because Latin American tropical forests represent the second largest carbon pool globally, only exceeded by African tropical forests (162 and 176 gigagrams carbon, respectively; Houghton et al. 1983). Carbon dioxide emission rates from the Neotropical region due to land use change may amount to 0.75 petagram carbon per year (Houghton 2003), although emissions from land use and cover change have large uncertainties in the global carbon cycle (Ramankutty et al. 2007). The integration of studies on the biogeochemical consequences of land use and land cover change with global models remains as one of the most interesting challenges for the understanding of anthropogenic impacts on global element cycles.

Acknowledgments

We thank Maribel Nava-Mendoza, Rodrigo Velázquez-Durán, Heberto Ferreira, and Alberto Valencia for their help in the preparation of this manuscript.

Chapter 11

Human Impacts on Pollination, Reproduction, and Breeding Systems in Tropical Forest Plants

Mauricio Quesada, Fernando Rosas, Ramiro Aguilar, Lorena Ashworth, Víctor M. Rosas-Guerrero, Roberto Sayago, Jorge A. Lobo, Yvonne Herrerías-Diego, and Gumersindo Sánchez-Montoya

Over the last two decades several studies have shown that plant species of contrasting life-forms ranging from small herbs to large trees may experience a decline in reproductive success following habitat fragmentation and population disruption (Bawa 1990; Aizen and Feinsinger 1994; Aguilar et al. 2006). Such outcome has been shown for many plants throughout the tropics, particularly trees, where human activities have resulted in elevated rates of habitat fragmentation and degradation (Ghazoul and Shaanker 2004; Quesada and Stoner 2004; Quesada et al. 2004). Because almost 90 percent of angiosperms (i.e., flowering plants) depend on animals for effective pollination and sexual reproduction (Buchmann and Nabhan 1996), it is of central concern to understand the capacity of pollinators for transferring pollen among individuals and its consequences on plant reproduction in newly created anthropogenic landscapes.

While evolutionary dependence of plants on animal mutualists for sexual reproduction has improved pollen transfer to stigmas, it has also prompted increased plant susceptibility to fragmentation and other forms of anthropogenic disturbance that characterize today's landscapes (e.g., Aizen et al. 2002; Ashworth et al. 2004). Changes in abundance, composition, and/or foraging behavior of pollinators as a consequence of habitat disturbance will have an effect on the amount and/or quality (autogamous vs. xenogamous) of pollen deposited on stigmas, thus affecting reproduction and the genetic structure of plants (Wilcock and Neiland 2002).

Much research has been conducted with regard to the effects of habitat loss and fragmentation on pollination, plant reproduction, and genetic diversity of plant populations over the past 20 years. Nevertheless, there has been certain research bias in the selected natural systems evaluated, where species from tropical forests represent only 16 percent of the entire studied species around the world (Aguilar et al. 2006; Aguilar et al. 2008). Moreover, no specific analysis of this subset of species has yet been conducted. This comparative underrepresentation of tropical plant species in fragmentation studies highlights the need to focus more thoroughly on population studies from these threatened and fragile habitats.

Some expected outcomes of habitat fragmentation include local extinction of plant and animal populations, the alteration of species richness and abundance, and changes in the trophic structure of communities. These negative effects of habitat fragmentation can be expressed at the landscape and population levels. At the landscape scale, fragmentation involves the transformation of a large area of habitat into several patches of smaller size, isolated from each other by surrounding anthropogenic habitats different from the original. Such loss and breaking apart of the habitat alters negatively the connectivity, functioning, and biodiversity within the matrix of the fragmented habitat (Fahrig 2003). At the population level, habitat fragmentation may reduce the effective population size and the magnitude and direction of gene flow, which in turn would produce negative changes in the population and genetic structure of plant species (Young et al. 1996; Aguilar et al. 2008). The reduction of both gene flow and effective population size by habitat fragmentation may cause inbreeding, genetic drift, and a consequent decline of genetic variation. Therefore, the loss of genetic diversity may limit the ability of local populations to respond selectively to varying local conditions, compromising their persistence and increasing their risk of extinction due to inbreeding depression.

Habitat fragmentation may not only lead to a reduction in population size and genetic variation but also disrupt key interactions of the plants with their pollinators and seed dispersers. The interaction between plants and pollinators can be disrupted by habitat loss, reduction of pollinator abundance, changes in floral resource availability and distribution, or competitive exclusion from floral resources by inefficient or exotic pollinators. Most of the plant-pollinator interactions may depend on the relative abundance of floral resources; thus, changes in plant abundances caused by forest disturbance may lead to modification in the composition, functioning, and maintenance of plant-pollinator webs (Aizen and Feinsinger 2003; Lopezaraiza et al. 2007). We should expect small isolated or fragmented plant populations to be less attractive to pollinators than large populations. As a result of this, rates of pollinator visitation and seed production may often be lower in small than in large populations of plants pollinated by animals.

The negative consequences of habitat fragmentation for plant populations could be exacerbated by the complex interactions of reproductive (sex expression) and mating systems (selfing vs. outcrossing or mixed strategies) in combination with population size and pollination and seed dispersal systems. Previous studies of seasonally dry tropical forests (SDTFs) indicate that the reproduction of plants is dependent on the presence of natural pollinators (Frankie et al. 1974; Bullock 1985). Therefore, changes in the abundance and activity patterns of pollinators induced by habitat fragmentation are expected to reduce gene flow between isolated plant populations. The negative effects of forest fragmentation on the viability of populations could be particularly noticeable in tropical tree species that posses self-incompatibility systems and depend on pollinators for sexual reproduction (Bawa 1974, 1990; Aguilar et al. 2006). Disturbances that impact animal vectors of pollen transfer may therefore affect the reproductive output of tropical trees. Pollination of tropical plants is mainly conducted by animal vectors such as bees, butterflies, flies, birds, and bats, and the natural populations of these animals inhabit and depend on the existence of forests.

The objectives of this chapter are to (1) evaluate the effects of forest fragmentation on plant-pollinator interactions, plant phenology, reproductive dynamics, and genetic parameters of tropical plants; (2) describe and compare plant life-history traits, pollination systems, and plant reproductive traits between tropical forests; and (3) predict vulnerability patterns to forest fragmentation based on ecological and reproductive traits of plants.

Habitat Fragmentation Effects on Pollinator and Reproductive Dynamics and Genetic Parameters

We conducted a quantitative synthesis of the published literature on the effects of habitat fragmentation on plant reproductive dynamics and population genetic parameters of tropical plant species. To accomplish this, we gathered data from two published databases (Aguilar et al. 2006; Aguilar et al. 2008).

To assess the effects of habitat fragmentation on the pollination process and on the sexual reproduction of tropical plant species, we used an extract of the database compiled by Aguilar et al. (2006). This meta-analysis used published data from the literature within the period of 1987–2006 to evaluate the effects of habitat fragmentation on pollination and reproduction of plant species from different habitats throughout the world (see Aguilar et al. 2006). This analysis included articles that evaluated directly or indirectly, explicitly or implicitly, the effects of habitat fragmentation on the reproductive dynamics of animal-pollinated plants. As response variables, we used pollinator visit frequency, pollen loads on stigmas, or pollen tubes in styles for assessing the effects on pollination. We used fruit or seed production to assess the effects on plant reproductive success as provided by the published studies.

To assess the effects of habitat fragmentation on population genetic parameters of tropical plant species, we used part of the data obtained by Aguilar et al. (2008) through a literature search in the Science Citation Index and Biological Abstracts databases using a combination of "fragment*" and "genet*" and "plant" as keywords within the period of 1989–2007. This quantitative review evaluated the effects of habitat fragmentation on the genetic variability and inbreeding parameters of plant populations in fragmented habitats around the world. As measures of genetic variability, we considered both expected heterozygosity and allelic richness and analyzed them separately. Inbreeding was measured through Wright's fixation index (f_{IS}).

With these reproductive and genetic variables provided by the published studies that evaluated the effects of fragmentation, we calculated the effect size using Hedges's *d*. This effect size is calculated using the mean values, sample sizes, and standard deviations of each parameter (from text, tables, or graphs) and represents a standardized measure of the magnitude and direction of fragmentation effects for each of the species included in the analysis (cf. Gurevitch and Hedges 2001). Positive values of the effect size (*d*) imply positive effects of habitat fragmentation on a given parameter, whereas negative *d* values imply negative effects of fragmentation on these parameters. The only exception is the inbreeding coefficient parameter, which has an opposite interpretation: positive *d* values imply higher inbreeding in fragmented conditions, whereas negative *d* values mean lower inbreeding in fragmented habitats.

Within both databases (for reproductive and genetic parameters), we selected exclusively the results for tropical plant species and ran separated meta-analyses with these species using the MetaWin 2.0 statistical program (Rosenberg et al. 2000). Confidence intervals of effect sizes were calculated using bootstrap resampling procedures as described in Adams et al.

(1997). An effect of habitat fragmentation was considered significant if the 95 percent bias-corrected bootstrap confidence intervals (CI) of the effect size (*d*) did not overlap zero (Rosenberg et al. 2000). Data were analyzed using random-effect models, which assume that differences among studies are due to both sampling error and random variation, which is usually the rule in ecological data (Gurevitch and Hedges 2001).

Habitat Fragmentation Effects on Pollinator and Reproductive Dynamics

The literature search conducted by Aguilar et al. (2006) gathered 54 published studies that evaluated the effects of fragmentation on pollination and/or plant reproduction in 89 unique plant species from different regions of the world. Within this sample, there were 17 species from tropical forests, most of them (70 percent) trees, which represent 19 percent of the total sample of species included in this quantitative review. We ran two separate meta-analyses: one assessing fragmentation effects on pollination on 11 data points from 10 unique species and another one assessing effects on sexual reproduction on 17 data points from 16 unique species (table 11-1). In both analyses we included a replicate of *Ceiba grandiflora* data, as this species was studied twice in two different regions (Quesada et al. 2003; Quesada et al. 2004).

The overall effect size of habitat fragmentation on the pollination process of tropical plants was negative, of a large magnitude ($d = -0.923$), and significantly different from zero, according to the 95 percent bias-corrected bootstrap confidence limits (fig. 11-1A). Similarly, the overall effect size of habitat fragmentation on the sexual reproduction of tropical plants was also negative, of a large magnitude ($d = -0.971$), and significantly different from zero (i.e., confidence limits do not overlap zero; fig. 11-1A). These response patterns are in agreement with the overall general effect size found in Aguilar et al. (2006), but with larger magnitude of effect sizes for tropical species.

One of the best examples showing these patterns was in the SDTF species *Samanea saman* (Cascante et al. 2002). This study demonstrated that fragmentation of SDTF changed pollination patterns, which in turn reduced the genetic variability of the progeny and seedling vigor of this tree species (Cascante et al. 2002). The study also showed higher genetic similarity in the progeny of isolated trees, both within and between fruits. Seeds produced by different fruits within isolated trees were more likely to

TABLE 11-1. Tropical plant species in the meta-analyses evaluating the effects of fragmentation on pollination, reproduction, heterozygosity, allelic richness, and/or inbreeding coefficient

Species	*Family*	*Life-form*	*Parameter evaluated*					*Geographic region*	*Source*
			P	*RS*	*He*	*AR*	*IC*		
Anacardium excelsum	Anacardiaceae	Tree	X	X				Central America	Ghazoul and McLeish 2001
Brongniartia vazquezii	Fabaceae	Shrub			X	X	X	North America	Gonzáles-Astorga and Núñez-Farfán 2001
Carapa guianensis	Meliaceae	Tree			X	X	X	Central America	Dayanandan et al. 1999
Caryocar brasiliense	Caryocaraceae	Tree			X	X	X	South America	Collevatti et al. 2001
Catasetum viridiflavum	Orchidaceae	Epiphyte	X	X	X	X		Central America	Murren 2002, 2003
Ceiba aesculifolia	Bombacaceae	Tree	X	X	X		X	Central America	Quesada et al. 2004
Ceiba grandiflora	Bombacaceae	Tree	X	X				Central America	Quesada et al. 2003, Quesada et al. 2004
Ceiba pentandra	Bombacaceae	Tree	X					Central America	Quesada et al. 2004
Dieffenbachia seguine	Araceae	Perennial herb			X	X	X	North America	Cuartas-Hernández and Núñez-Farfán 2006
Dinizia excelsa	Fabaceae	Tree		X		X		South America	Dick 2001
Dombeya acutangula	Sterculiaceae	Tree		X				Asia	Gigord et al. 1999
Dyospiros montana	Ebenaceae	Tree	X	X				Asia	Somanathan and Borges 2000
Elaeocarpus williamsianus	Elaeocarpaceae	Tree		X				Oceania	Rossetto et al. 2004

TABLE II-1. *(continued)*

Species	*Family*	*Life-form*	*Parameter evaluated*					*Geographic region*	*Source*
			P	*RS*	*He*	*AR*	*IC*		
Enterolobium cyclocarpum	Fabaceae	Tree	X	X				Central America	Rocha and Aguilar 2001
Heliconia acuminata	Heliconiaceae	Perennial herb		X				South America	Bruna and Kress 2002
Oncidium ascendens	Orchidaceae	Epiphyte		X				Central America	Parra-Tabla et al. 2000
Pachira quinata	Bombacaceae	Tree		X	X			Central America	Fuchs et al. 2003
Pentadethra macroloba	Fabaceae	Tree			X	X		Central America	Hall, Chase et al. 1994
Pithecellobium elegans	Fabaceae	Tree			X	X	X	Central America	Hall et al. 1996
Psychotria tenuinervis	Rubiaceae	Shrub	X	X				South America	Ramos and Santos 2006
Samanea saman	Mimosaceae	Tree	X	X				Central America	Cascante et al. 2002
Shorea siamensis	Dipterocar-paceae	Tree	X	X				Asia	Ghazoul et al. 1998
Spondias mombin	Anacardiaceae	Tree		X				Central America	Nason and Hamrick 1997
Swietenia humilis	Meliaceae	Tree			X	X	X	Central America	G.M. White et al. 1999
Swietenia macrophylla	Meliaceae	Tree			X	X	X	Central America	Novick et al. 2003
Symphonia globurifera	Clusiaceae	Tree			X	X	X	Central America	Aldrich et al. 1998

For each species we show the botanical family, the parameter evaluated by the authors, geographic region where the study was conducted, and the source publication. For parameters, P = pollination, RS = reproductive success, He = heterozygosity, AR = allelic richness, IC = inbreeding coefficient.

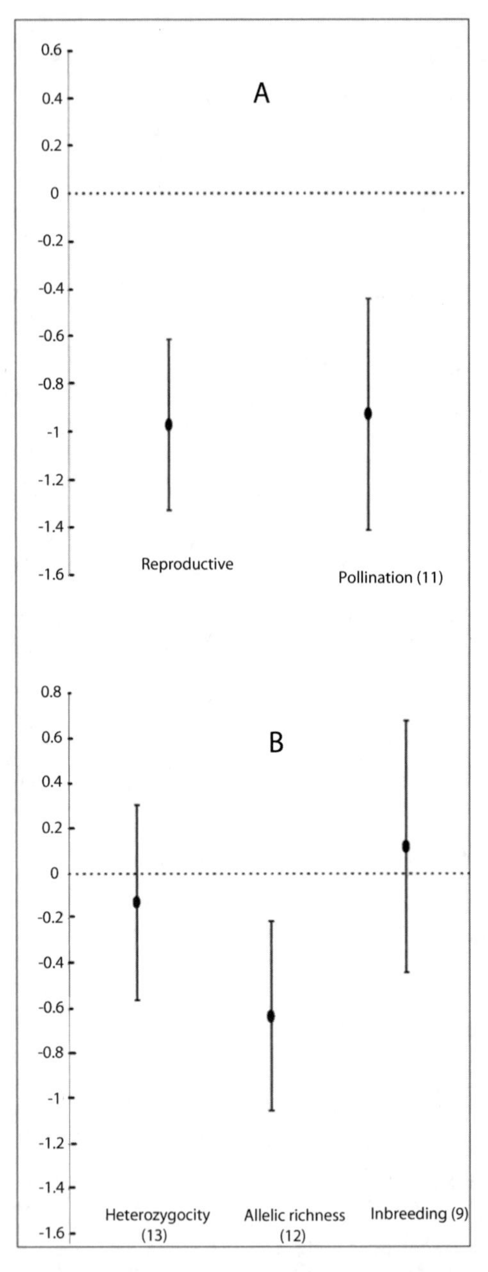

FIGURE 11-1. Overall effect of habitat fragmentation on pollination and reproduction (A) and on genetic parameters (B) of tropical plant species. Weighted-mean effect sizes and 95 percent bias-corrected confidence intervals are shown. Sample sizes are given in parentheses. Dotted line shows Hedges's $d = 0$.

be related than seeds from trees in continuous populations. Seeds produced by trees from continuous populations were more likely to germinate and to produce greater leaf area and biomass as seedlings than progeny from isolated trees. However, isolated trees showed high reproductive capacity in spite of their habitat condition. Few other studies recognize the importance of quantifying fragmentation effects on pollinator activity and its consequences for the reproductive success and genetic variation of tropical trees (Quesada et al. 2003; Quesada et al. 2004; Cuartas-Hernández and Núñez-Farfán 2006).

Habitat Fragmentation Effects on Population Genetic Parameters

Our previous literature survey (Aguilar et al. 2008) measured the effects of habitat fragmentation on the genetic variability of 102 unique plant species from many different regions around the world. From this list, 13 species (nearly 13 percent) belonged to tropical forests, and most of them (77 percent) were trees (table 11-1). We ran three separate meta-analyses evaluating the fragmentation effects on three genetic parameters: heterozygosity (13 data points from 12 unique species), allelic richness (12 data points from 11 unique species), and inbreeding coefficient (9 data points from 9 unique species). In each of the first two meta-analyses, we included a replicate of *Carapa guianensis* data, as two different authors studied this species in different regions (Hall, Orrell et al. 1994; Dayanandan et al. 1999).

The overall effect size of habitat fragmentation on expected heterozygosity of tropical plants was negative and of a small magnitude (d = –0.129). However, this value was not significantly different from zero, given that the 95 percent bias-corrected bootstrap confidence limits overlap zero (fig. 11-1B). Thus, fragmentation is not having a significant effect on the expected heterozygosity of this sample of tropical plants. The overall effect size of fragmentation on allelic richness was also negative but of a larger magnitude (d = –0.634) and significantly different from zero (fig. 11-1B), which implies fragmentation is decreasing allelic richness in these tropical plant species. This result indicates that forest fragmentation at a regional scale decreases the genetic diversity of remnant populations. For example, in *Carapa guianensis*, allelic richness of the cohort in forest fragments was lower than of the cohort in undisturbed habitats. This difference could be caused by changes in local mating patterns of this species.

Finally, the overall effect size of fragmentation on inbreeding coefficient was positive (meaning increased inbreeding in fragmented habitats compared with continuous forests), of a small magnitude ($d = +0.117$), but not significantly different from zero, as confidence intervals overlap zero (fig. 11-1B). This result indicates fragmentation is not significantly affecting the inbreeding coefficient of these fragmented tropical plant populations.

Habitat Fragmentation Effects on Phenological Patterns

In this section we review the literature and analyze the possible consequences of forest fragmentation on the phenology of tropical plants.

Few studies have indicated that forest fragmentation affects climate or several environmental factors in forest remnants (Laurance, Lovejoy et al. 2002). Remnant forest fragments tend to show increased average temperatures, higher rates of evapotranspiration, and higher exposure to winds, which result in decreased soil moisture compared with continuous forests (Wright 1996; Kapos et al. 1997; Laurance, Albernaz et al. 2002). Several of these environmental factors, such as changes in water content stored by plants (Reich and Borchert 1984; but see Wright and Cornejo 1990; Wright 1991), seasonal variations in rainfall (Opler et al. 1976), changes in temperature (Ashton et al. 1988; Williams-Linera 1997), photoperiod (Leopold 1951; Tallak and Muller 1981), irradiance (Wright and van Schaik 1994), and sporadic climatic events (Sakai et al. 1999), have been shown to trigger phenological events in tropical plants. However, very few studies have analyzed the effects of forest fragmentation on plant phenology.

Many studies in the tropics have found that leaf abscission is highly synchronized with dry conditions that are related to soil water content and tree water status (Reich and Borchert 1984; Borchert 1994b). In tropical wet forests, most tree species are evergreen with a relatively continuous pattern of leaf production, but the amount of leaf fall is correlated with the intensity of a dry season. In SDTF, the dry deciduous community of plants drops its leaves at the beginning of the dry season. Leaf flushing appears to be different between habitats (Reich 1995). In wet forests, leaves tend to be produced during the driest period (Frankie et al. 1974), but foliar development is apparently controlled more by internal than by environmental factors (Reich 1995). In contrast, dry forests flush predominantly at the beginning of the wet season, but there is also a small peak during the beginning of the dry season in riparian habitat plant communities (Frankie et al. 1974). In dry forests, primordial leaf buds and leaf expansion take place before the initiation of the rainy season. There is also intra- and interspecific variation in leaf production's response to soil water content and stored stem water availability in dry tropical species (Borchert 1994). Leaf production of trees of the same species varies according to habitat soil water availability. Different species of trees vary in their stem water storage capacity, and this in turn is related to the timing of leaf production. Given the understanding of the environmental factors that affect leaf phenology in tropical plants, we predict that SDTF will be more sensitive to environmental changes caused by forest fragmentation. A greater increase in temperature and evapotranspiration and a decrease in soil moisture in forest fragments are likely to reduce leaf life span, possibly affecting carbon uptake of trees more in dry forests than in wet forests.

Certain plant-herbivore interactions are affected by the timing of leaf flushing because many herbivores depend on leaves to complete part of their life cycle (Janzen 1970; Marquis 1988; Aide 1993). This is particularly important in SDTF. Changes in the timing of leaf flushing and leaf life span provoked by habitat fragmentation's effects on environmental factors may negatively affect herbivore population dynamics. Enviromental changes caused by habitat fragmentation are also likely to trigger changes in the phenology of flowering and fruiting. Synchronization of flowering seems to be partially controlled by physical abiotic factors; in wet forests a flowering peak usually occurs at the beginning of the wet and dry seasons, whereas in dry forests most plant species flower during the dry season. Little is known about the physiological processes that control flower and fruit production in tropical plants (Chapotin et al. 2003). Apparently, a certain threshold level of drought is required to trigger flowering in some tropical plant species (Alvim 1960; Wright et al. 1999). One of the few studies that have evaluated this phenomenon found that the frequency of flowering was similar in populations of the tree *Ceiba aesculifolia* in disturbed and undisturbed conditions, but flowering initiation date and flowering peak occurred between 2 and 3 weeks earlier in disturbed populations than in undisturbed habitats during 3 consecutive years (Herrerías-Diego et al. 2006). Tree populations in disturbed areas may be experiencing drier soil conditions and greater temperatures that trigger their flowering period earlier than the flowering of trees in undisturbed populations.

Changes in flowering phenology caused by habitat loss will also disrupt the pollination patterns of many long-distance pollinators and trap-liners such as some large bees, hawkmoths, nectarivorous bats, and hummingbirds that follow the flowering sequential phenology of plant communities (Stiles 1977; Fleming et al. 1993; Haber and Stevenson 2004; Lobo et al. 2003; Quesada et al. 2003; Quesada et al. 2004).

Variation in the synchrony of flowering also has been proposed as an important factor that affects the reproduction, genetic structure, and mating patterns of tropical plant populations in disturbed habitats (Stephenson 1982; Murawski and Hamrick 1992; Chase et al. 1996; Nason and Hamrick 1997; Fuchs et al. 2003). For example, Fuchs et al. (2003) found for the SDTF tree *Paquira quinata* that trees in fragmented habitats consistently presented either early or late flowering peaks. This resulted in higher selfing rates within trees through geitonogamy and the production of single-sired fruits. In contrast, trees in undisturbed natural forests had higher outcrossing rates and multiple paternity of fruits.

Several biotic factors, such as pollinator attraction and competition for pollinators, have been proposed as important evolutionary forces responsible for phenological patterns in tropical plants (Janzen 1967; Stiles 1975; Appanah 1985; Zimmerman et al. 1989; Sakai et al. 1999; Lobo et al. 2003). Changes in flowering phenological patterns caused by forest fragmentation are likely to affect the behavior and visitation rate of pollinators. If the flowering pattern of plant species that share pollinators of the same guild is displaced over time (Frankie et al. 1974; Stiles 1975; Lobo et al. 2003), competition for the same pollinators will occur, resulting in negative consequences for the reproductive success of the plants and the ability of the pollinators to obtain resources over time. For example, in a SDTF in Mexico, trees of the family Bombacaceae provided the main resource to nectarivorous bats during 8 months of the year (Lobo et al. 2003; Stoner et al. 2003). The sequential use of bombacaceous species by these bats was coupled with the flowering phenology of the tree species. Changes in flowering phenology caused by habitat fragmentation changed the pollination patterns of bats and negatively affected the reproductive output and mating patterns of some of these trees species (Quesada et al. 2004).

Fruiting phenology may also be altered by environmental changes associated with habitat disturbance, but this has remained unexplored. Timing of fruit and seed production is key to understand dispersal, regeneration, and establishment of natural populations in disturbed habitats. Most species of tropical trees fruit after they flower; thus, delays in the flowering patterns will directly affect fruiting and seed dispersal patterns. In SDTF, most tree species are wind dispersed and depend on high temperature and low relative humidity for abscission and dispersal (Greene et al. 2008). Changes of such environmental conditions in disturbed habitats will change fruit maturation and seed dispersal patterns. Displacement in time of fruiting phenology of tropical tree species that provide keystone resources could have negative consequences on populations of birds and mammals that disperse their seeds and, ultimately, negative effects on recruitment of the species they disperse (Howe 1984). Seed dispersal by animals is negatively affected by deforestation and results in lower recruitment in forest fragments (Cordeiro and Howe 2001).

Comparison of Plant Life History and Reproductive Traits and Pollination Systems between Forests

To predict the possible vulnerability of tropical plants to forest fragmentation, we first compiled basic information on plant life-history traits of species from several SDTFs from different published and unpublished databases. Specifically, we gathered information on life-form, sexual expression, compatibility systems, and pollination and seed dispersal vectors of SDTF plant species from Brazil (Caatinga: Machado and Lopes 2004; Machado et al. 2006; Cerrado: Oliveira and Gibbs 2000), Venezuela (Colinas de Bello Monte: Jaimes and Ramírez 1999), Mexico (Chamela: Bullock 1985; Lott 2002; our own database), and Costa Rica (Guanacaste: Bawa 1974; Bawa and Opler 1975; our own database). Also, we obtained the same information for plant species from different regions of tropical rain forest, such as Brazil (Atlantic forest: Silva et al. 1997), Costa Rica (La Selva: Janzen 1983; Bawa et al. 1985; Chazdon et al. 2003), Mexico (Los Tuxtlas: Ibarra-Manríquez and Oyama 1992), Panama (Barro Colorado: Croat 1979), and Venezuela (montane tropical forest: Sobrevila and Arroyo 1982). We made a complete list of plant species from each type of forest and carefully checked it to avoid repetition of species taxonomic identities. With this information we were able to determine the incidence of the different plant life-history traits in each type of forest and also to compare the frequency distribution of these traits between SDTF and tropical rain forest plant species.

Patterns of Plant Life-History Traits, Pollination Systems, and Plant Reproductive Traits

The available published and unpublished databases allowed us to obtain information on a total of 1364 unique plant species from SDTFs and 668 unique plant species from tropical rain forests (TRFs). We were able to assign at least one of the five life-history traits (namely, life-form, sexual expression, compatibility system, and pollen and seed dispersal vectors) to each one of these species. Our analysis contains the largest and most comprehensive database of tropical plant reproductive traits compiled to this date.

Herbs represented nearly 33 percent of life-forms, while shrubs and trees characterized 24 percent and 23 percent of sampled species from SDTF, respectively. Vines made up 18 percent of the species, and nearly 4 percent were epiphytes. The sexual expression of plants was obtained for 1310 species of SDTF and 443 species of TRF. Hermaphroditism was by far the most represented type of sexual expression throughout all life-forms: about 70 percent of the species were hermaphrodite (fig. 11-2). Monoecious and dioecious species were less frequent, and both were similarly represented within this sample of species (about 15 percent of the species; fig. 11-2). The incidence of monoecy was comparable (between 3 and 5 percent) in herbs, shrubs, trees, and vines. Dioecy was mainly represented in trees and shrubs (5 percent and 2.5 percent, respectively) and less represented in vines, epiphytes, and herbs (1.7 percent, 0.8 percent, and 0.6 percent, respectively).

The overall percentages of occurrence of sexual expressions were very similar between SDTF and TRF, and there was a significantly higher pro-

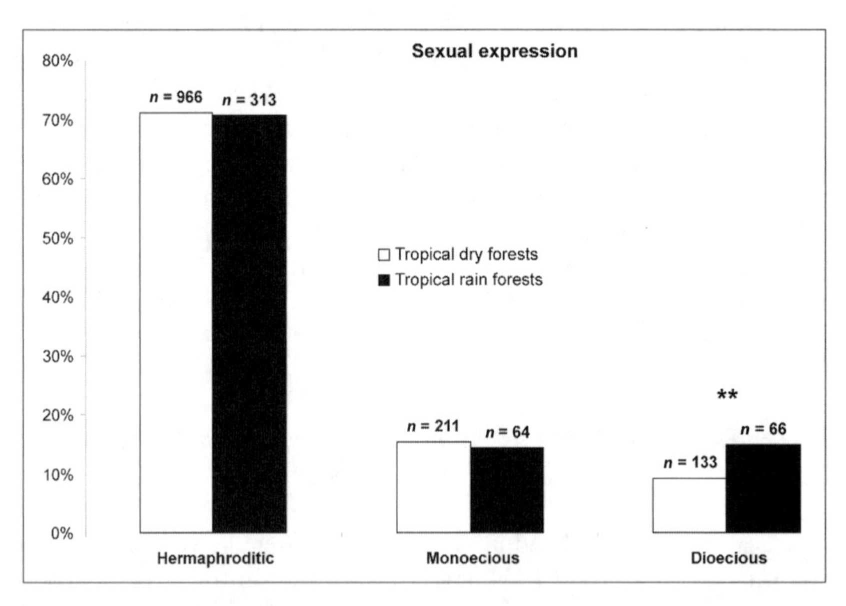

FIGURE 11-2. Frequency distribution patterns of sexual expression in plant species from tropical dry and tropical rain forests. Numbers of species are displayed on the tops of the bars. ** $P < 0.05$ for chi-square test.

portion of dioecious species in TRF compared with SDTF (chi-square = 5.86; p = 0.015; fig. 11-2). Furthermore, in a comparison of the distribution patterns of sexual expression using the proportions previously calculated by several authors (synthesized in table 2 in Machado et al. 2006) from four SDTFs and five TRFs from different countries, the results were comparable to ours. On average, hermaphrodite species represent 75 percent and 70 percent of the species from these SDTFs and TRFs, respectively. Monoecious species are on average 10 percent and 11 percent of the species, while dioecious species represent 14 percent and 20 percent of the species from SDTFs and TRFs, respectively. In this case, however, the larger proportion of dioecious species in TRFs was not significantly different from the proportion found in SDTFs (Mann-Withney U = 6.12; p = 0.327).

Information on compatibility systems was obtained for 171 SDTF species and 104 TRF species. A great majority of the sampled species from SDTF, nearly 75 percent, was self-incompatible. Most of the long-lived, woody species such as trees (85.7 percent), shrubs (64.5 percent), and vines (70.6 percent) were self-incompatible, as were 50 percent of the herbaceous species from SDTF. The distribution pattern of compatibility systems between the two types of forests was similar. Although there was a higher proportion of self-incompatible species in SDTF and of self-compatible species in TRF, these differences were not statistically significant (chi-square = 0.77, p = 0.380 and chi-square = 2.23, p = 0.135, respectively). The higher proportion of self-compatible species in TRF is mostly due to the higher incidence (72 percent) of self-compatibility among the herbaceous species of the sample. Furthermore, we evaluated the incidence of obligate outcrossing species by adding the proportion of self-incompatible and dioecious species in each type of forest. Overall there was a higher proportion of obligate outcross species in SDTF compared with TRF, and this difference was statistically significant (chi-square = 6.93, p = 0.01).

We gathered information on the type of pollination vector for 585 SDTF species and 516 TRF species. We grouped the pollinator vectors in three main categories: wind, insect, and vertebrate. Clearly, insect-pollinated species are the vast majority of species in both tropical forests. In SDTF, insects are the pollination vectors of more than 83 percent of the species, vertebrates pollinate almost 12 percent of the species, and wind is responsible for the pollination of nearly 3.5 percent of the species. These proportions remain similar among species sharing the same life-form, with the few wind-pollinated species being herbs and shrubs. Regarding the pollination vector and the sexual expression of plants, we found that almost 95 percent of the vertebrate pollinators interact with hermaphrodite plants,

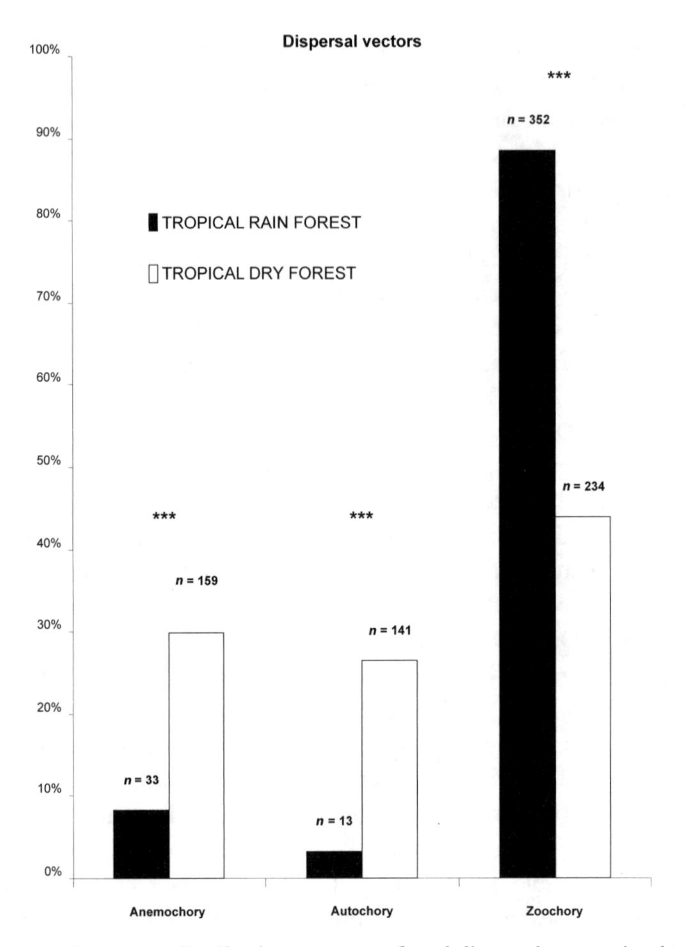

FIGURE 11-3. Frequency distribution patterns of seed dispersal vectors in plant species from tropical dry and tropical rain forests. Numbers of species are displayed on the top of the bars. *** $P < 0.01$ for chi-square test.

whereas 80 percent of the wind-pollinated species were monoecious plants. Insects are the pollinators of almost all (98 percent) dioecious species. The overall distribution of pollinator vectors between both SDTF and TRF is quite comparable. A rather different pattern, however, was observed in TRF species when pollinator vectors, life-form, and sexual expression were combined. Unlike SDTF species, most wind-pollinated species in TRF (nearly 70 percent) were dioecious trees.

It is important to notice that most of the information on pollination systems, from our databases as well as many others from the literature, is based mainly on indirect observation of floral visitors. Nevertheless, floral visitors do not necessarily always function as legitimate pollinators, because not all floral visitors transport pollen from anthers to stigmas (e.g., Schemske and Horvitz 1984; Bawa et al. 1985; Armbruster et al. 1989). For example, in a study to discriminate visitors from legitimate pollinators in 22 sympatric species of *Ipomoea* that occur in a SDTF in Mexico, we found that 12 plant species appeared to be generalists when all flower visitors were included as pollinators, whereas only 8 plant species appeared to be generalists when just legitimate pollinators were assigned to each species (Rosas-Guerrero et al., in prep.). Therefore, it is likely that many studies have overestimated the frequency of legitimate pollinators by interpreting the pollination syndrome from flower morphology or by observations of flower visitors.

Finally, we characterized the seed dispersal vector for 534 SDTF species and 398 TRF species. To do this, we summarized the dispersal vectors in three main categories: anemochory, autochory, and zoochory. For species of SDTF, there was a higher proportion of animal-dispersed seeds (zoochory), followed by quite similar proportions of anemochorous and autochorous plants (fig. 11-3). In fact, these two nonanimal dispersal types make up over 56 percent of the total sample of species in SDTF. The three dispersal types were compared among the different life-forms and sexual expressions, and the following patterns were observed: anemochory was present in more than 90 percent of epiphytes, whereas zoochory was more frequent in trees and shrubs (60.4 percent and 54.2 percent, respectively), and autochory was more common in herbs (52.3 percent). Nearly 70 percent of hermaphrodite plants were either anemochorous (41 percent) or zoochorous (38 percent), whereas 94 percent of monoecious plants were autochorous (47 percent) or zoochorous (47 percent), and almost 60 percent of dioecious species were zoochorous.

We found very different distribution patterns when comparing types of dispersal vectors between species from SDTF and TRF (fig. 11-3). Anemochory and autochory were significantly more frequent in SDTF than in TRF (chi-square = 43.8, $p < 0.001$ and chi-square = 65.9, $p < 0.001$, respectively; fig. 11-3). Zoochory, on the contrary, was significantly more represented in TRF compared with SDTF (chi-square = 43.4, $p < 0.001$; fig. 11-3). In species of TRF, zoochory is the main dispersal type for all the different life-forms (frequency range of 79 percent and 100 percent of occurrence), and sexual expressions (frequency range of 83 percent and 100 percent of occurrence). Anemochorous and autochorous species in TRF were mostly present in hermaphrodite trees.

Vulnerability of Plants to Anthropogenic Disturbance in SDTFs: Conclusions

The literature review analyzed in this chapter has shown that forest fragmentation affects important life-history components of tropical plants, particularly biotic pollination, plant phenology, plant reproductive success, and genetic diversity. A meta-analysis showed that tropical plant populations in fragmented forests show a decrease in pollination, decrease in reproductive output, and loss of genetic diversity. Given these results, we should expect a reduction in plant populations with greater probability of local extinction because of demographic, environmental, and genetic stochasticity (Young et al. 1996; Aguilar et al. 2008).

Mutualisms between Plants and Pollinators

A quantitative analysis shows that most tropical plants are animal pollinated, with a high incidence of outcrossing, mediated by either self-incompatibility systems or dioecy. Flowering plants of tropical ecosystems are highly dependent on animals to move their pollen to receptive compatible plants of the same species to accomplish sexual reproduction. Such dependence of tropical plants on mutualistic relationships with pollinators for reproduction is the result of a long history of evolutionary changes. Hence, disruption of such long-term plant-pollinator interactions will make tropical plants particularly vulnerable to forest fragmentation, more so than temperate systems that are mainly wind pollinated.

Our results showed that pollination and plant reproductive success are negatively affected by fragmentation (fig. 11-1). There were negative effects of fragmentation on pollination, interpreted as pollination visitation frequency, deposition of pollen on stigmas, and number of pollen tubes in styles. Because native small insects and native bees are the main pollinators of the species considered in the meta-analysis, we predict that this guild of insect pollinators will be the most susceptible to forest fragmentation. At the community level we expect their abundance and species richness to be reduced in remnant fragments, compared with continuous forest. We expect a reduction of the pollinators' population sizes due to reduced availability of flower resources and nesting sites. Moreover, small native insects and native bees will be more susceptible to fragmentation because they have a limited ability to fly between remnant fragments. Such restriction on pollinator movement may translate into a reduction of plant fitness due to pollen limitation, biparental inbreeding, and geitonogamy.

Unlike native insect pollinators, exotic pollinators such as feral honey bees (*Apis mellifera*) are able to dominate flower visitor assemblages on fragmented habitats (Aizen and Feinsinger 1994). Thus, there is a potential that native pollinators will be replaced by exotic ones following forest fragmentation. The success of these exotic bees has been related to their social and generalized pollinator behavior; thus, it is expected that these bees will prefer common plants with massive synchronous flowering rather than plants that flower asynchronously or in lower density. Therefore vulnerability of specialized pollination systems will increase because of disappearance of legitimate pollinators. Furthermore, the foraging behavior of *Apis mellifera*, typically characterized by long visitation time to many flowers of the same individual before movement to another plant, increases the deposition of geitonogamous pollen. While this may represent a short-term rescue effect in self-compatible species in fragmented habitats (Dick 2001; Aguilar et al. 2006), it increases the chances of reproductive failure among outcrossing, self-incompatible plants, which, as shown here, represent the majority of species in tropical systems.

A general prediction is that large-bodied pollinators such as birds, bats, large bees, and hawkmoths can fly long distances and are less likely than smaller pollinators to be affected by the increased distance between flower resources in remnant fragments (Ghazoul and Shaanker 2004). However, Quesada et al. (2004) showed that the effects of forest fragmentation on bat pollination in SDTF are plant-pollinator specific. Based on the evidence presented in this study, we predict that highly specialized pollination systems involving large pollinators of self-incompatible plants with prolonged flowering patterns will be especially susceptible to forest fragmentation (e.g., the *Ceiba grandiflora*–bat pollination system).

Sexual Expression and Mating Systems

The extent of the effect of fragmentation on biotic pollination and its translation into seed production is closely related to the degree of dependence of plants on cross-pollination. The breeding system is a relevant trait of plants, determining both the dependence of plants' reproductive success on the availability of pollinators and the extent of reproductive vulnerability of plants to forest fragmentation. It is expected that automatic self-pollination through reproductive assurance will be the most successful reproductive mechanism for survival in fragmented habitats, but it may suffer problems associated with inbreeding in the long term. Conversely, obligate hermaphroditic outcrossing and dioecious plants will show complete dependence, not only on pollinators, but also on the presence of other reproductive individuals of the same species for mating and seed production. Our results showed that 76 percent of the species are self-incompatible or dioecious. These obligate outcrossing species will be particularly vulnerable to the disruption of pollination mutualisms caused by forest fragmentation, particularly dioecious species that are essentially pollinated by small insects with restricted flying movement. In addition, because our analysis showed a higher proportion of obligate outcrossing species in SDTF than in TRF, we may predict more susceptibility of SDTF to forest fragmentation.

The presence of mixed mating systems has been suggested for few tropical tree species (Bullock 1995). Such variation in breeding systems has been associated with changes in pollinators between different habitats or along altitudinal gradients (Murawski and Hamrick 1992; Lobo et al. 2005). For example, Lobo et al. (2005) showed that differences in pollinators between two habitats were correlated with the breeding system and the levels of relatedness of the progeny produced in *Ceiba pentandra* in Costa Rica. High levels of outcrossing were found in the SDTF where bats were the predominant pollinators, whereas a mixed mating system was found in wet seasonal forests where bats were not common pollinators. Genetic relatedness of seeds was greater in the region where bats were absent, indicating higher probability of selfing (Lobo et al. 2005). Other studies on temperate regions have shown that self-incompatibility can be flexible. These studies have proposed different genetic mechanisms to explain such flexibility; ability to switch to partial self-incompatibility has particularly been related to polyploidy or gene duplication (de Nettancourt 2001), pleiotropic effects of modifying genes (Levin 1996; Good-Avila and Stephenson 2002, 2003), or temporal plasticity of self-incompatibility proteins (Richardson et al. 1990; Vogler et al. 1998). These mechanisms can be present in tropical plants, but this field remains completely unexplored.

Changes in mating systems have also been associated with forest fragmentation and habitat disturbance. For example, in the SDTF tree *Pachira quinata*, populations from continuous forest presented high levels of outcrossing rates, whereas populations from fragments experienced a mixed mating system (Quesada et al. 2001; Fuchs et al. 2003). Disturbance and habitat fragmentation may change mating patterns and gene flow of natural populations of tropical plants, with possible consequences for the genetic diversity of these systems.

In an ongoing study in the SDTF of Mexico, we found changes in the composition of plant communities under different successional stages. The number of plant families declined almost linearly, from 35 families in mature forest to 28, 10, and 7 families in sites that were 8 to 12 years old, 3 to 5 years old, and grassland, respectively. The number of tree species showed an evident reduction in grassland and early successional stages, and this is also related to a significant reduction of monoecious and dioecious reproductive systems (fig. 11-4).

Hermaphroditic and monoecious systems are maintained in all the successional stages, whereas dioecy disappears in grassland. Therefore, it is expected that mature and intermediate successional stages are more susceptible to habitat fragmentation.

Finally, obligate outcrossing species will be particularly vulnerable to the disruption of pollination mutualisms caused by forest fragmentation, particularly dioecious species that are essentially pollinated by small insects with restricted flying movement. We predict more susceptibility of SDTF to forest fragmentation because of a higher proportion of obligate outcrossing species in SDTF than TRF.

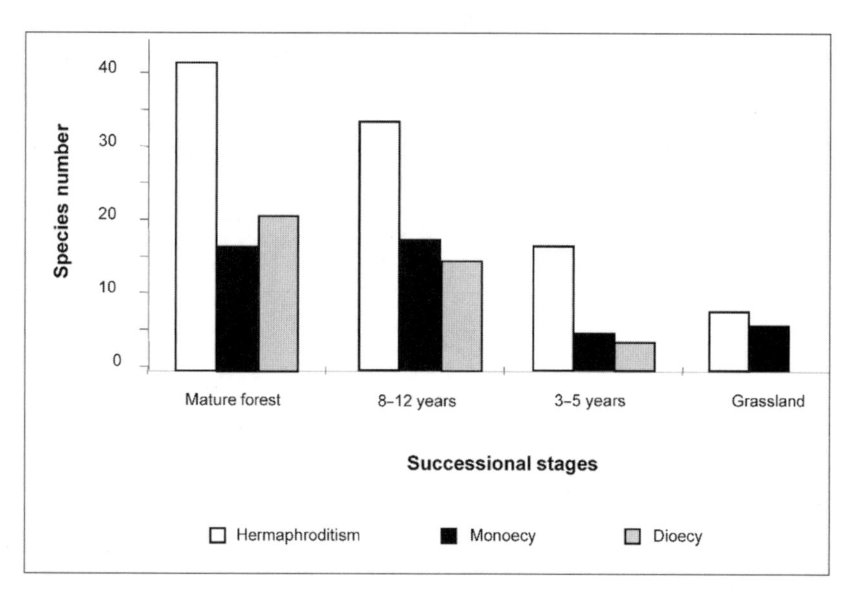

FIGURE 11-4. Frequency of plant reproductive systems in different successional stages of a SDTF of Mexico.

Conclusions

In this review we were able to show general patterns of the effects of forest fragmentation on plant-pollinator interactions, plant phenology, reproductive dynamics, and genetic parameters of tropical plants. However, it is clear that our conclusions are limited to fewer than 30 species of plants that are biased by trees. Other life-forms with contrasting pollination systems, sexual expression, and breeding systems need to be studied in this context. Particular attention is needed in the case of insect pollinators and their interactions with dioecious plants.

Acknowledgments

This work was funded by Consejo Nacional de Ciencia y Tecnología (CONACyT) Mexico (grants SEP-CONACyT 2005-C01-51043, 2005-C01 50863, 2009-C01-131008), Sabbatical grant (91527), the Dirección General de Asuntos del Personal Académico, UNAM (grant PAPIIT IN221305 and IN224108), the Inter-American Institute for Global Change Research, (IAI) Project CRN2-21, and the Agencia Nacional de Promoción Científica y Tecnológica, Argentina (PICT06-132 & PICT04-20341). This research was conducted while L.A. and R.A. were Postdoctoral Associates at Centro de Investigaciones en Ecosistemas (CIEco, UNAM). We also would like to thank Heberto Ferreira and Alberto Valencia for technical support. Three anonymous referees made constructive criticisms which led to improvements in the final version of this manuscript.

Chapter 12

Seasonally Dry Tropical Forest Biodiversity and Conservation Value in Agricultural Landscapes of Mesoamerica

Robin L. Chazdon, Celia A. Harvey, Miguel Martínez-Ramos, Patricia Balvanera, Kathryn E. Stoner, Jorge E. Schondube, Luis Daniel Avila Cabadilla, and Mónica Flores-Hidalgo

The Mesoamerican region is blessed with tremendous biological richness, a high level of species endemism, and a diverse cultural heritage. Yet in 2000, only 30 percent of the region's forest cover remained. Overall, 37.4 percent of the land area of Mesoamerica is used for agriculture (CCAD 2002); much of this agriculture is concentrated in the more seasonal areas previously occupied by seasonally dry tropical forest (SDTF). Pasture is the predominant agricultural land use in Central America, constituting 61.1 percent of all agricultural land in 2000 (FAO 2005). Rice, sugarcane, maize, and beans constitute other major agricultural products grown in dry forest zones of Mesoamerica (Donald 2004; Harvey, Alpizar et al. 2005).

Only 5.7 percent of SDTFs in Central America have protected area status (Miles et al. 2006). As formerly continuous blocks of forest are reduced and fragmented throughout Mesoamerica, remaining forested refuges and protected areas become increasingly embedded within agricultural landscapes (DeClerck et al. 2010). Conservation of biodiversity (including forest specialist species) in these areas therefore requires active management at the landscape level (Janzen 1986a; Daily et al. 2003; McNeely and Schroth 2006; Vandermeer and Perfecto 2007; Harvey, Komar et al. 2008; Chazdon, Harvey et al. 2009) and partnerships with farmers who are practicing *circa situm* (on-farm) conservation (Boshier et al. 2004).

Here, we review our present knowledge of plant and animal biodiversity at the landscape level in seasonally dry forest zones of Mesoamerica. We describe how much and what kind of biodiversity is found within agricultural landscapes, and in what kinds of habitats. We highlight the conservation value of agricultural landscapes for protecting forest specialists and endangered species. To examine the potential for secondary forests on abandoned pastures in SDTF regions to support biodiversity, we include a detailed case study of plant, bird, and bat diversity associated with secondary forests of Chamela-Cuixmala in Mexico. Finally, we present recommendations to enhance biodiversity within agricultural landscapes and enumerate research priorities for deepening our knowledge of species, populations, and communities and their interactions in these landscapes.

The Nature of Biodiversity in Agricultural Landscapes

When concern over biodiversity loss in tropical forests became widespread in the 1980s, virtually no detailed information was available on biodiversity status in areas outside the boundaries of intact forest. Janzen (1988) speculated that "when dry forest habitat is replaced by fencerows, ditchsides, unkempt pastures, and woodlots, the species richness of the breeding fauna and flora is reduced by 90 to 95 percent." We now know that unkempt pastures, woodlots, and other agricultural habitats can contain significant levels of biodiversity. In the agricultural landscape of Las Cruces, Costa Rica, researchers have detected at least 45 percent of the native bird species and 54 percent of the native mammal species within agricultural and pastoral habitats (Daily et al. 2001; Daily et al. 2003; Ranganathan and Daily 2008). Studies in the fragmented landscape of Los Tuxtlas, in Mexico, have detected 226 bird species, 39 bat species, 39 nonflying mammal species, and 36 dung beetle species, which represent 68 percent of the original bird fauna, 80 percent of the original bat fauna, and 65 percent of the original nonflying mammal species (Estrada 2008).

What is the nature of the biodiversity within tropical agricultural landscapes? Janzen (1986a) classified three groups of species that live and interact in agricultural habitats. The first group is the crop species, which are planted (in the case of crops) or raised (in the case of cattle) and depend upon humans for their well-being and survival. Second are species that can thrive and reproduce in managed systems, disturbance-adapted species, or species associated with human activities. Third are remnant species originating from forested ecosystems that have persisted in the transformed landscape but are not breeding or regenerating. This group has been termed the "living dead," as they are not expected to persist beyond the remnant generation (Janzen 1986a).

Many species have been incorrectly labeled as "living dead," as one-time surveys may not clearly reveal whether breeding and reproduction are occurring within the disturbed landscape. Studies of gene flow (particularly pollen flow) within pastures and between remnant trees and trees in forest fragments show that trees in pastures are not reproductively "dead" but typically form part of the breeding population, despite being physically isolated from other trees (Aldrich and Hamrick 1998; G.M. White et al. 1999; Boshier et al. 2004). Moreover, many tree species are able to reproduce within actively grazed pastures. Esquivel et al. (2008) found that 37 of the 85 tree species present in grazed pastures in Muy Muy, Nicaragua (an area of transition between dry and humid forest), regenerated under the current management conditions, suggesting that these species may be able to maintain their populations over the long term. In other cases, management of heterogeneous landscapes, changes in land use, and restoration practices have the potential to transform the living dead into biological legacies (Bengtsson et al. 2003; Lamb et al. 2005).

Janzen's initial categorization also omitted species that are considered to be restricted to mature forests but are able to persist and reproduce in types of forest cover within agricultural landscapes. A subset of mature forest species can survive within highly modified landscapes (Chazdon, Peres et al. 2009). In the dry forest pasture landscape of Rivas, Nicaragua, riparian forests, secondary forests, and forest fallows were characterized by tree species typical of SDTF (Harvey et al. 2006). Although most species of birds, butterflies, beetles, and bats were generalist species, 27 bird and 4 bat species classified as forest dependent were observed at low abundance (Harvey et al. 2006). Some bird species typical of forest habitats are capable of persisting and breeding successfully in agricultural areas (Sekercioglu et al. 2007). Clearly, there is still much to learn about the nature of biodiversity and the types of habitats that support forest-dependent species in agricultural landscapes (Chazdon, Peres et al. 2009; Gardner et al. 2009).

The Nature of Agricultural Landscapes in Dry Forest Zones

Biodiversity is most diverse in agricultural landscapes with heterogeneous and abundant vegetation cover (Kindt et al. 2004; Schroth et al. 2004; Bennett et al. 2006). Remnant old-growth forest fragments and riparian strips provide the highest quality tree cover, with vegetation structure and composition most similar to intact forest areas. Live fences are widely used in Mesoamerica and provide corridors for animal movement and breeding/nesting sites (Budowski 1987; Estrada et al. 2000; Estrada and Coates-Estrada 2001; Harvey, Villanueva et al. 2005; Chacón and Harvey 2006). Fallow fields and secondary forest provide sheltered areas and resources for animal species (Dunn 2004) and are often composed of a high diversity of plant species (Finegan and Nasi 2004; Gordon et al. 2004). Remnant trees in pastures provide stepping-stones for animal movement as well as food, shelter, and shade (Harvey and Haber 1998). Finally, there is the agrobiodiversity itself, in the form of annual crops, perennial crops, tree crops, agroforestry systems, and associated flora and fauna (Power 1996; Perfecto and Vandermeer 2002; Schroth et al. 2004; Jarvis et al. 2007).

Remnant tree cover, tree crops, agroforestry, and tree plantations provide a benign and permeable matrix within agricultural landscapes (Harvey and Haber 1998). The type, size, and spatial configuration of habitat patches influence species composition and movement within the landscape (Medina et al. 2007). Remnant tree cover is not randomly distributed across the landscape; remnant patches are usually associated with steep topography, stream drainages, and river basins or located in areas with limited access (i.e., away from roads). Thus, species utilization of remnant patches of vegetation will be affected by topography, the composition and structure of the surrounding landscape mosaic, and connectivity with other habitats within the landscape.

Vegetation Cover and Diversity in Pastoral Dry-Forest Landscapes

The abundance, diversity, and spatial configuration of tree cover within pasture landscapes strongly determine overall patterns of biodiversity. Tree cover types within these landscapes include riparian forests, old- or second-growth forest fragments, live fences, and remnant trees. The diversity and abundance of remnant tree cover within pastures reflect farmers' management decisions (Barrance et al. 2003; Boshier et al. 2004). When farmers clear an area for pasture or crop production, they sometimes leave forest patches and remnant trees as sources of future timber, fruits, as shade for cattle or workers, or as protection for steep slopes or riparian areas (Harvey and Haber 1998; Muñoz et al. 2003). Over time, additional trees regenerate within the area, arising from underground stems, seed banks, and/or incoming seed rain from adjacent land uses. Resprouting from root suckers is a particularly important mechanism for tree recruitment in SDTF of central Brazil (Vieira et al. 2006). Even after the area has been cleared and converted to pastures, farmers continue to shape the tree cover within the landscapes by selectively removing some tree species and allowing others (usually species of commercial value as timber species or fruit trees) to grow into adults (Muñoz et al. 2003; Villanueva et al. 2003). These decisions are driven by a variety of socioeconomic factors, particularly the need for timber, fence posts for fencing pastures, and firewood (Villanueva et al. 2003).

Detailed inventories of on-farm tree cover clearly illustrate the extent to which farmers shape patterns of tree cover within agricultural landscapes (table 12-1). A study of isolated trees in pastures in three dry forest regions (Cañas, in Costa Rica, and Rivas and Matiguás in Nicaragua) reported high overall tree species richness within pastures, with 71 to 101 species found in pastures in each landscape (Harvey, Villanueva et al. 2008). In all three landscapes, a handful of species dominated; these species were either common timber species or important forage species for cattle. In dry forest zones of southern Mexico, Nicaragua, Cuba, Colombia, and Bolivia, silvopastoral systems often involve leguminous species in the genera *Acacia* and *Prosopis*. Pods of these and other legume species, including *A. pennatula*, *Samanea saman*, *Caesalpinia coriaria*, and *Senna atomaria*, are used as supplemental forage for cattle and are locally sold to supplement farmer income (Rice and Greenberg 2004). Management typically reduces the tree diversity considerably, favoring species that provide commercial products or services to farmers (Michon et al. 2007).

Live fences are common and conspicuous elements of agricultural landscapes throughout Mesoamerica (Harvey et al. 2004). In three dry forest pasture landscapes in Costa Rica and Nicaragua, live fences were found on 49 to 89 percent of the farms (Harvey, Villanueva et al. 2005). These trees contributed to forest connectivity within the landscape; 3.4 to 14.1 percent of the live fences joined directly to forest vegetation. Farmers plant one or two species within the live fences, usually choosing species that resprout easily and can be established from stakes. Consequently, tree diversity within live fences is usually quite low, with most live fences in SDTF regions of Costa Rica and Nicaragua dominated by one of the following species: *Bursera simaruba*, *Pachira quinata*, or *Gliricidia sepium*. Live fences do not contribute greatly to tree biodiversity within individual farms, as they are dominated by a small number of species, but at the

TABLE 12-1. Ten most common trees occuring as dispersed trees and in live fences in three tropical dry forest ecosystems in Costa Rica and Nicaragua

Cañas, Costa Rica		Rivas, Nicaragua		Matiguás, Nicaragua	
Dispersed trees	**Live fences**	**Dispersed trees**	**Live fences**	**Dispersed trees**	**Live fences**
Tabebuia rosea	*Bursera simaruba*	*Cordia alliodora*	*Cordia dentata*	*Guazuma ulmifolia*	*Busera simaruba*
Guazuma ulmifolia	*Pachira quinata*	*Guazuma ulmifolia*	*Guazuma ulmifolia*	*Cordia alliodora*	*Guazuma ulmifolia*
Cordia alliodora	*Spondias purpurea*	*Tabebuia rosea*	*Myrospermum frutescens*	*Tabebuia rosea*	*Pachira quinata*
Acrocomia aculeata	*Ficus werckleana*	*Byrsonima crassifolia*	*Acacia collinsii*	*Enterolobium cyclocarpum*	*Gliricidia sepium*
Byrsonima crassifolia	*Tabebuia rosea*	*Gliricidia sepium*	*Erythrina* spp.	*Albizia saman*	*Erythrina* spp.
Tabebuia ochracea	*Gliricidia sepium*	*Cordia dentata*	*Simarouba amara*	*Platymiscium parviflorum*	*Cordia alliodora*
Pachira quinata	*Guazuma ulmifolia*	*Myrospermun frutescens*	*Gliricidia sepium*	*Gliricidia sepium*	*Tabebuia rosea*
Andira inermis	*Caesalpinia eriostachys*	*Acrocomia vinifera*	*Cordia alliodora*	*Lonchocarpus minimiflorus*	*Spondias mombin*
Piscidia carthagenensis	*Tabebuia ochracea*	*Enterolobium cyclocarpum*	*Caesalpinia violacea*	*Cordia collococca*	*Enterolobium cyclocarpum*
Acosmium panamensis	*Byrsonima crassifolia*	*Swietenia humilis*	*Tabebuia rosea*	*Tabebuia ochracea*	*Spondias* spp.

Data are based on complete inventories of all pastures occurring on 15 farms in each landscape. Tree species are ordered in decreasing order of abundance.
Sources are Esquivel et al. 2003 (Cañas), López et al. 2004 (Rivas), and Ruíz-Alemán et al. 2005 (Matiguás).

landscape level, from 72 to 85 species were found within each region because of the presence of remnant trees or pioneer species within the live fences (Harvey, Villanueva et al. 2005).

To date, only one study has examined nonarboreal vegetation within agricultural landscapes in Mesoamerica. In three wet forest regions of southern Costa Rica, 37 to 42 percent of the species of herbs and shrubs were growing within an agricultural matrix of pastures, coffee plantations, and small forest fragments (Mayfield and Daily 2005). Similar studies have yet to be conducted in dry-forest agricultural landscapes.

Animal Diversity in Pastoral Dry-Forest Landscapes

The different types of vegetation cover within pasture landscapes support substantial levels of animal diversity, by providing key resources, habitat, and connectivity to sustain animal populations (e.g., Daily et al. 2001, Daily et al. 2003; Harvey et al. 2006; Manning et al. 2006). Numerous studies have documented bird diversity within pastoral dry-forest landscapes. For example, in the inter-Andean Cauca Valley of northwestern Colombia, a preliminary survey of avifauna in a landscape of silvopastures, citrus groves, sugarcane fields under organic management, and remnant forest fragments of El Hatico Ranch and Nature Reserve revealed 135 of the region's 141 bird species, including several species not previously recorded from the valley (Cárdenas et al. 2000). Of these, 66 percent were found within agroecosystems, and 51 species were breeding or feeding fledglings within agricultural or silvopastoral habitats and remnants of natural vegetation. In comparison, for open rangeland in this region, Naranjo (1992) recorded only 42 species in pastureland, and only 14 of them regularly used this habitat. Thus, the presence of forest remnant vegetation, live fences, and silvopastoral management in El Hatico enabled persistence of forest specialist birds.

Studies of bird diversity in agricultural landscapes in the dry region of Nicaragua reported significant numbers of bird species using different types of tree cover (Harvey et al. 2006; Vílchez et al. 2008). In the pastoral landscape of Rivas, 83 bird species were reported within all types of tree cover. Similarly, in Matiguás, 137 bird species were observed in the agricultural landscape, with the greatest species richness occurring in the riparian forests, secondary forests, and forest fallows (table 12-2). In both landscapes, species composition varied greatly across different tree cover types. Forest-dependent birds and birds of conservation concern were generally more abundant in the forestlike types of tree cover than in more open pasture habitats.

Bats, like birds, use a variety of tree cover types within the agricultural matrix, including riparian forests, secondary forests, forest fallows, live fences, and even dispersed trees in pastures (Harvey et al. 2006; Medina et al. 2007). A total of 24 bat species were found in the agricultural landscape of Rivas, Nicaragua (Harvey et al. 2006), whereas 39 species were found

TABLE 12-2. Total species richness of birds, bats, and dung beetles in two agricultural landscapes dominated by pastures in Matiguás and Rivas, Nicaragua

Tree cover type	*Birds (point counts)*		*Bats*		*Dung beetles*	
	Matiguás	*Rivas*	*Matiguás*	*Rivas*	*Matiguás*	*Rivas*
Riparian forests	73	42	24	19	20	23
Secondary forests	71	49	21	14	26	29
Forest fallows	63	42	20	14	20	28
Live fences	47	32	20	18	17	24
Pastures with high tree cover	53	41	22	15	23	24
Pastures with low tree cover	51	35	20	15	10	20
Total	**137**	**83**	**39**	**24**	**33**	**32**

Sampling efforts and methods were identical across the two landscapes. Birds were sampled using point counts, bats were sampled using mist nets, and dung beetles were sampled used baited pit-fall traps. Eight plots of each tree cover type were sampled in each landscape. For additional details on methods see the sources (Harvey et al. 2006; Harvey et al. in press)

in the agricultural landscape of Matiguás, Nicaragua (Medina et al. 2007). Within the Matiguás landscape, bats move readily among different types of tree cover within the landscape, easily traveling large distances (more than 10 kilometers) within the landscape (Medina et al. 2007). Riparian forests (and to a lesser degree, live fences) appear to be key travel routes for bats as they cross agricultural landscapes. These linear features help bats orient their flight as they cross the agricultural landscape (as has been reported elsewhere by Verboom and Huitema [1997], Law and Lean [1999], and Galindo-González and Sosa [2003]) and fly over areas of open pasture (Estrada and Coates-Estrada 2001; Medellín et al. 2000; Medina et al. 2007). Moreover, bats feed on fruiting trees in pastures and other habitats within agricultural landscapes and disperse seeds of many tree species important for forest regeneration (Medellín and Gaona 1999; Melo et al. 2009).

Ants are another important indicator group for assessing the biodiversity value of agricultural habitats and landscapes (Power 1996; Perfecto and Vandermeer 2002). Ramírez and Enriquez (2003) sampled ant diversity in two silvopastoral systems (*Prosopis* and *Leucena*) and in remnant dry forest at El Hatico Ranch and Nature Reserve in Colombia. The forest showed the highest species richness, whereas *Prosopis* and *Leucena* silvopastoral systems supported 62 and 97 percent of the species found in the dry forest fragment, respectively, despite considerably lower vegetation cover (20 to 71 percent). For ants, higher vegetation cover did not correspond with higher species richness across these habitats (Ramírez and Enriquez 2003). In the dry forest region of Veracruz, Mexico, Gove et al. (2005) found that ant species richness was higher in isolated trees in pasture and secondary growth than in pastures lacking trees. Isolated trees also provided unique habitat for some arboreal species such as *Cephalotes*.

Dung beetle diversity within agricultural landscapes can also be high, particularly in landscapes that retain a significant level of tree cover. In many dry forest regions converted to cattle production, the availability of large quantities of cattle dung can provide a plentiful food source for some dung beetle species, and if the landscape still retains sufficient tree cover and shade, the conditions may be quite favorable for certain (but not all) species. In the Rivas landscape of Nicaragua, a total of 32 dung beetles were found, while in the Matiguás landscape, a total of 33 dung beetle species were reported (Harvey et al. in press). In both landscapes, dung beetles were reported across all types of tree cover studied (from forest patches to live fences to pastures with trees), although species richness was generally higher in habitats with greatest tree cover and lowest in pastures with low tree cover.

Conservation of Vulnerable and Endangered Species in Agricultural Landscapes

Agricultural landscapes in dry forest regions support a considerable proportion of the original biodiversity but fewer species than found in intact forest. For most taxa that have been examined, the most abundant and frequent species within agricultural landscapes are generalist species typical of open or disturbed habitats (Boshier et al. 2004; Harvey et al. 2006). Small forest fragments within the agricultural matrix can have high conservation value, however (Sekercioglu et al. 2007). In the Rivas landscape of southwestern Nicaragua, 14 endangered bird species were observed in forest fragments (Harvey et al. 2006).

Conservation value is ranked highest for habitats utilized by species that are highly sensitive to human disturbance. Petit and Petit (2003) studied bird communities in 11 distinct habitat types in Panama, ranging from extensive tracts of lowland humid forest to forest fragments and a variety of agricultural land uses. Bird species were classified into five broad habitat association guilds and into three classes of "vulnerability," representing sensitivity of species to human disturbance. Agricultural habitats, such as sugarcane and rice fields, grazed and fallow pasture, and pine plantation, showed no forest-specialist bird species and no species of high vulnerability (Petit and Petit 2003). Yet these habitats did support species with moderate levels of vulnerability. Shade coffee plantations, riparian forest, and lowland forest fragments supported some forest specialists and forest generalists, as well as species of high and moderate vulnerability. These habitats serve as refugia within the broader agricultural landscape in which they are found (Griffith 2000). In regions where most of the forest cover has been lost, shade coffee plantations provide a particularly important habitat for forest-dwelling bird species (Perfecto et al. 1996; Wunderle and Latta 1996; Greenberg et al. 1997; Petit and Petit 2003).

The conservation value of agricultural landscapes is greatly enhanced by linear forms of vegetation that serve as corridors connecting forest fragments. Williams and Vaughan (2001) studied habitat use by white-faced monkeys (*Cebus capucinus*) in an agricultural landscape in Curu Wildlife Refuge of northwestern Costa Rica. Live fences, palm canals, and riparian forest were the most utilized habitats. Management of vegetation corridors in agricultural landscapes is therefore essential to support populations of primates, birds, bats, butterflies, and other animal taxa (Estrada and Coates-Estrada 1996; Sorensen and Fedigan 2000). Forest fragments in Oaxaca, Mexico, are often embedded in an agricultural matrix of land uses, including older secondary forests. These secondary forests and forest fragments have considerably high conservation importance (Gordon et al. 2004).

Restoration and Succession: A Brighter Future for Biodiversity?

In addition to their current value as habitat and resources for certain species, agricultural landscapes also hold the potential to enhance biodiversity through natural and assisted regeneration processes. Increases in vegetation complexity and forest cover will likely benefit biodiversity, potentially ameliorating at least some of the negative effects of forest conversion to agriculture. In the Chorotega dry forest region of northwestern Costa Rica, declining beef prices from 1985 to 1989 caused progressive abandonment of cattle ranches (Arroyo-Mora et al. 2005). In this highly deforested region, forest cover increased from 1979 to 1986 (at a rate of 1.63 percent per year) and from 1986 to 2000 (4.91 percent per year). Secondary forests now cover large areas of this dry forest region (Kalacska et al. 2004; Arroyo-Mora et al. 2005). Within Guanacaste National Park (700 square kilometers), pasture declined by 28 percent from 1979 to 1985 and is being replaced by successional deciduous or evergreen dry-forest vegetation (Janzen 1988d; Kramer 1997; Kalacska et al. 2004). On a smaller scale, SDTFs are resurging in many areas of the Neotropics (Sánchez-Azofeifa et al. 2005; Wright and Muller-Landau 2006; Hecht and Saatchi 2007; Lebrija-Trejos et al. 2008).

As with their wet-forest counterparts, SDTFs show a high capacity to recover vegetation structure and biodiversity through succession (Ruiz et al. 2005; Vieira and Scariot 2006; Chazdon et al. 2007; Lebrija-Trejos et al. 2008). In Mexico, dry-forest succession differs from humid-forest succession in the low species richness of pioneers and lack of long-lived pioneer species, leading to more rapid recovery of tree species composition (Ewel 1980; Lebrija-Trejos et al. 2008).

Case Study: Biodiversity in Secondary Forests of Chamela-Cuixmala, Mexico

Secondary forests within agricultural landscapes provide critical habitats for biodiversity, but few data are available for SDTF species. To explore the potential value of secondary forests as catalysts for the conservation of plant and animal species, we describe changes in woody vegetation, bat, and bird diversity in successional habitats within one agricultural landscape in a SDTF region surrounding the Chamela reserve in western Mexico. We assess changes in abundance and biodiversity using a chronosequence composed of three pasture sites 0–1 years since abandonment, three sites 3-5 years postabandonment, three sites 8–12 years postabandonment, and three old-growth forest sites representing a late successional stage from the Chamela reserve. In July 2003, one 1-hectare site was delimited for study for each of these sites.

The Chamela-Cuixmala region was opened to human colonization in the late 1960s. Slash-and-burn practices initially established agricultural fields during the first 2 years, and pastures were later established for raising cattle. About 70 percent of the SDTF in the local communal *ejidos* surrounding the biosphere reserve has been converted to pastures (Maass et al. 2005). Lack of economic or human resources to maintain pastures along with water scarcity (various years of drought) have resulted in the abandonment of pastures in this region (Burgos and Maass 2004; J. Trilleras-Motha and P. Balvanera, unpublished data). Abandoned pastures are rapidly colonized by *Acacia* and *Mimosa* species, which form monospecific-vegetation thick carpets (Ortiz 2001; Burgos and Maass 2004).

Woody Plants

Within each of the 12 study sites, a 30 by 60 meter permanent plot was established using a nested design. All stems of trees, shrubs, and lianas greater than or equal to 1 centimeter in diameter at breast height (or 1.3 meters aboveground level) were tagged, measured, and taxonomically identified within 10 by 50 meters, those greater than or equal to 2.5 centimeters in diameter within 20 by 50 meters, and those greater than or equal to 10 centimeters in diameter within 30 by 60 meters. Stem density increases slowly during the first 5 years of succession, followed by rapid increases thereafter; sites with fallow ages of 8–12 years were similar in stem density to old-growth forest sites (fig. 12-1A). This trend was paralleled by changes in alpha diversity (i.e., number of species per plot) observed along the chronosequence. Pasture sites with less than 6 years of abandonment had, on average, a third or less of the number of species recorded at the old-growth forest sites, whereas sites with fallow ages of 8–12 years had a similar alpha diversity (about 80 species in 1800 square meters) compared with the old-growth forest sites (fig. 12-1B).

Facilitation appears to operate at the early successional stages, where few colonizing species can establish and grow under the harsh conditions of the abandoned pastures. Among the colonizing species, *Cnidosculus spinosus* and the legumes *Acacia farmesiana*, *Caesalpinia caladenia*, *Mimosa arenosa*, and *Bahuinia subrotundifolia* were practically restricted to the youngest successional sites (less than 6 years old), accounting for 40 percent of total stem density in such sites. Other relatively abundant colonizing species reached their maximum densities in the 8–12-year-old sites. Among these, *Croton pseudoniveus*, the legumes *Lonchocarpus constrictus* and *Piptadenia constricta*, and *Casearia corymbosa* accounted for 14 percent of total stem density recorded at the youngest successional sites. Yet other colonizing species were among the 15 most abundant trees found in the old-growth forest sites. Among these, *Caesalpinia eriostachys*, *Lysiloma microphylla*, *Caesalpinia tremula*, *Cordia alliodora*, and *Heliocarpus pallidus* constituted 12

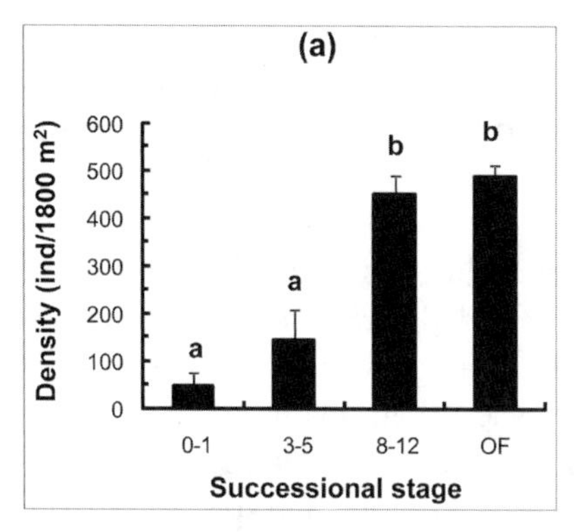

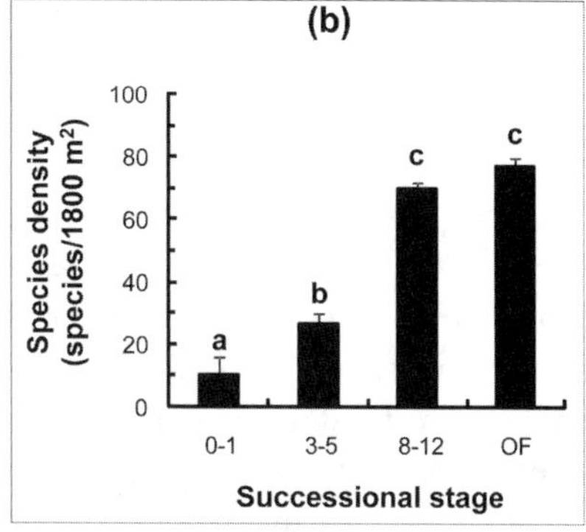

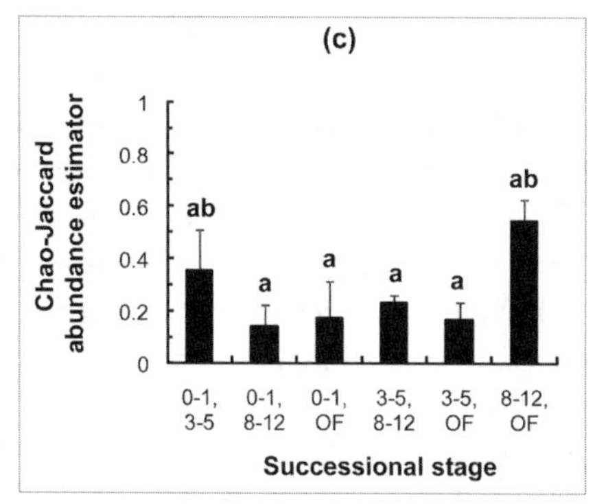

FIGURE 12-1. Structural and species diversity changes of woody plants along a chronosequence of abandoned pasture fields and old-growth forest sites in the Chamela-Cuixmala region, western Mexico. (a) Changes in stem density of trees, shrub, and lianas with diameter at breast height greater than 1 centimeter. (b) Changes in alpha diversity, number of species of trees, shrubs, and lianas per sampled area. (c) Changes in beta diversity, proportion of shared species among sites of different successional ages. In all cases, bars represent means plus 1 standard error considering three sites per successional stage. Numbers on the bottom axis indicate intervals of years since field abandonment, and "OF" indicates old-growth forest sites. Different letters on bars indicate significant differences between successional stage classes ($P < 0.05$).

percent of total stem density recorded at the youngest successional sites. Thus, early secondary vegetation incorporates tree species that are structurally and functionally important in middle-aged secondary forest as well as in the old-growth forest.

Overall, 229 woody species were recorded in the 12 studied plots (total sampled area = 1.2 hectares). Combining sites from the same successional age category together, species density in old-growth forest accounted for 65 percent of the gamma diversity value, species density in 8–12-year-old secondary forest accounted for 49 percent, species density in the 3–5-year-old secondary forest accounted for 37 percent, and species density in the 0–1-year-old abandoned pasture fields accounted for only 11 percent. Thus, the secondary forest sites of different ages held a substantial component (about 80 species) of the recorded gamma diversity. Furthermore, on average, species composition of old-growth forest sites showed low similarity with secondary forest sites less than 6 years old and higher similarity with the 8–12-year-old secondary forest sites (fig. 12-1C).

Current agricultural practices appear to promote high species diversity of woody vegetation within the landscape. The small size of clearings, reduced use of heavy machinery, low use of agrochemicals, presence of a large fragment of old-growth forest (the Chamela Biological Station Reserve), low frequency of fire use, low cattle loads for short periods, and generally moderate to low soil degradation in the region facilitate forest regeneration there (J. Trilleras-Motha and P. Balvanera, unpublished data). Contrary to the notion that shrubby legume species of *Acacia* and *Mimosa* arrest succession in abandoned agricultural fields (Ortiz 2001; Burgos and Maass 2004), our results suggest that they constitute a transient successional stage that facilitates the incorporation of a diverse array of woody species after only 6 years since field abandonment. In 30-year-old secondary forests dominated by *Mimosa arenosa* in the region, regenerating primary forest species were found among the youngest saplings. The relative dominance of regenerating primary species was clearly related to the intensity of management (Romero-Duque et al. 2007).

Birds

Effects of the process of ecological succession from pastures to mature forest on resident bird species were studied using both nonradius point counts and mist nets. Point counts were used to estimate bird species richness and abundances, while mist netting was used to complete the inventory of species and determine the reproductive and molting status of birds. Eighty point counts were established in each of the four successional stages, for a total of 320 independent sampling points. Point counts were randomly located inside the different successional stages and at least 250 meters apart to ensure independence of the data collected (Bibby et al. 2000). Sampling was conducted in, and around, two of the three sites used to represent each successional state in the vegetation study, for a total of eight sampling areas. Mist-netting stations were established inside the eight vegetation plots discussed above. Each sampling area was visited two times during the dry season and two times during the rainy season in 2005, allowing for a better understanding of the effect of agricultural landscapes on the bird species breeding in the area.

Bird species richness and abundances were clearly positively related to increasing complexity of vegetation structure and plant diversity during succession. Bird communities in pastures were very distinct from those of the other successional stages. Pasture communities had low species richness and were dominated by a small number of granivorous species. Bird species richness was similar among secondary and old-growth forests, suggesting that once shrubs establish in abandoned pasture, bird communities can recover their alpha diversity and structure within 8 years (fig. 12-2).

Bird species richness increased rapidly during the first 3 to 5 years of succession and then stabilized, with 8–12-year-old secondary forests being very similar to old-growth forest sites (fig. 12-2A). However, bird abundances increased slowly along the successional gradient, reaching their highest values in the old-growth forest sites (fig. 12-2B).

Patterns of species richness and abundance along the successional gradient seem to be the result of three complementary processes: (1) the loss of forest species when the native vegetation is turned into pastures, (2) the invasion of pastures by granivorous and/or grassland-specialist species, and (3) the recolonization of secondary forest habitats by insectivorous, frugivorous, and nectarivorous species. Pasture sites presented a higher number of granivorous species compared with forest sites, and usually their communities were dominated by one species (*Aimophila ruficauda*). Granivore species richness and abundances decreased during succession, while the number of species and individuals of fruit- and nectar-eating birds increased. Although the number of insectivorous species increased with the time of abandonment (from 16 species in pastures to 30 in old-growth forest), their abundances remained constant (15 plus or minus 2.5 individuals per hectare), increasing the complexity and evenness of insectivorous bird communities. Because insect abundance did not differ significantly among the different sites (C.A. Chávez-Zichinelli, unpublished data), this result suggests that an increase in the complexity of vegetation structure associated with successional age could provide a larger number of feeding niches for insectivorous birds.

In total, 110 bird species were recorded in the 320 sampling points and eight mist-netting stations (gamma diversity). Among these, 109 were terrestrial birds, and 1 was an aquatic species that uses terrestrial habitats to nest (*Dendrocygna autumnalis*). This represents 88 percent of the terrestrial

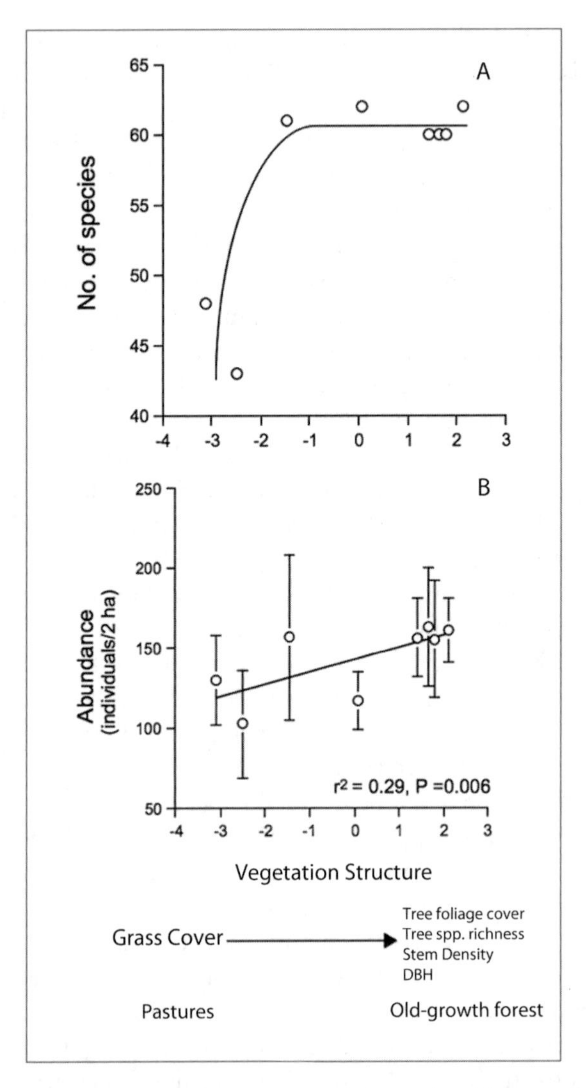

FIGURE 12-2. Effects of vegetation structure on bird species richness and abundance. Vegetation structure was determined by using a principal component analysis that included grass cover, tree foliage cover, density of stems, tree species richness, and diameter at breast height (DBH). The first principal axis (used here as the horizontal axis) comprised 83 percent of the total variation. The analysis ordered the eight sampling sites (each represented by a data point) as a chronosequence, with pasture sites on the left and old-growth forest sites on the right. The number of bird species increases rapidly with succession, reaching a saturation point and staying constant after a few years of pasture abandonment. Bird abundances increase along the successional gradient, reaching their highest values in old-growth forests.

resident species reported for the region (Arizmendi et al. 2002). Lumping data for all sampling points of each successional age category together, species density in old-growth forest accounted for 60.9 percent (67 species) of the gamma diversity, while species density in 8–12-year-old forest, 3–5-year-old forest, and 0–1-year-old abandoned pastures accounted for 61.8 percent (68 species), 67.3 percent (74 species), and 56.3 percent (62 species) of the gamma diversity, respectively. Thus, pastures and secondary forests supported a large fraction of the bird diversity in the landscape (102 species), with only 8 species being restricted to old-growth forest. These old-growth species include one large-sized parrot (*Amazona oratrix*), one large-sized woodpecker (*Campephilus guatemalensis*), a hummingbird (*Chlorostilbon canivetis*), and several species of flycatchers (*Contopus pertinax*, *Myiarchus cinerascens*, *Myiopagis viridicata*, and *Tityra semifasciata*). Old-growth forest sites shared 62 percent of their species with secondary forest sites younger than 6 years old and 93 percent with the 8–12-year-old secondary forest sites. Surprisingly, 13 percent of all species (12 species) were shared by all successional stages. Among these species are 8 species endemic to Mexico that are considered to be dry forest specialists (*Cacicus melanicterus*, *Deltarhynchus flammulatus*, *Granatellus venustus*, *Melanerpes chrysogenys*, *Passserina lechlancherii*, *Polioptila nigriceps*, *Thryothorus felix*, and *T. sinaloa*).

These results suggest that birds respond quickly to changes in vegetation structure associated with succession. Agricultural landscapes increase bird diversity by allowing grassland species to invade previously forested habitats. Some forest species adapt to pasture conditions, or they start to return to early successional stages as soon as shrubs/young trees get established. Whereas secondary forests showed bird diversity and community structure similar to those found in old-growth forest, the population-level mechanisms that allow these patterns to occur are not yet understood. The patchy configuration of the agricultural landscape in the Chamela-Cuixmala region allows birds to move easily among different successional stages and old-growth forest, promoting the rapid recovery of bird communities despite large-scale land use transformations.

Bats

The structure and composition of bat communities in pasture and different successional stages was investigated by mist net sampling. This method focused on sampling the leaf-nosed family (Phyllostomidae), which includes the most important Neotropical frugivorous and nectarivorous species.

Since most aerial insectivores are adept at detecting and avoiding mist nets, their capture in mist nets is not a reliable method for assessing their abundance or distribution, and data for these species were excluded from this analysis. Every 6 weeks from June 2004 to August 2006, five mist nets (two 6 meters, two 9 meters, and one 12 meters) were placed in each of the 12 sites representing the chronosequence composed of three pasture sites 0–1 years since abandonment, three sites 3–5 years postabandonment, three sites 8–12 years postabandonment, and three old-growth forest sites representing late successional stage. Because of very low capture rate, the three pasture sites were sampled approximately half as much as the other sites. Sampling was conducted from sunset for 5 hours.

During 142 sampling nights, 606 phyllostomid bats were captured, representing 16 species (L.D. Avila Cabadilla and K. Stoner, unpublished data). We captured 87.5 percent of the phyllostomid species reported for this region (Stoner 2002). As with bird communities, bat species richness and abundances were clearly related to successional increases in vegetation structure and plant diversity. However, unlike bird communities, bat communities were less diverse in the youngest successional stages (fig. 12-3).

Sixteen species were captured in mature forest, 9 in late and early successional stages, and only 4 in pastures. Furthermore, a clear change in the

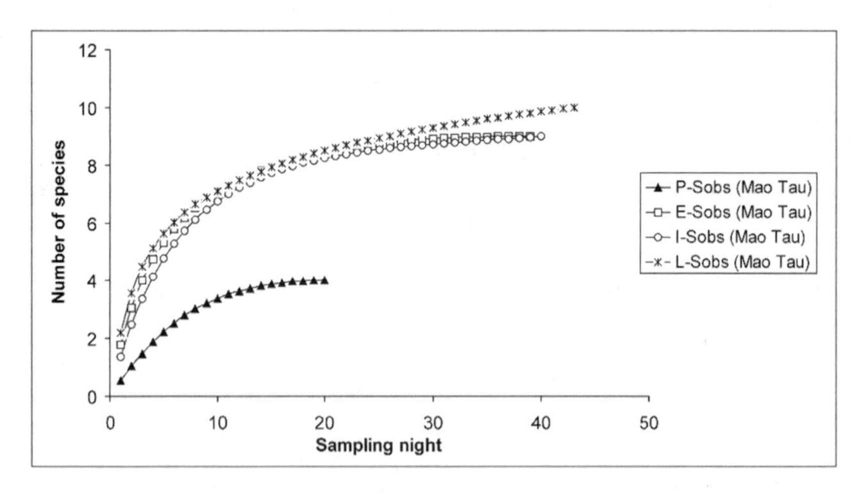

FIGURE 12-3. Species accumulation curve for bats in pastures (triangles), secondary vegetation 3–5 years old (squares), secondary vegetation 8–12 years old (circles), and mature forest representing a late successional stage (asterisks) in the Chamela-Cuixmala region, western Mexico.

number of species within foraging guilds was observed for bats (fig. 12-4). In particular, more species of frugivores and nectarivores were found in mature forest than in either successional stage. The one species of gleaning insectivore (*Micronycteris megalotis*, Phyllostominae) was found only in mature forest.

Although these trends suggest that bird communities recover more quickly than bat communities during succession, more studies in SDTF are needed. Given the importance of bats in both seed dispersal (Geiselman et al. 2002; Muscarella and Fleming 2007) and pollination (Winter and von Helversen 2001; Stoner et al. 2003) in Neotropical ecosystems, the lag time in the recovery of this important group of animals may have negative consequences for the regeneration of SDTF ecosystems.

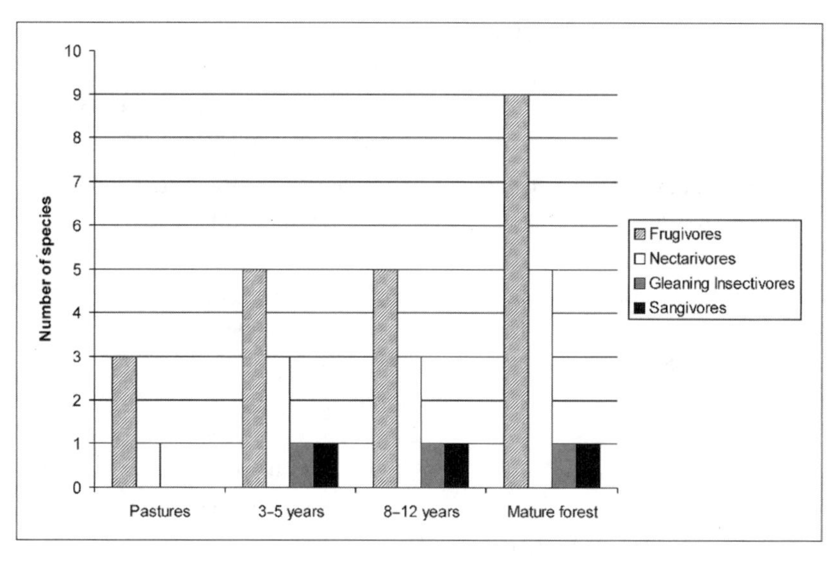

FIGURE 12-4. The number of phyllostomid species observed in pasture, successional forest, and mature forest in the Chamela-Cuixmala region, western Mexico, classified by foraging guilds.

Steps to Enhance Biodiversity and Conservation Value of Agricultural Landscapes

Under favorable conditions, abandonment of pastures in SDTF regions can lead to substantial recovery of biodiversity through forest succession. The negative effects of agricultural land use for biodiversity can be further reduced through a range of specific farm management strategies and agricultural policies (Harvey et al. 2008a; Chazdon, Harvey et al. 2009; table 12-3).

The most fundamental approach to enhancing biodiversity in agricul-

TABLE 12-3. Examples of farm management activities that influence the biodiversity value of agricultural landscapes

Activities that increase the conservation value of agricultural landscapes	*Activities that decrease the conservation value of agricultural landscapes*
Controlled burning (avoiding escape of fire into intact forests by establishing firebreaks)	Indiscriminant use of fire (and escape of fires into adjacent forests)
Rotational and controlled grazing systems	Overgrazing and unsustainable stocking rates
Maintenance of natural regeneration within pastures	Elimination of natural regeneration in pastures and crop fields
Establishment of diversified forest plantations and agroforestry systems	Establishment of monocultures and exotic species
Planting of multistory, diverse live fences (preferably connected to forest patches)	Removal of existing live fences or simplification of live fence structure and diversity
Sustainable management of forests	Unsustainable harvesting of timber, firewood, and/or nontimber forest products from forest patches, riparian areas, fallow areas, and pastures
Establishment of fallows within agricultural areas	Uncontrolled grazing and trampling of riparian areas by cattle
Restoration of degraded areas to native forest cover	Conversion of forests or fallow areas to pasture or crop land
Use of sustainable land use practices (e.g., organic agriculture, agroforestry)	Unsustainable hunting or collection (for pet trade) of wildlife within forests and fallow areas
Use of integrated pest management strategies	Pollution or contamination of soils and water from agrochemicals (fertilizers, pesticides, etc.)

Sources are Harvey et al. 2005 and Harvey et al. in press.

tural landscapes is to increase the amount, diversity, and connectivity of tree cover. Protecting existing forest fragments, riparian forests, and remnants of native habitat within agricultural landscapes should be a high priority (Bengtsson et al. 2003; Daily et al. 2001; Daily et al. 2003; Harvey, Villanueva et al. 2005; Harvey et al. 2006). Different taxa and guilds vary in their response to the types of on-farm tree cover present in agricultural landscapes. Highly degraded, unproductive areas within these landscapes should be restored through reforestation, natural regeneration, and/or enrichment planting. The establishment and management of diverse forms of tree cover will ensure heterogeneous habitats for wildlife and resources for migratory species of birds and butterflies. Linear forms of tree cover, such as riparian strips and live fences, can increase habitat connectivity and enhance biodiversity within agricultural landscapes (Chacón and Harvey 2006). Diversification of agroforestry systems offers another mechanism for enhancing tree cover, particularly in buffer zones surrounding existing protected areas and in biological corridors linking protected areas within the Mesoamerican Biological Corridor (Kaiser 2001; Miller et al. 2001).

Tree cover within agricultural landscapes is highly dynamic, largely due to changes in the way farmers design and manage their farms (Boshier et al. 2004; Harvey et al. 2004; Harvey et al. 2005; Komar 2008). Conservation biologists need to actively engage farmers in seeking long-term, landscape-level management plans that address both conservation and production concerns. Tree diversity within pastures can be enhanced by encouraging farmers to retain a greater diversity of adult trees within pastures and to select (and retain) a greater number of regenerating trees within pastures. Diversity within live fences can be enhanced by providing farmers with a greater selection of tree species to plant (Zahawi 2005). Tree species that show limited regeneration in pastures could also potentially be promoted through specific management strategies such as retention of adult trees as seed sources, protection of saplings and seedlings from weeding and grazing, and enrichment planting. Farm management practices are thought to be responsible for saving *Leucena salvadorensis* from extinction in Mesoamerica, through maintenance of genetic diversity on farms (Chamberlain et al. 1996; Hughes 1998).

Remnant trees in pastures can serve as seed sources to promote regeneration of many species within the surrounding landscape. Unfortunately, many forestry laws do not restrict harvesting of trees from pastures or regulate their densities or use, due to the misconception that these trees have little or no conservation value. On-farm tree cover also provides critical resources for migratory species. Many dry forest butterfly species migrate seasonally to wetter lowland forests or to higher-elevation cloud forests (Haber and Stevenson 2004) and require tree cover along their migratory routes.

Challenges to Biodiversity Conservation within Agricultural Landscapes

Any efforts to conserve biodiversity within agroecosystems in the SDTF of Mesoamerica will require strong alliances with the farmers who own and manage agricultural land (Vandermeer and Perfecto 2005, 2007; Harvey, Komar et al. 2008). Decisions about tree cover within these landscapes may have negative or positive impacts on the value of this land for biodiversity conservation by changing the amount or type of tree cover, its structural and floristic complexity, or its arrangement in the landscape (Harvey, Alpizar et al. 2005).

Common management practices such as the indiscriminant use of fire, cattle grazing in riparian areas and forest patches, hunting, conversion of fallow areas to pastures, or conversion of forests to pastures reduce the quality of the agricultural landscape for biodiversity conservation. Fire is used widely throughout the region to clear new areas for shifting cultivation, prepare land for the planting of crops, eliminate weeds and stimulate regrowth in existing pastures, or eliminate weeds and pests (snakes, rats) from pastures and crop fields prior to planting grass or crops, and also prior to harvesting sugarcane (another key land use in the dry forest region). Although the information on the number and extent of fires is scarce, each year larger areas of land are burned—particularly in the dry season when farmers use fire to prepare lands for cultivation (Middleton et al. 1997; Billings and Schmidtke 2002). For example, it is estimated that 1.1 million hectares of land were burned in Central America during 1998 (of which 653.3 hectares were in agricultural landscapes and the remainder in forest areas), while an estimated 1.5 million hectares were burned in Central America and Mexico during 1999 (SCBD 2001). A significant proportion of these fires occur in dry forest ecosystems. A study in 1998 estimated that 42,486 wildfires occurred in Central America, burning 1.1 million hectares of land.

Harvesting of firewood presents a second important threat to on-farm tree cover. Particularly in rural areas where firewood is the key fuel, indiscriminant harvesting of firewood from forests and fallow areas can severely reduce tree diversity and negatively impact wildlife. In Masaya, Nicaragua, at least 64 tree species were being commercialized in the fuelwood market by vendors (McCrary et al. 2005). The majority of the fuelwood comes from on-farm tree cover, such as live fences, windbreaks, and fallow areas, yet as much as one-third to one-half of the fuelwood was harvested from natural forests (including protected areas). Harvesting of fuelwood from natural forests is particularly prevalent during the dry season, when these forests are most easily accessed.

Decisions by farmers to increase tree cover within their farms by retaining forest patches, fencing off forests and riparian areas to prevent cattle entry, planting new live fences, establishing forest plantations, or converting agricultural systems to agroforestry systems will rarely be made without financial incentives. Therefore, payments to subsidize reforestation efforts or to cover fencing costs are required to encourage practices that enhance on-farm tree cover and increase the diversity and quality of forested habitats. For example, in projects where payments for environmental services are available to farmers to help offset the costs of establishing on-farm tree cover, farmers have readily increased the number of live fences in pastures and allowed greater natural regeneration of trees (e.g., Pagiola et al. 2005).

A further challenge in biodiversity conservation within agricultural landscapes is that land use and landscape composition are subject to rapid change with minimal regulation. Changes in agrarian and economic policies, such as the adoption of the Central American Free Trade Agreement or sudden increases in prices for cattle or other commodities, can rapidly change the composition of the landscape, with certain impacts on biodiversity (Harvey, Alpizar et al. 2005). The 1996 Forestry Law of Costa Rica prohibits clearing of existing forest for agricultural land use—but farmers are now hesitant to abandon agricultural lands, as they will lose their rights to use these lands for agriculture in the future. Similarly, if sugarcane becomes more profitable because of changes in the Central American Free Trade Agreement, many pasture areas may be converted to sugarcane, further decreasing the potential for forest regeneration.

Research Priorities for Assessing Biodiversity in Agricultural Landscapes

Our synthesis clearly reveals the need for further research to assess the biodiversity of species utilizing agricultural areas and to promote application of adaptive management to changing agricultural systems and landscapes, changing socioeconomic contexts, and emerging threats to biodiversity.

Landscape-level studies should be conducted in a broad range of different types of agricultural landscapes in dry forest zones encompassing different countries and different historic and current patterns of land use and agricultural management (Chazdon, Harvey et al. 2009). These studies will provide critical information needed to maximize biodiversity conservation potential in agricultural landscapes of dry tropical forest regions:

- investigations of demography, breeding behavior, and population genetics of wild flora and fauna in agricultural landscapes
- studies of habitat use, resource use, species interactions, range sizes, and movement patterns of animals within the agricultural landscape
- assessments of the effectiveness of buffer zones and corridors for the conservation of target species of conservation concern
- assessments of the effects of agrochemicals (and pesticide drift), human disturbance, fire, and harvesting of natural products on plant and animal communities
- characterization of relationships between biodiversity, farm productivity, and ecosystem services to provide a more rigorous scientific basis for environmental payment schemes and other incentives for conservation
- assessment of the economic value (and costs) of biodiversity to agriculture (e.g., Ricketts et al. 2004) to evaluate the gains, losses, and trade-offs of encouraging wildlife presence on farms
- investigations of long-term dynamics of natural forest regeneration and succession in agricultural landscapes, including studies of vegetation, vertebrate and invertebrate taxa, and comparisons with wet-forest successional dynamics

Conclusions

Because of their great extent and proximity to protected areas, agricultural landscapes of the tropics are critically important for biodiversity conservation. Within Mesoamerica today, agricultural landscapes still contain significant levels of biodiversity (DeClerck et al. 2010). Urgent action is needed to promote sustainable farming practices that enhance conservation of wild species within these landscapes (Harvey, komar et al. 2008). At the same time, basic ecological knowledge regarding the habitat utilization, movements, breeding biology, and diversity of species within these landscapes is urgently needed to provide a sound basis for conservation policies and action. As illustrated in chapter 11, SDTF fragmentation represents a significant threat to the maintenance of species interactions, particularly those relevant to plant reproductive success and genetic diversity. Given that agricultural landscapes bring about forest fragmentation, work is urgently needed to define to what extent agroscapes can support significant proportions not only of the local biodiversity but also of the ecological processes that generate and maintain such biodiversity. Conservation biologists need to extend the scope of their research activities to include studies within agricultural areas (Chazdon, Harvey et al. 2009). Ultimately, the fate of tropical biodiversity will be determined by management decisions and government policies that affect entire landscapes and regions. Despite many challenges, agricultural landscapes in dry forest regions offer the potential to cultivate a more secure future for biodiversity.

Acknowledgments

Our synthesis was stimulated by a working group on Biodiversity and Conservation Value in Agricultural Landscapes of Mesoamerica, supported by the National Center for Ecological Analysis and Synthesis (NCEAS) Center funded by the NSF (Grant #DEB-0072909), the University of California, and the Santa Barbara campus. Robin L. Chazdon's research was supported by NSF DEB-0424767 and NSF DEB 0639393. Research by Celia A. Harvey was supported by the FRAGMENT and CORRIDOR projects, funded by the European Union (INCO-DEV ICA4-CT-2001-10099 and INCO-CT-2005-517644). Research by Kathryn E. Stoner and Luis Daniel Avila Cabadilla was partially supported by the Inter-American Institute for Global Change Research (IAI CRN2-21). Funding for the Chamela successional research was provided by a grant from SEMARNAT-CONACYT (CB-2005-01, No. 24843) and SEP-CONACYT (CB-2005-01-51043). We thank Lorena Morales-Pérez for use of unpublished data.

Chapter 13

Pasture Recolonization by a Tropical Oak and the Regeneration Ecology of Seasonally Dry Tropical Forests

Jeffrey A. Klemens, Nicholas J. Deacon, and Jeannine Cavender-Bares

Fragmentation and habitat destruction of tropical forests is nowhere more apparent than in the seasonally dry tropical forests (SDTFs) of Central America (Janzen 1988b; chap. 1). In Central America, old-growth tropical dry forest had been reduced to less than 20 percent of its original extent by the mid 1980s (Trejo and Dirzo 2000), largely as a result of disproportionately high human population density and intensive agricultural activity within this habitat zone (Murphy and Lugo 1986a). Although rates of deforestation in Central America peaked in the twentieth century, palynology data indicate that humans have been using fire to manipulate forest cover in Central American SDTF for thousands of years (Janzen 1988b; Piperno 2006).

Although preserving the few remaining stands of old-growth SDTF in Central America is of critical importance to biodiversity conservation in the region, the remaining stands of intact, mature forest may be too small and isolated from one another to preserve this system (Bierregaard et al. 1992; Laurance, Lovejoy et al. 2002; Laurance et al. 2006; chap. 12). The conservation biology of Central American SDTF, therefore, must be in large part restoration biology (Janzen 1987).

The challenges facing restoration efforts in SDTF are manifold. Much of the tropical conservation literature has focused on the status and importance of tropical rain forest (Quesada, Sánchez-Azofeifa et al. 2009).

This may be due to the fact that old-growth SDTF is becoming rare in the Neotropics (chap. 3). The very rarity of SDTF hinders scientific work on forest regeneration, as there is very little high-quality SDTF to study. Without an understanding of how ecological processes relevant to regeneration operate in intact forests, it is challenging to determine how they are altered by fragmentation. Another challenge is that SDTFs are dominated by secondary growth (Quesada, Sánchez-Azofeifa et al. 2009). Abandoned agricultural fields face particular restoration challenges, such as the presence of exotic ungulates or a history of management by intense and frequent fire. Historical land use has resulted in great variation in species composition and forest characteristics of current SDTF, from closed-canopy evergreen forests to scrub savannas dominated by exotic pasture grasses. Variation in climate and soil properties also contributes to structural variation in SDTF. Even relatively undisturbed SDTF will show a high degree of variation across sites within regions due to the amount and pattern of rainfall and soil properties (Powers et al. 2009).

Despite the rarity and continued loss of SDTF, restoration of SDTF has rarely been attempted on a large scale. Worldwide, most restoration efforts have been passive and undocumented in the scientific literature (e.g., Fajardo et al. 2005). However, several regions in Latin America are well described. Colón and Lugo (2006) demonstrated that the outcome of 50 years of passive restoration of SDTF in Puerto Rico was highly dependent on past land use patterns. Calvo-Alvarado et al. (2009) found evidence for major recovery of forested area since the 1980s in Guanacaste province, Costa Rica, although they concluded that this was mostly a result of economic and social changes in the region and had little to do with conservation policy. Nevertheless, deforestation of SDTF continues outside park boundaries. Sánchez-Azofeifa, Daily et al. (2003) showed that outside of the network of protected areas in Costa Rica, deforestation has not declined. Similarly, substantial forest regrowth has occurred within the Chamela-Cuixmala Biosphere Reserve, but beyond the boundaries a high degree of fragmentation continues (Sánchez-Azofeifa et al. 2009).

Although few studies document restoration of SDTF, many studies have identified factors that limit forest regeneration These include ungulate exclusion (e.g., Cabin et al. 2002), a complex issue because ungulates may either facilitate regeneration (by reducing fuel loads in fire-prone areas; Janzen 1986a) or retard it (by grazing or trampling establishing seedlings; Cabin et al. 2002). The use of "framework" or "nuclear" trees to reduce grass cover and provide habitat for seed dispersers has also been advocated, but species selection requires an understanding of both the growth of the framework species and the seedling community that establishes beneath it (Elliot et al. 2003). The role of seed rain and the distribution and germination of seeds outside of fragments (Holl 1999) as well as appropriate timing of reseeding efforts (e.g., Mascia Vieira et al. 2008) have also been identified as critical factors in regeneration of SDTF. One common thread among all of these studies, however, is that the parameters for restoration discovered from this bottom-up approach depend largely on local conditions. As a result, it is unlikely that general solutions to SDTF restoration will be discovered. In an instructive example, Griscom et al. (2009) showed that while removal of cattle was beneficial to restoration, herbicide application to control exotic grasses only improved outcomes in sites that were far from remnant forest fragments. Inappropriate restoration techniques may even have a negative impact on "natural" regeneration. Sampaio, Holl et al. (2007) showed that in Brazilian cerrado, passive restoration (preventing disturbance) is superior to some active restoration techniques (those that involve mowing or plowing before planting nursery-grown seedlings). As a result, they argued that reseeding efforts should be targeted and limited to species not naturally regenerating in the area of interest.

SDTFs in Guanacaste and the Area de Conservación Guanacaste

The Area de Conservación Guanacaste (ACG) is a 110,000-hectare conservation area located in northwestern Costa Rica. About 50,000 hectares of its area are SDTFs, averaging about 1500 millimeters of rain per year. It experiences a 6-month dry season running from December to May of each year. Unpredictable weather events such as hurricanes cause a high degree of year-to-year variation in rainfall, although it is unusual for more than a few millimeters of rain to fall during the dry season.

Throughout Guanacaste there has been widespread conversion of SDTF to pastureland for the grazing of cattle and selective logging for valuable species such as lignum vitae (*Guayacan sanctum*) and mahogany (*Swietenia humilis*). Tropical dry forest in the ACG now exists as a complex mosaic of SDTF and pastures in varying states of regeneration. True old-growth SDTF is restricted to a few very small patches, which have been selectively logged. There also exist extensive stands of SDTF that are more than 100 years old that support high tree and animal diversity.

The core area of the ACG, Parque Nacional Santa Rosa, was declared a national park in 1971 to protect high-quality tropical dry forest habitat (Janzen 1986b). Throughout the 1970s, 80s, and 90s, the main external threat to this conserved wildland and the remnant forests surrounding it was anthropogenic fire used to maintain cattle pastures. These fires were devastating to remnant forest patches because of high fuel load from the introduced pasture grass species jaragua (*Hyparrhenia rufa*). Following the establishment of the national park, fires set by neighboring landowners encroached on the protected area, causing persistent degradation of remnant forest stands.

Since the mid 1980s a fire control program within the ACG has reduced the impact of fire on SDTF regeneration tremendously, limiting the area burned to fewer than 2000 hectares per year. Since active fire suppression began in the 1980s, rapid recolonization of abandoned pastures by woody vegetation has been observed in much of the seasonal ACG (Janzen 1986a, 1988d).

Efforts to actively manage regeneration of SDTF in the ACG have been minimal. However, Guanacaste SDTF appears to be amenable to passive restoration. SDTF soils are more fertile than many tropical soils, and many species grow well even in soils taken from beneath pasture vegetation (Klemens 2003). Also, the major introduced species, jaragua (*Hyparrhenia rufa*), does not outcompete woody vegetation directly but persists because it tolerates fire better than woody species (Daubenmire 1972a). Many SDTF species are therefore able to establish in the middle of a jaragua pasture, including many "old-growth" species (Janzen 1988b; Gerhardt 1993), although subsequent survival in the pasture environment seems to be highly species specific (Gerhardt 1993) and has not been monitored in a comprehensive way.

Quercus oleoides in Guanacaste

Our work focuses on the regeneration of the tropical live oak, *Quercus oleoides*, in the ACG. Although the ACG is one of the "success stories" of SDTF restoration, the regeneration of *Q. oleoides* forests is thought to be severely "limited" under the current regime of passive regeneration, compared with other taxa in the regional species pool. These mono-dominant forests formed a unique SDTF ecosystem that once covered tens of thousands of hectares of Guanacaste Province prior to the severe forest fragmentation of the last century (Boucher 1981; Janzen 1987). Structurally and ecologically important species that do not respond to a particular regeneration regime are a major challenge to restoration in SDTF. Despite being formerly widespread, *Q. oleoides* is something of a biological enigma here: the local population is geographically disjunct, physiologically differentiated, and genetically distinct from conspecifics and similar species (J. Cavender-Bares, A. Pahlich, A. Gonzalez-Rodriguez, and N. Deacon, unpublished data; Cavender-Bares 2007). Unlike many other dry forest species, *Q. oleoides* is geographically quite restricted within Costa Rica. Moreover, *Q. oleoides* is ectomycorrhizal in a habitat dominated by vesicular arbuscular mycorrhizal associations, possesses an atypical developmental process with regard to germination and emergence system (a fruit type and associated dispersal syndrome that is extremely rare in the tropics), is wind pollinated in a habitat dominated by insect-pollinated species, and is evergreen in a habitat where most species are deciduous or semideciduous. Finally, a large proportion of trees in the population show a reproductive phenology that seems largely mismatched to the seasonally dry environment of Guanacaste (Cavender-Bares et al. 2010), producing crops of dessication-sensitive acorns at the beginning of a lengthy dry season that is more severe than anywhere else in its range.

Despite this seeming mismatch between traits and environment, *Q. oleoides* is the most common large tree wherever it occurs and may represent 80 percent or more of individuals present at a site (Boucher 1983). The persistence of leaves in the dry season is probably explained by drought adaptations common to many oaks (e.g., Abrams 1990; Pallardy and Rhoads 1993; Goulden 1996; Cavender-Bares et al. 2004). The evergreen leaves are very resistant to wilting, compared with co-occurring species (Brodribb et al. 2003; Brodribb and Holbrook 2005; Cavender-Bares et al. 2007). Continued gas-exchange rates and high predawn water potentials of mature trees during the dry season suggest that they are deep rooted as adults, although seedlings are considerably more vulnerable (Cavender-Bares, unpublished data). The abundance of *Q. oleoides* in the region and its unusual combination of functional traits make it an extremely important species in Guanacaste dry forest communities. The seeds are consumed by a wide range of mammalian and avian seed predators, and its evergreen habit affects the abiotic environment experienced by many dry forest organisms.

Scale of the Study

In the ACG, *Q. oleoides* extends from the dry forest of sectors Santa Rosa and Santa Elena, at approximately 280 meters elevation, to the much wetter forests at approximately 800 meters on the Pacific slope of Volcán Rincón de la Vieja.

Because our study was designed to examine the range of habitat types present within the ACG, six study sites were chosen that exhibited an abrupt ecotone between mono-dominant oak forest and abandoned pasture: three low-elevation (approximately 280 meters) dry forest sites in sector Santa Rosa and three high-elevation (approximately 800 meters) wet forest sites on the Pacific slope of Volcán Rincón de la Vieja. The particular sites were selected for the presence of a clear forest-pasture ecotone across which to establish transects and to be generally representative of the biophysical environments present at the two elevations (fig. 13-1).

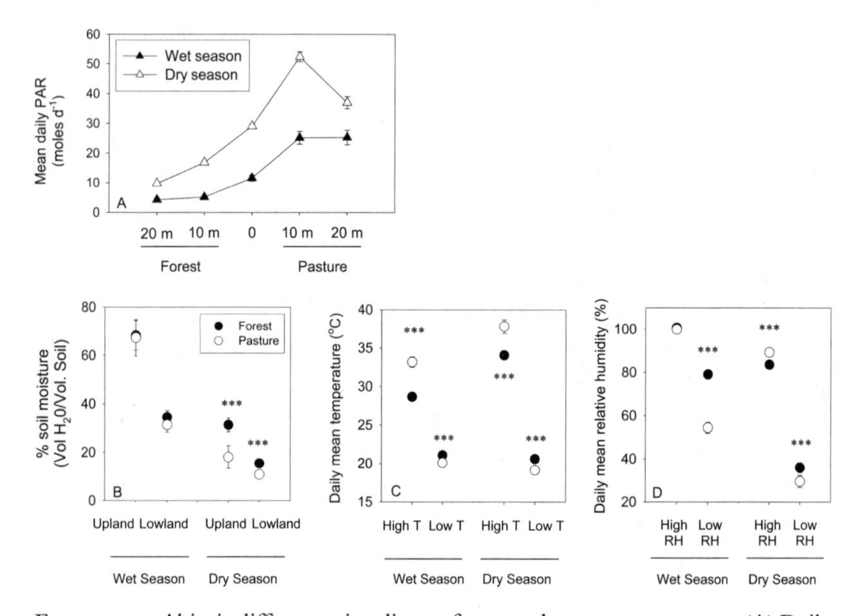

FIGURE 13-1. Abiotic differences in adjacent forest and pasture ecosystems. (A) Daily integrated photosynthetic active radiation (PAR) across the forest-pasture ecotone during the wet and dry seasons averaged for six sites. The x-axis indicates the distance from the edge (0) into the forest or the pasture. (B) Estimated values of volumetric soil water content θ to 30 centimeters depth) averaged for 10 meters and 20 meters into the forest or the pasture for the three upland (800 meters) and three lowland (300 meters) sites. θ measurements were made using time domain reflectometry and calibrated with the equation $\theta = (67.345*x) - 149.74$ for highlands and $\theta = (32.643*x) - 67.678$ for lowlands, where x is flight time of the electromagnetic pulse in arbitrary units, following Cavender-Bares and Holbrook (2001). (C) Seasonal high and low daily means for temperature. (D) Relative humidity at 20 meters into the forest or pasture for the same sites. For B–D, significant differences between means in the forest and the pasture, based on two-way ANOVA with habitat and site as main effects, are indicated as follows: $***P < 0.001$, $**P < 0.01$, $*P < 0.05$.

SEED PREDATION

The only previous empirical work focused on regeneration of *Q. oleoides* is the work of Boucher (1981). Boucher concluded that for small stands of oaks, at least, seed production was insufficient to satiate the local predator community and suggested that this explained the failure of *Q. oleoides* forests to regenerate on the landscape. A large number of vertebrate seed predators do consume acorns in large numbers (table 13-1). There are, however, several reasons to doubt whether seed predation is sufficient as a general explanation for regeneration failure in Guanacaste. First, as shown above, oak seedlings exist at relatively high density within oak patches and on the forest pasture ecotone (fig. 13-2). Second, viable acorns can be collected from oak stands throughout the wet season and can be observed germinating in place. Third, Boucher assumed a long time interval during which acorns were exposed to predators, and this interval was used to project seedling germination and survival rates based on seed predation rates observed over the short term. However, in all cases where we

TABLE 13-1. Vertebrate seed predators known to consume *Quercus oleoides* acorns and details of the spatial patterns of the interaction

Consumer/disperser	*Interaction with* Quercus oleoides	*Distance*	*Movement in pasture*
Variegated squirrel (*Sciurus variegatoides*)	Predator (1)	Close	No, highly arboreal
Agouti (*Dasyprocta punctata*)	Predator (1), scatterhoarder (3)	> 30 m	Limited, cross pastures, do not cache there
Deer (*Odocoileus virginianus*)	Predator (1)	Close	Yes
Collared peccary (*Tayassu tajacu*)	Predator (1)	None	Cross, limited foraging
White-faced monkey (*Cebus capucinus*)	Rare predator* (1)	Close	No
Spiny pocket mouse (*Liomys salvini*)	Predator (1)	Close	Short distances up to 50m**
Magpie jay (*Calocitta formosa*)	Predator (1), cacher (?)	Far	Forage in pastures
Parrots (*Amazona* sp., *Pionopsitta* sp.)	Predator (2)	Close (drop)	Forage in pastures

*infrequently in oak forest, effect mostly on edge oaks in contact with more diverse forest
**in pasture to forest direction (Janzen 1986c)
Sources are (1) Boucher 1981, (2) Boucher 1983, (3) Hallwachs 1994.

observed mature acorns falling from trees, they germinated and extended the primary radicle within days of landing on the soil surface given even moderate levels of soil moisture. An experiment conducted on seeds from a number of different parental trees indicated that 78 plus or minus 23 percent of seeds collected from the ground would germinate immediately if placed in a damp environment. However, even mild desiccation (i.e., air drying during the humid wet season in the absence of direct sunlight) reduced germination rates to 18 plus or minus 19 percent after 5 days of desiccation. Germination was reduced to 0 percent after 15 or 29 days of desiccation (data not shown).

Finally, direct measurements of seed removal rates indicated that seed predators are not removing seeds at a high enough rate to prevent seedling emergence, given the rapidity of germination. Furthermore, at sites where predation rates varied between the forest and the pasture, seed survival was higher in the pasture than the forest (fig. 13-3).

These results, taken together with the observations of naturally occur-

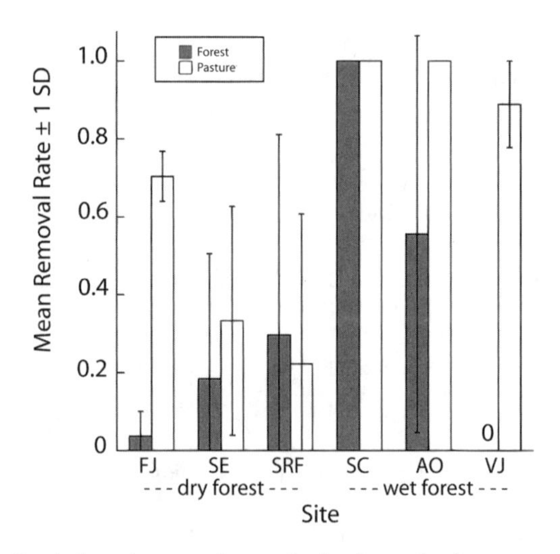

FIGURE 13-3. Pooled seed removal rates in the forest in the pasture habitats for the six study sites. Values are the averages of the per cage removal rates in forest and pasture points on each transect. A randomized two-way nested ANOVA of short-term removal rate shows a trend in forest type over site (df = 1, MS = 1.777, F = 7.20, P = 0.0586). There was a significant effect of site nested within forest (df = 4, MS = 0.247, F = 3.78, P = 0.0172) and the interaction of habitat by site (df = 4, MS = 0.257, F = 7.86, P = 0.0010). Habitat and the interaction of habitat by site were not significant.

ring seedling establishment, indicate that while seed predation undoubtedly plays a role in local regeneration dynamics, it is unlikely to serve as a general explanation for regeneration failure.

Survival and Growth of Seedlings

The pasture is significantly hotter than the forest understory, and during the dry season it is significantly drier, in terms of both atmospheric relative humidity and soil moisture (fig. 13-1). It is not well understood to what extent the seedlings tolerate the dry season, particularly in the harsh and variable conditions of the pasture environment. In general, oak seedlings are less resistant to drought than mature trees because of shallower rooting and less desiccation-tolerant leaves (Cavender-Bares and Bazzaz 2000). Potential competitive interactions with the exotic pasture grass, jaragua, which already has established root systems, provide further reason to consider that seedlings may be particularly vulnerable in the pasture.

To test seedling performance in pastures relative to the forest understory, seeds were planted into the predator exclusion cages on the transects at each site in early July 2005. Seeds were monitored monthly or semimonthly over the next year for emergence and survival. A subset of these seedlings was harvested after 6 months to assess the degree of mycorrhizal infection on the roots. All remaining seedlings were harvested after one year and root systems excavated to assess the rate of mycorrhizal infection.

Emergence

Overall emergence rates varied among sites and among transect positions; however, there was only a moderate drop in emergence rate in the pasture habitat (fig. 13-4A).

Survival and Growth

Survival of emerged seedlings was uniformly high, with the median survival higher than 80 percent for all distances (fig. 13-4B). There was no significant effect of transect position on survival. For growth, measured as total aboveground dry mass for each harvested plant, there was a significant effect of transect position. Plant growth was highest at 10 meters into the pasture (fig. 13-4C) and was generally higher in the pastures.

While additional experiments are ongoing, these data lead us to

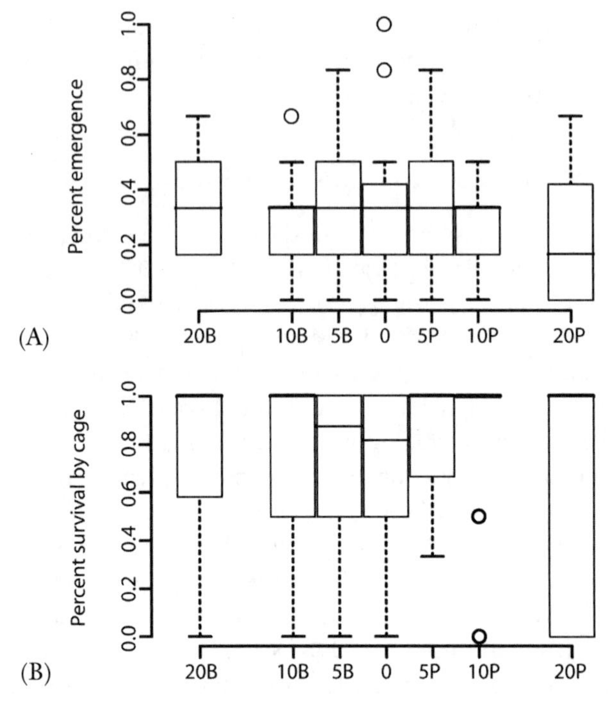

FIGURE 13-4. (A) Effect of transect position on emergence rate per cage. A two-way nested ANOVA comparing emergence across the forest-pasture transition shows significant effects of distance (df = 6, MS = 0.230, F = 8.018, $P < 0.0001$), site within forest type (df = 2, MS = 3.036, F = 177, $P < 0.0001$), and the interaction of site by distance (df = 12, MS = 0.181, F = 6.336, $P < 0.0001$). Forest type, transect nested within site, the forest by distance interaction, and the transect by distance interaction were not significant. (B) Survival to greater than 1 year of emerged seedlings as a function of transect position. A two-way nested ANOVA comparing survival across the forest-pasture transition shows a significant effect of site nested within forest type (df = 2, MS = 0.735, F = 7.470, P = 0.001). Distance, forest type, the interactions of forest by distance, and site by distance were not significant. For all box plots, heavy lines are medians, boxes represent middle quartiles of the distribution, whiskers are set to 1.5 times box length, and open circles are outliers.

conclude that seeds and young seedlings of *Q. oleoides* are robust in the face of the biotic and abiotic conditions of the pasture environment. Competition with grasses does not appear to inhibit seedling performance once seedlings have emerged and may even facilitate it. Gerhardt (1993) observed that the dry season is the time at which most seedling mortality oc-

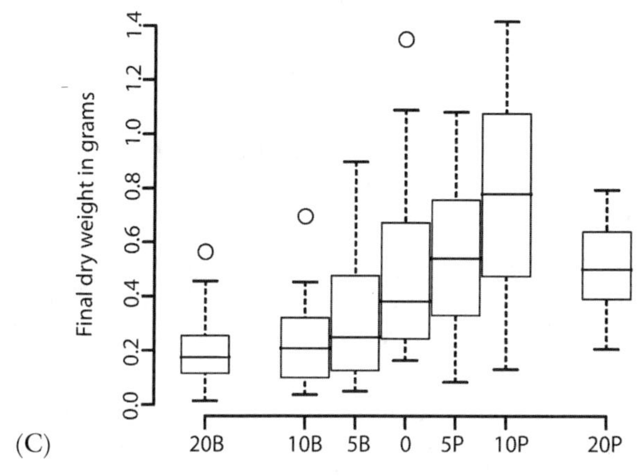

FIGURE 13-4. (C) Dry mass in grams of surviving seedlings as a function of transect position at final harvest in July 2006. A two-way nested ANOVA comparing final seedling size across the forest-pasture transition shows a significant effect of distance (df = 6, MS = 0.675, F = 8.426, P = 0.001). There is no significance of forest type or site nested within forest type. The interactions of forest by distance and site by distance were also not significant.

curs in SDTF. However, in our study seedling mortality often occurred in the wet season, perhaps due to submersion by flooding during the wet season, fungal infection, or light limitation in forested areas.

Ectomycorrhizal Infection

In December 2005 (5 months postplanting) a small sample of plants was harvested from each position on one transect at one wet forest and two dry forest sites in order to assess mycorrhizal infection. For each plant, a target sample of 300 to 400 root tips was examined under a dissecting microscope. For many plants fewer root tips were recovered, as oak root tips are very delicate and have a high tendency to shear when removed from clay soils, despite soaking of the soil with water and cutting through the soil with a sharp knife before extraction. As a result, complete samples could not always be collected, and the wet forest site examined is missing the data from the edge transect position. Fungal morphotypes were identified, photographed, and counted. From these data, the number of morphotypes and the percentage of root tips that were infected with mycorrhizae were calculated following Kennedy et al. (2003).

Among our samples, infection was uniformly high (fig. 13-5A).

ANOVAs of both morphotype diversity and percent colonization indicated that there was a significant or marginally significant effect of site on both measures of fungal colonization but no effect of transect position on either variable. Also notable is that some of the highest ectomycorrhizal diversity observed among all samples occurred in the pasture environment (fig. 13-5B). Species richness of morphotypes was plotted against the number of tips successfully collected from each sample in order to see if differences in richness might simply be a function of the number of tips recovered. There was no relationship between number of tips collected and number of morphotypes encountered ($R^2 = 0.04, p = 0.38$) and sites tended to cluster in terms of both morphotype diversity and tips collected, meaning that relative comparisons of transect positions within sites are appropriate.

These data, taken together with the data presented earlier on robust plant growth in the pastures, suggest that a lack of fungal symbionts is not likely to be important in limiting regeneration in this species. This is sur-

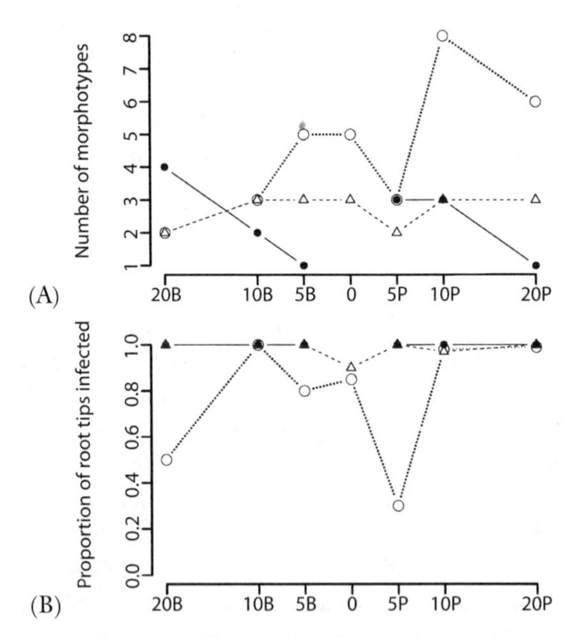

FIGURE 13-5. Mycorrhizal infection of plants grown from seed along three transects, each at a different site, at the early harvest (December 2005). (A) Number of distinct morphotypes present at each transect position. (B) Proportion of root tips colonized for the same samples. Dotted lines with open circles are for site SRF, dashed lines with open triangles for FJ, and solid circles with unbroken line for VJ.

prising, given the paucity of ectomycorrhizal species in this system and the important role that ectomycorrhizal fungi may play in maintaining monodominance in tropical forest systems (McGuire 2006). Studies in tropical vesicular arbuscular mycorrhizae systems have shown, however, that pasture soils often contain as much or more innoculum potential as forest soils (Fischer et al. 1994), and a study in SDTF showed that innoculum from early successional stages (2 years postburning) promoted seedling growth more effectively than fungal innoculum drawn from later successional stages (Allen et al. 2003).

It also contrasts somewhat with recent studies of forest-grassland systems in the temperate zone (Dickie and Reich 2005; Dickie et al. 2007). It is possible that at some point beyond our sampling scale (20 meters into pasture) infection would decrease, and at that scale it may limit regeneration.

Dispersal Limitation

Growth and survival of seedlings seem to be robust in the pasture environment over the timescale of our study. Therefore, at least in favorable years, it seems clear that *Q. oleoides* seedlings will not be prevented from establishing in the pasture environment. Furthermore, there does not appear to be a generalized reproductive failure, based on the abundant recruitment of seedlings within oak patches (fig. 13-2).

The remaining plausible explanation is that acorns do not arrive in the pasture environment in the first place. One major difference in the oak forests of Guanacaste compared with the oak forests of temperate America is the absence of squirrels that exhibit caching behavior. North American squirrels are behaviorally flexible and highly mobile dispersers that serve as effective dispersers of acorns in the temperate zone (Steele and Koprowski 2001). The variegated squirrel, the only squirrel native to most of the range of *Q. oleoides* in Guanacaste, is almost completely arboreal and is not known to exhibit caching behavior (Harris 1937; Boucher 1981).

Most vertebrates that interact with *Q. oleoides* in Guanacaste probably act exclusively as seed predators (Boucher 1983) (table 13-1). The mammal most likely to play a large role in oak seed dispersal is the Central American agouti (*Dasyprocta punctata*), which has been shown to play a disproportionate role in the dispersal of other large-seeded SDTF species (Hallwachs 1986). Although agoutis will cross open pasture, they do not cache seeds in pastures (Hallwachs 1994) or even in large forest gaps (Hallwachs 1986). It has been shown that in fragmented oak forests in the United

States, fragmentation-tolerant squirrel species (*Sciurus niger*) can make up for the absence of more fragmentation-sensitive species (*S. carolinensis*) in maintaining substantial dispersal (Moore and Swihart 2007). However, no such secondary disperser occurs in Costa Rican SDTF.

Preliminary data on seed dispersal in this system (Klemens and M.A. Steele, unpublished data) indicate that most seeds are killed within a few meters of the presentation point, and we have so far failed to detect any movement of seeds from the forest environment into the pastures. In fact, when seeds are moved, they seem to be moved deeper into oak patches, away from the pasture. Many unanswered questions remain, however, and this is an important area for future study.

Conclusions and Implications for Restoration

Our data indicate that acorns germinate and survive in abandoned pastures. Once they are established, growth appears to be promoted by the high-light environment of the pastures. However, given the lack of effective dispersal agents, recolonization is likely to occur only very gradually outward from the edge of the forest. The combined results of these experiments help provide guidelines for active restoration of tropical live oak forest ecosystems. Given the plausible role of dispersal limitation and the high performance of established oak seedlings in the pasture, a direct-planting strategy is likely to work well. Seeds or seedlings can be planted into abandoned pastures during the wet season to give their roots time to establish before the dry season. As seedling nurseries to facilitate such efforts can be readily established at low cost and with minimal maintenance, our findings thus far are optimistic with respect to the regeneration potential of the live oaks in Guanacaste. Given the complexity of the tropical dry forest agroscape, this study demonstrates the utility of basic research in deciding among alternative restoration strategies. Our understanding of this system remains preliminary, but it reinforces the notion that any understanding of SDTF regeneration will depend on understanding how highly local, and even species-specific, factors, both "natural" and anthropogenically caused, interact to promote or retard regeneration. This may be the major challenge to restoration ecologists working within a particular system. As there is no reason to expect that many general solutions to the problem of SDTF restoration will emerge beyond "stop the disturbance," for any restoration effort there is probably no substitute for immersion in the natural history of the organisms involved.

Acknowledgments

This work could not have been conducted without the tireless efforts of our field assistant, Marileth Leitón Briceño. Meredith Sheperd assisted with data collection. Dan Janzen and Winnie Hallwachs provided helpful discussion and advice at all stages of the project. Sal Agosta, Mike Steele, Jack Bradbury, Ian Dickie, Jennifer Powers, and N. Michele Holbrook all provided helpful discussions relevant to the project. We would also like to thank the entire staff of the ACG for making this research possible and, in particular, Roger Blanco, María Marta Chavarría, and Felipe Chavarría. JAK was supported by an NSF postdoctoral fellowship awarded in 2004. NJD was supported by several grants from the University of Minnesota. Grants from the University of Minnesota and NSF (IOS-0843665) to JCB supported multiple aspects of the study. JCB would like to thank R. Dirzo and H. Mooney for the invitation to contribute this chapter.

Chapter 14

Economic Botany and Management Potential of Neotropical Seasonally Dry Forests

CHARLES M. PETERS

Seasonally dry tropical forests (SDTFs) are preferred habitats for people who live in the Neotropics. The reasons for this are straightforward. There are fewer big trees in dry forests than in moist or wet forests, and the land is easier to clear for farming. It is also easier to burn the felled trees and slash after clearing, because of the seasonal drought. Once the tree cover has been removed, the underlying soils are frequently more fertile than those found in regions of higher rainfall, because nutrient leaching is less extreme. Finally, the pronounced dry season keeps pest levels down in agricultural fields and reduces the incidence of mosquito-borne and fungal pathogens that cause health problems. It is not surprising that in many Central American countries the population density in dry forest areas is three to four times higher than that found in wet forest areas (Tosi and Voertman 1964).

Long-term human presence in Neotropical dry forests has produced both positive and negative results. On the positive side, communities living in dry forest areas have amassed a wealth of ethnobotanical information about local plant species. They have learned the properties, uses, and ecological requirements of hundreds of native species. For example, one recent study in Pernambuco, Brazil, found that more than 80 percent of the species recorded had local uses (Albuquerque et al. 2005), more than 400 species of medicinal plants were recorded in a survey of four municipalities of northeastern Brazil (Victor 1990), and the Tarahumara of Chihuahua, Mexico, use 176 different species for food alone (Bye 1989). Clearly, the people who opted to settle in Neotropical SDTF learned a lot about the local flora.

The negative result of human occupation of SDTF areas is the ever-increasing rate at which the forest is being disturbed, depleted, or converted to other forms of land use. Dry forests have been overexploited for nontimber forest products, subjected to uncontrolled logging, cleared for agriculture, and turned into tourist facilities (Almeida and Albuquerque 2002; Quesada and Stoner 2004; Gordon et al. 2006). Hundreds of hectares of dry forest have been cut and burned to make cattle pastures in Latin America (Toledo 1992). The overall extent of SDTF in the Neotropics is currently a small fraction of its original distribution. Obviously, clearing a large tract of dry forest eliminates all of the plant resources that were growing there, many of which may have been quite valuable.

Here, I apply the promise of the first result of the human-SDTF interaction, that is, the vast knowledge about useful forest species, to the critical problem engendered by the second, that is, forest destruction. Can the resource richness of Neotropical SDTF be used in a way that promotes forest conservation? Are there plant species that could be harvested sustainably to provide local communities with a source of income—and a reason to maintain the forest? Are there any examples from the Neotropics where this is actually occurring?

To start addressing these questions, the economic botany of Neotropical dry forests is first briefly reviewed. Given the diversity of species involved and the paucity of literature about the useful flora in dry forests, this is only a very selective overview. The potential for managing dry forest resources on a sustained-yield basis is then assessed by focusing on the ecology of Neotropical SDTF from a forestry perspective. I conclude by offering two examples from Mexico where local communities are attempting to conserve their SDTF by using them sustainably.

A Survey of Plant Resources in Neotropical Seasonally Dry Forests

Although Neotropical SDTFs extend from the tip of the Baja peninsula in Mexico, through Central America and the Caribbean, and down to northern Argentina (Pennington, Lewis et al. 2006), much of the available literature on the useful flora of these forests focuses on either Mexico or northeastern Brazil. In general, we know more about the medicinal plants in these two areas than about food or fiber resources, there tends to be more small-scale community studies than regional surveys, and more attention is given to the subsistence use, rather than the commercial trade, of SDTF resources. Relative to Neotropical moist or wet forests, quantitative research on the economic botany of dry forests is sorely lacking.

A decidedly subjective selection of 30 useful plants from Neotropical SDTF is presented in table 14-1. Species were chosen to represent a variety of different life-forms and resource types and to showcase the ethnobotany of particularly interesting plants. Categories of uses are defined ecologically based on the particular plant tissue exploited rather than the type of product used. Many of the species shown in table 14-1 have multiple uses involving several different parts of the plant. The fresh fruits of *Guazuma ulmifolia* are edible, and the toasted-and-ground seeds are used to prepare a coffeelike beverage. The fibrous inner bark is used for cordage; the wood is easily worked and used for making boxes, cabinets, barrels, doors, and windows; and the flowers, bark, leaves, buds, and roots are all used medicinally (Little and Wadsworth 1964; Vallejo and Oveido 1994). The thorny bark of *Ceiba acuminata* is used for carving handicrafts, the roots are an important source of starch in times of drought, and the puffy seed fibers are used to stuff pillows and to add cushioning to cribs (Yetman et al. 2000). *Hymenea courbaril* produces resources from all three categories.

Neotropical SDTFs contain hundreds of species of medicinal plants (e.g., Argueta et al. 1994; Matos 1999; Silva and Albuquerque 2005), and while the use of many of these plant remedies is limited to specific cultural groups or regions, several of the species listed in table 14-1 are also traded in national and international markets. The bark of *Amphipterygium adstringens* ("cuachalalate"), for instance, is one of the most valuable and widely used medicinal plants in Mexico (Olivera 1998), and there is also a consistent and growing demand for the species in foreign markets (FAO 2003). Infusions made from *Amphipterygium* bark are reported to be effective in the treatment of more than 30 ailments, including gastric ulcers, high cholesterol, malaria, and stomach cancer (Solares 1995).

The yellow oleoresin obtained from *Copaifera officinalis* is also a common and commercially valuable remedy. The oil is used as a topical analgesic for wounds and skin irritations, and it is reported to have antibacterial properties useful for treating sore throats, urinary tract infections, and stomach ulcers (Basile et al. 1988; Veiga et al. 2001). Prior to the discovery

TABLE 14-1. Selective list of plant resources from Neotropical SDTF

Nomenclature	Life-form	Uses
		Vegetative tissues
Anacardiaceae *Amphipterygium adstringens* (Schltdl.) Standl. Cuachalalate	Tree	Bark used medicinally to treat ulcers, wounds, and gastritis
Fabaceae *Anadenanthera colubrina* (Vell.) Brenan Angico	Tree	Bark source of tannins; used medicinally to treat respiratory problems
Arecaceae *Astrocaryum vulgare* Mart. Tucum	Palm	Leaf fibers used to make fishing nets, bags, and rope
Arecacae *Brahea dulcis* (HBK) Mart. Palma soyate	Palm	Leaves used to make mats, baskets, and hats; thatch
Fabaceae *Bauhinia cheilantha* (Bong.) Steud. Mororó	Shrub	
Burseraceae *Bursera glabrifolia* (H.B.K.) Engl. Copal	Tree	Wood used for carving handicrafts; wood oil used in soaps and perfumes
Malpighiaceae *Byrsonima crassifolia* L. Kunth. Nanche	Tree	Bark used for tanning
Rubiaceae *Calycophyllum candidissimum* (Vahl) DC Palo camarón, Madroño	Tree	Wood good source of sawtimber; bark used medicinally to treat stomach ulcers
Capparaceae *Capparis jacobinae* Moric. Ex Eichler Icó	Shrub	
Bombacaeae *Ceiba acuminata* (S. Watson) Rose Pachote, Kapok	Tree	Bark used for carving handicrafts; roots a source of starch
Fabaceae *Copaifera officinalis* (Jacq.) L. Copaiba, Palo de aceite	Tree	
Arecaceae *Copernica cerifera* (Arruda) Mart Carnaúba	Palm	Wax from the leaves used in cosmetics, ointments, and industrial products
Bignoniaceae *Crescentia alata* Kunth	Tree	
Fouquieriaceae *Fouquieria macdougalii* Nash Jaboncillo	Shrub	Bark a source of shampoo
Sterculiaceae *Guazuma ulmifolia* Lam. Guásima	Tree	Inner bark used to treat liver ailments; outer bark a purgative; wood used for making furniture
Hippocrateaceae *Hemiangium excelsum* (Kunth) A.C. Smith Cancerina	Liana	Bark used to treat wounds and gastric ulcers

Uses (continued)		Sources
Reproductive propagules	Exudates	
		Martínez 1969; Argueta et al. 1994; Hersch Martínez 1996
		Sampaio 2002; Monteiro et al. 2006
Edible fruit		Sampaio 2002
Edible fruits		Aguilar et al. 1997; Acosta et al. 1998; Illsley et al. 2001
Edible fruits; used medicinally for treatment of diabetes		Albuquerque et al. 2005
	Resin used for incense	Chibnick 2003; Peters et al. 2003
Edible fruit; juice used as a dye		Bye 1995; Casas and Caballero 1996
		Sabogal 1992
Fruits eaten as a vegetable		Albuquerque and Andrade 2002
Seed fibers used for stuffing pillows		Robichaux and Yetman 2000
	Oleoresin used medicinally for stomach ulcers, respiratory problems, and skin disorders	Lawless 1995; Fajardo et al. 2005
		Bayma 1958; Sampaio 2002
Fruits used to make handicrafts and household utensils; edible seeds		Bye 1995
		Robichaux and Yetman 2000
Dried fruits used to make hot beverages; used medicinally to treat kidney ailments		Martínez 1969; Pennington and Sarukhán 1998
		Hersch Martínez 1996

(table continues)

TABLE 14-1. *(continued)*

Nomenclature	*Life-form*	*Uses*
		Vegetative tissues
Fabaceae *Hymenaea courbaril* L. Jatobá	Tree	Bark used medicinally
Fabaceae *Leucaena esculenta* (Moc. and Sessé ex A. DC.) Benth. Guaje	Tree	
Chrysobalanaceae *Licania rigida* Benth. Oiticica	Tree	
Celastraceae *Maytenus rigida* Mart. Bon-nome	Tree	Leaves used medicinally to treat ulcers; bark infusion used for coughs
Bromeliaceae *Neoglaziovia variegata* (Arruda) Mez Caroá	Herb	Leaves yield a strong white fiber
Fabaceae *Pithecellobium dulce* (Roxb) Bent. Guamúchil	Tree	Tonic from bark used medicinally; bark used for tanning
Asteraceae *Porophyllum ruderale* ssp. *macrocephalum* (DC.) R.R. Johnson Papaloquelite	Herb	Leaves eaten as a vegetable
Euphorbiaceaee *Sebastiana pavoniana* (Müll. Arg.) Müll. Arg.Brincador	Shrub	
Selaginellaceae *Selaginella lepidophylla* (Hook and Grev.) Spring. Doradilla		Whole plant used medicinally
Anacardiaceae *Spondias tuberosa* Arruda Umbú	Tree	
Cactaceae *Stenocereus stellatus* (Pfeiff.) Riccob. Pitaya	Cactus	
Arecaceae *Syagrus coronata* (Mart.) Beccari Licuri	Palm	Wax from leaves; palm heart
Asteraceae *Vanillosmopsis arborea* (Gardner) Baker Candeia	Shrub	Bark oil used medicinally
Rhamnaceae *Ziziphus joazeiro* Mart. Joazeiro	Small tree	Bark and leaves used medicinally; inner bark a source of toothpaste

Uses (continued)		Sources
Reproductive propagules	*Exudates*	
Seedpods contain edible pulp	Resin used as incense and medicine	Langenheim 1981; Albuquerque and Andrade 2002
Edible pods and seeds		Bye 1995; Casas and Caballero 1996
Oil seeds		Prance 1972; Sampaio 2002
		Albuquerque and Andrade 2002; Silva and Albuquerque 2005
		Sampaio 2002
Edible arils		Bye 1995; Casas et al. 2006
		Bye 1981; Vázquez 1991
Triangular seeds used as toys (Mexican jumping beans)		Robichaux and Yetman 2000
		Martínez 1969; Hersch Martínez 1996
Edible fruits		Silva 1986; Santos 1999; Sampaio 2002
Edible fruits		Pimienta Barrios and Nobel 1994; Casas et al. 1997
Edible fruits; oil seed		Noblick 1986; Silva and Tassara 1996
		Viera 1999
Edible fruit		Lorenzi and Matos 2002; Sampaio 2002

of sulfa drugs and penicillin, copaiba oil was the principal treatment for gonorrhea (Dwyer 1951). The oil is marketed throughout the world, and a quick Internet search revealed that a 2-ounce bottle of *Copaifera officinalis* oil currently sells for about US$22. To put this price in context, individual *Copaifera* trees may yield up to 40 liters of oleoresin per year (Wood and Osol 1943), and many Venezuelan SDTFs are dominated by this species (Fajardo et al. 2005).

Medicinal plants are not the only commercial resources found in Neotropical SDTF. Carnauba wax, a basic ingredient in car polish, cosmetics, and the coating used on dental floss, is collected from the leaves of *Copernicia prunifera* (Johnson 1972), a common species in caatinga forest in northeastern Brazil but especially common in forests subject to periodic flooding (Henderson et al. 1995). The collection and processing of carnauba wax is a major industry in Brazil. In 2005 more than 3000 tons of wax and 19,000 tons of wax powder valued at approximately US$7 million and US$24 million, respectively, were produced (IBGE 2005). Although it is estimated that there are more than 200,000 hectares of caatinga forest containing *Copernicia prunifera* in the Brazilian states of Ceará, Piauí, and Rio Grande do Norte, much of the carnauba wax currently entering the market is produced in large plantations.

Oiticica seeds from *Licania rigida* are another important plant resource from SDTF regions of Brazil. A high-quality drying oil similar to tung oil (*Aleurites* spp.) is extracted from the seeds and used to make paint, varnish, and printing inks and for improving rubber elasticity (Linskens and Jackson 1991). The production of oiticica seeds in 1995 was estimated at 13,000 tons (FAO 2000), while more recent harvest data suggest that collection rates have dropped to 1500 tons (IBGE 2005).

A final example from Brazil is *Spondias tuberosa*. The tart fruits from this SDTF tree are highly esteemed by the residents of northeastern Brazil, and during harvest season the collection and sale of the fruits is a major source of income for rural communities. More than 9000 tons of fruit were harvested in 2005 (IBGE 2005), with an estimated market value of US$2.3 million. The fruit is eaten fresh or used to make jellies or desserts, and one of the most famous delicacies from the region, *imbuzada*, is prepared by adding the juice from *S. tuberosa* fruit to boiled, sweetened milk.

Factoring in the 33 tons of *Anadenathera* bark that are collected and sold for medicinal purposes, the 77 tons of *Astrocaryum* leaves harvested for fiber, the 206 tons of *Neoglaziovia* cordage, the 13,000 tons of *Copernicia* straw, the 329 tons of *Syagrus* palm hearts, and the 6440 tons of *Syagrus* leaves, the total value of the plant resources harvested from the dry forests of Brazil alone is somewhere on the order of US$50 million per year (Sampaio 2002). Adjusting this figure to account for the annual harvest of medicinal plants from the SDTF of Mexico and the sale of handicrafts made from *Crescentia* fruits, *Sebastiana* seeds, *Bursera* wood, and *Brahea* leaves, it becomes clear that the useful flora of Neotropical SDTF is of considerable economic importance—even when only the 30 taxa of table 14-1 are considered.

It is important to note that these plant resources have acquired economic value with little promotion, limited market development and advertising, and virtually no investment in management to ensure a continual supply of product. When SDTF are cleared to make pastures or build hotels, these revenue streams, many of which may have been quite important to local communities, are lost. In the absence of any type of management or harvest control, keeping the forest and continuing to exploit it commercially may be only marginally better in the long term. It is very easy to overexploit and deplete a valuable forest resource, and once this resource is gone, the perceived value of the forest drops to zero.

The missing link here is management. There are certainly enough useful products in Neotropical SDTF to warrant a program of controlled exploitation, and even the limited existing markets would seem to provide adequate economic incentive for promoting sustainable forest use. In terms of conservation, the basic idea is to use resource management to increase the value of intact forest relative to more-destructive forms of land use. If the forest is viewed as a valuable source of revenue that must be protected and stewarded, local communities, regional development agencies, and governments might be less inclined to do something else with the land. Current efforts to manage the wetter forest types in the Neotropics, however, have met with mixed success (Zarin et al. 2004). Is there any reason to think that we might do better in dry forests?

Forester's Guide to the Ecology of Neotropical Seasonally Dry Forests

Several ecological characteristics of tropical moist and wet forests exert a controlling influence on the nature and intensity of resource exploitation. The greatest bottlenecks to sustainable forest use are (1) the high species diversity and low density of conspecific individuals, (2) the irregularity of flowering and fruiting, and (3) the importance of animals for pollination and seed dispersal (Peters 1994). Low densities of useful species mean low yields per unit area, long search and travel times, and greater harvest costs. Low-density populations are also more prone to overexploitation. Unpredictability in flowering and fruiting makes it hard to schedule harvests and guarantee product delivery to buyers, and the critical role of animals in the reproduction of many tropical forest species—as pollinators, dispersers, or seed predators—requires that management activities extend to both the harvest plant and its animal coterie.

A review of recent literature on the ecology of tropical dry forests suggests that these limitations may not be as stringent in dry forests as they are in wet forests. This is not to say that dry forest are not species rich, or that there is no variability in flowering times, or that only a few species involve animals in the reproductive cycles. There are, however, notable structural, phenological, and reproductive features of Neotropical SDTF that seem to increase the potential of sustainable resource management as a viable land use option.

Density of Conspecific Trees

One of the most frequently cited and widely appreciated features of tropical forests is the high level of species diversity. Tropical wet forests have been reported to contain up to 300 species of trees greater than or equal to 10 centimeters in diameter at breast height per hectare (Gentry 1988), with dry forests exhibiting from 80 to 90 tree species greater than or equal to 2.5 centimeters in diameter in 0.1-hectare samples (Gentry 1995; Gillespie et al. 2000; Lott and Atkinson 2006). From a management standpoint, an unfortunate correlate with high species diversity is that each species is usually represented by only a few individuals. This trend is most apparent in tropical wet forests, where the majority of tree species occur in densities of only two to three individuals per hectare (e.g., Campbell et al. 1986; Gentry and Terborgh 1990).

As is shown in figure 14-1A, the mean density of conspecific individuals in Neotropical SDTF is considerably higher than that exhibited by moist or wet forests. Using plot data from 40 sites reported by Gentry (1988, 1995) and calculating mean conspecific density as simply the quotient of the total number of trees divided by the number of species (diameter greater than or equal to 10 centimeters in both), the mean number of trees per species is 2.55 and 7.75 for Neotropical wet and dry forests, respectively.

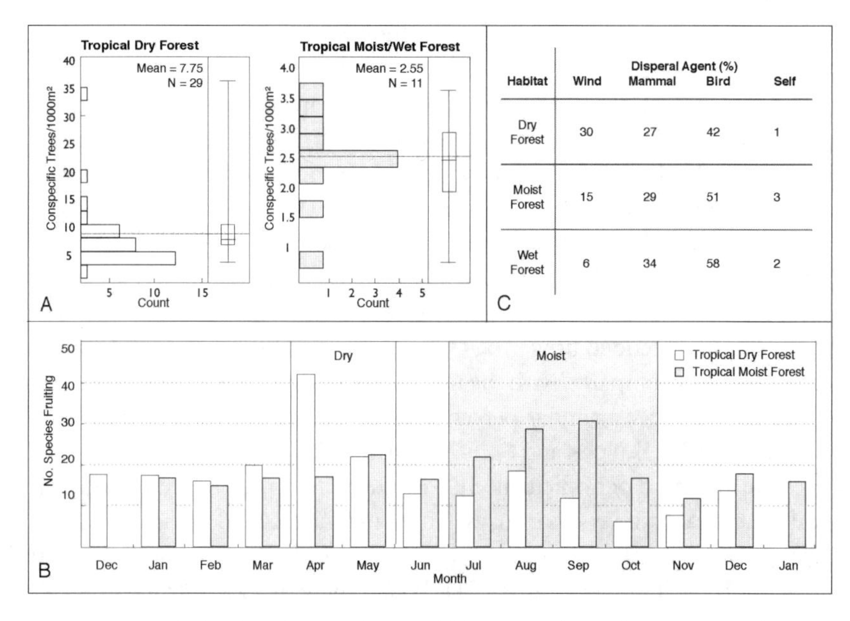

FIGURE 14-1. Ecological characteristics of Neotropical dry forests of special relevance to resource management. (A) Density of conspecific trees in tropical dry and tropical moist/wet forests. Horizontal line shows mean value; median value, 95 percent confidence interval, and range are shown in box plots. Adapted from data in Gentry 1988 and Gentry 1995. (B) Fruiting phenology of trees in tropical dry and tropical moist forests. Vertical lines (dry forest) and shaded polygon (moist forest) indicate periods required for half of the species in each habitat to fruit. Data from Frankie et al. 1974. (C) Dispersal syndromes exhibited by trees in tropical dry, moist, and wet forests. Adapted from Gentry 1982b.

These estimated densities are comparable to, and frequently considerably less than, other values reported in the literature for populations of SDTF trees. In terms of the species listed in table 14-1, Gentry (1982b) reported densities of 12 and 22 trees per 500 square meters for *Copaifera officinalis* and *Guazuma ulmifolia*, respectively, in Venezula. Gillespie et al. (2000) recorded 53 individuals of *Calycophyllum candidissimum* (greater than or equal to 2.5 centimeters in diameter) in 0.1-hectare plots in Nicaragua, and 101 individuals of *Anadenanthera colubrina* greater than or equal to 3.0 centimeters basal diameter were reported by Monteiro et al. (2006) from a 1.0-hectare plot in Brazil. The high density of conspecific trees found in Neotropical SDTF and its implication for resource management have been previously noted by several authors (e.g., Killeen et al. 1998; Lugo et. al 2006).

Phenology of Flowering and Fruiting

The synchrony and duration of reproduction appear to be slightly more predictable in Neotropical SDTF than in moist or wet forests. As summarized by Bullock (1995), synchronous flowering, at both the population and community levels, is more common in SDTF, especially among species that flower during the dry season or at the dry-wet season transition. The duration of the flowering events exhibited by different species is also compressed. Almost half of the tree species (47 percent) in Costa Rican wet forest exhibit extended flowering cycles more than 20 weeks long (Frankie et al. 1974), as compared with the more seasonal, 6 weeks or less, flowering cycles exhibited by a similar percentage (41 percent) of canopy species in Mexican SDTF (Bullock and Solís Magallanes 1990).

The increased periodicity in SDTF flowering is manifested as a notable peak in annual fruit production, as shown in figure 14-1B, which shows the fruiting behavior of selected tree species from tropical moist and dry forests in Costa Rica (Frankie et al. 1974). While there are some fruits available in every month in both forest types, the dry forest exhibits a pronounced peak in April, when 41 tree species are recorded with fruit. As indicated by the vertical lines in figure 14-1B, within a span of 2 months more than half of the tree species at the dry forest site have produced fruit; 4 months are needed for a similar proportion of species to fruit in the moist forest site (fig. 14-1B, shaded polygon).

Seed Dispersal

A second aspect of the reproductive ecology of Neotropical dry forests relevant to management is the reduced importance of animals for seed dispersal (fig. 14-1C). As has been documented in numerous studies (e.g., Gottsberger and Silberbauer-Gottsberger 1983; Wikander 1984; Oliveira and Moreira 1992; Killeen et al. 1998), there is a greater prevalence of wind-dispersed species in SDTF than in moist/wet forests, where most of the species are dispersed by birds or mammals (Gentry 1982b). Wind dispersal is especially common among climbing plants in dry forests, and almost all of the liana species in this habitat are anemochorous (Gentry 1991).

Finally, there are several other characteristics of Neotropical dry forests that would seem to simplify the task of forest management. For example, a large percentage of the tree species in dry forest have the ability to resprout after cutting (Murphy and Lugo 1986a; McDonald and McLaren 2003), which greatly facilitates the coppicing and regeneration of certain types of vegetative tissues. The larger root system possessed by sprouts gives them a considerable advantage over competing seedlings, especially during the dry season (Miller and Kauffman 1998; Kennard et al. 2002). It has also been suggested that the seasonal openness of the canopy in deciduous forests permits a less restrictive pattern of regeneration than that dependent on the stochastic occurrence of treefall gaps (Quigley and Platt 2003).

From an operational standpoint, there are numerous reasons to be optimistic about the management potential of Neotropical dry forests. These forests contain lots of valuable plant species, and many of them occur naturally in high-density populations, produce flowers and fruits at predictable intervals, don't require the use of animals to move their seeds around, and resprout if cut. From a conservation standpoint, it would seem that there are also many reasons to move away from the current pattern of overexploitation and resource depletion and to start managing these forests to provide a continual flow of products to the market and a continual flow of income to the communities involved.

Within the context of this chapter, sustainable resource management seems like a reasonable thing to experiment with in SDTF. But what happens in the real world of conflicting user groups, byzantine government regulations, questionable land tenure, skimpy markets, and uncontrollable development agendas? Taking a look at a few cases where local communities are successfully managing dry forest resources is probably the most useful way to address this question.

Management of Neotropical Seasonally Dry Forests

Mexico is an appropriate place to start looking for examples of sustainable resource management. The dry forests of western Mexico are some of the most species-rich in the Neotropics (Gentry 1995; Lott and Atkinson 2006), and there is a long history of resource use by indigenous populations (Codex Mendoza 1978). The practice of community forestry is also more established here than in any other country in the Neotropics, and as much as 80 percent of Mexico's forests are under the control of indigenous communities (Bray et al. 2003; Bray et al. 2005). It has been estimated that there are currently 500 to 600 community forest enterprises operating in Mexico (Alatorre Frenk 2000).

For the purposes of this chapter, two of the Mexican species listed in table 14-1, *Brahea dulcis* and *Bursera glabrifolia*, have been chosen as informative examples of community resource management in Neotropical dry forests. The species represent different life-forms, the tissues exploited and uses are different, the products supply different markets, and the pattern, intensity, and historical development of management are also notably distinct.

Brahea dulcis *in Guerrero*

Brahea dulcis (*palma soyate*) is a small, clonal palm (fig. 14-2A) that grows in Mexican SDTF. The species is particularly abundant in the Montaña Region of Guerrero (fig. 14-2 map), where it has been used since pre-Hispanic times to make mats and baskets (fig. 14-2B) and as a source of thatch. The technique of using the tender young leaves of *Brahea* to make sombreros was introduced by the Spanish, and this use, together with the production of handicrafts from the young leaves, has continued to the present day and currently represents an important source of revenue for local communities (Mastache et al. 1982).

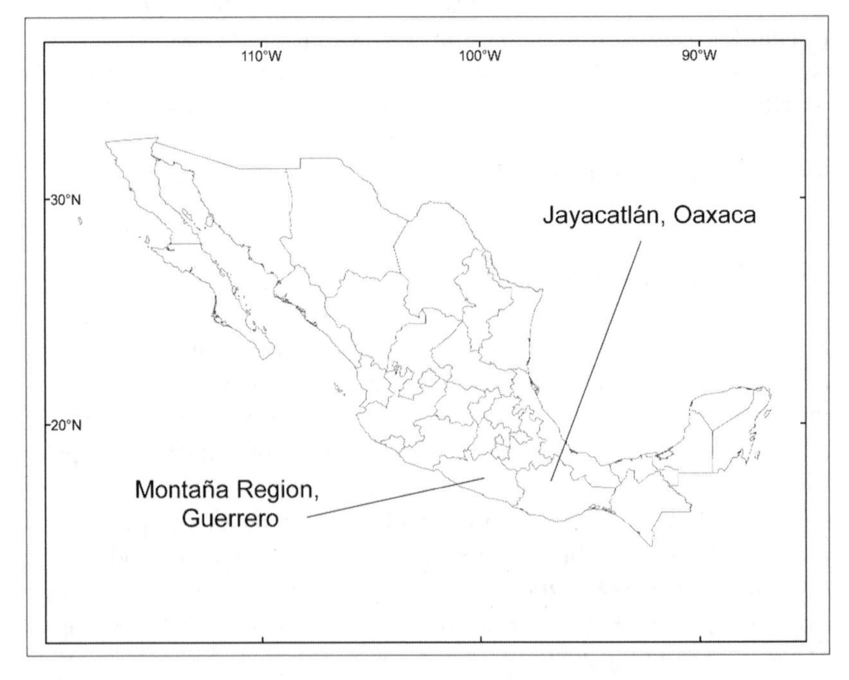

FIGURE 14-2. Locations of case studies for *Brahea dulcis* (Montaña Region, Guerrero) and *Bursera glabrifolia* (Jayacatlán, Oaxaca).

(A)

(B)

FIGURE 14-2. (A) *Brahea* palm exhibiting young leaf used for weaving. (B) Making a basket from *Brahea* leaves for harvesting corn.

FIGURE 14-2. (C) *Bursera glabrifolia* tree. (D) Copal wood to be used for carving. (E) Carved and painted alebrije made from *B. glabrifolia* wood harvested at Jayacatlán.

Two different forms of the palm are known to occur (Miranda 1947; Castillo 1993). In forests near communities that exploit the dry leaves of *Brahea* for thatch, the palm forms small groves of individuals (*soyacahuiteras*) that attain heights of 6 meters or more. Near communities where the tender young leaves are cut to make sombreros, the palm grows in dense clumps of stems (*manchoneras*) that are less than 1.5 meters tall. The occurrence of these two different growth forms is thought to be the result of centuries of management and conscious selection by local Nahua communities (Illsley et al. 2001).

The ability of the communities in Guerrero to affect such notable changes in the *Brahea* populations they manage seems to result from three interrelated factors. First, they understand an impressive amount of the ecology of *Brahea dulcis*. For example, they clearly distinguish the difference between ramets and genets, appreciate the trade-offs between sexual and asexual reproduction, and know how management activities affect the rate of leaf replacement. Second, they have developed a variety of management strategies to increase the density of *Brahea* in different habitats and to enhance the production of specific plant tissues. Finally, the communities exhibit a level of social organization sufficient to regulate access to the resource, to control rates of harvest, and to enforce consensus norms concerning its use and sale (Aguilar et al. 1997; Illsley et al. 1998).

Currently, the overall management trend in the region is towards the formation and maintenance of the smaller *Brahea* morph, largely in response to the demand for young leaves to make sombreros and handicrafts. In the past, the greater subsistence demand for roofing material drove the management of *Brahea* in the other direction, that is, the production of the taller morph in *soyacahuiteras*. In spite of these demand swings, local communities continue to maintain considerable densities of both forms of *Brahea* in the region. Illsley et al. (2001) report densities of *Brahea* in managed *manchoneras* (small morph) and *soyacahuiteras* (tall morph) of 3233 and 1450 individuals per hectare, respectively.

Bursera glabrifolia *in Oaxaca*

The second example of dry forest management in Mexico differs markedly from *Brahea* in that both the product and the management system are quite recent. *Bursera glabrifolia* is a small, deciduous tree that occurs in SDTF of western Mexico (fig. 14-2C). Traditionally, the most widespread use of this species, and other related species of *Bursera*, was for its aromatic resin. Its local name, copal, is derived from the Nahuatl word for the exudate produced by certain members of the Burseraceae (Standley 1923).

In more recent times, *B. glabrifolia* has gained economic importance as the raw material (fig. 14-2D) used to carve the small, painted figures known as alebrijes. These carvings, which usually portray fanciful animals, dragons (fig. 14-2E), mermaids, or human-animal hybrids, are unique relative to other Mexican handicrafts in that they are not a traditional craft item nor are they produced only by artisans from a particular ethnic group. Alebrijes started to appear on the Mexican handicraft scene in the early 1960s, with most of the production coming from three villages: San Antonio Arrazola, San Martín Tilcajete, and La Unión Tejalapam, in the Central Valley of Oaxaca. The demand for these whimsical figures in national and international craft markets has increased steadily since then (Chibnick 2003).

The production of alebrijes requires a considerable amount of *B. glabrifolia* wood (López 2001). When the market first started, copal wood was harvested from the forests surrounding each carving village. This material was rapidly depleted, however, and the artisans were forced to go farther and farther to find harvestable *B. glabrifolia* trees. A vigorous clandestine trade in copal wood soon developed, most of this material coming from the extensive tracts of SDTF located several hours north of Oaxaca City.

In response to this situation, management activities focused on the sustainable production of *B. glabrifolia* wood were initiated in the village of San Juan Bautista Jayacatlán (fig. 14-2 map) in early 2000. The municipality contains about 5000 hectares of dry forest, each hectare of forest containing an average of 31.5 adult *B. glabrifolia* trees (Peters et al. 2003). Over the next 2 years, 300 hectares of forest were inventoried, quantitative growth studies were conducted, a yield table was constructed, and regeneration surveys were initiated to monitor the ecological impact of harvesting. Much of this research was conducted by local villagers who had been trained in the basic methodologies of forest mensuration. Once the baseline data had been collected, a management plan for the sustainable exploitation of *B. glabrifolia* wood at Jayacatlán was written and submitted to SEMARNAT (Mexican Secretary of the Environment and Natural Resources) in 2003 to obtain the necessary harvest permits. Selected carving communities were also approached about the possibility of purchasing copal wood from Jayacatlán.

Several positive things have happened since then. The management plan was approved by SEMARNAT, and Jayacatlán can now legally harvest and manage its forests. It seems that this was the first time that such a management permit had ever been granted for dry forests in Mexico—which may go a long way in explaining the problem. Subsequent projection matrix simulations revealed that the harvest intensities prescribed in the plan (about 4.3 square meters per hectare) are indeed sustainable (Hernández-Apolinar et al. 2006). Several of the carving communities near Oaxaca City have started buying "legal" copal wood from Jayacatlán, and as might be expected, a new type of handicraft, the eco-alebrije (fig. 14-2E) made from sustainably harvested copal wood, has now appeared in local craft markets.

Conclusions

Based on the preceding discussion, it seems that there are several benefits that could result from the sustainable management of Neotropical SDTF. From an ecological standpoint, management provides a clear incentive to maintain the forest, rather than convert it to some other form of land use. Species, resources, and forest functions are conserved, not discarded. A greater supply of managed, sustainably harvested dry forest resources may also reduce the incidence of illegal harvesting that is rapidly degrading the original resource base in these habitats.

From a social standpoint, communities involved in resource management programs develop a greater sense of forest stewardship and cooperation, and management operations may also serve to strengthen the organizational capacity of the group. Participatory forest management is a conservation activity that includes and empowers local communities, rather than excludes them.

Finally, sustainable forest exploitation provides a much-needed source of income to rural populations. And this, in the final analysis, may be the most fitting retribution for the wealth of ethnobotanical knowledge that these dry forest residents have compiled through hundreds of years of trial and error. But the clock is ticking, and indigenous plant knowledge is of little value to anyone if the plants themselves have already disappeared.

Acknowledgments

The author wishes to thank Silvia Purata and Catarina Illsley for kindly sharing information, insights, and photos from their groundbreaking research in Mexico, Rodolfo Dirzo and Hal Mooney for providing the motivation to write this chapter, and the Overbrook Foundation for its continual support of community forestry in the Neotropics.

Chapter 15

Ecosystem Services in Seasonally Dry Tropical Forests

Patricia Balvanera, Alicia Castillo, and María José Martínez-Harms

Human populations depend for their survival and well-being on benefits derived from ecosystems, also called ecosystem services (MA 2003). The nature, magnitude, and reliability of the services provided by a certain ecosystem depend on its particular characteristics, the human group that interacts with it, and the nature of their interaction (MA 2003; Maass et al. 2005). The ability of an ecosystem to provide services depends on its processes—that is, the interactions between its physical and biotic components—and the rates and variability over time and space of such processes and components (Kremen 2005). The benefits people obtain from ecosystems also depend on demographic, economic, political, cultural, scientific, and technological characteristics of the human groups that interact with the ecosystem (MA 2003; Castillo et al. 2005). Human groups determine which services they demand, extract, or expect from ecosystems and thus drive decisions about how to manage them (MA 2003; Bennet and Balvanera 2007). Given the tight relationship between the delivery of ecosystem services and human well-being, the long-term maintenance of the capacity of ecosystems to provide services is essential to ensure a promising future for humanity. Indeed, technical and social interventions need to be designed to foster the maintenance of the services to ensure human well-being (MA 2003).

The above statements are applicable to all types of ecosystem services and human societies. In this case we will use them as a framework for understanding the services seasonally dry tropical forests (SDTFs) provide to human societies. We will first analyze how their ecological characteristics, the history of the human-SDTF interaction, and the results from that interaction determine the type of ecosystem services they provide. We will then describe the different services provided, the ecosystem processes that result in such services, and how human-SDTF interactions modify an ecosystem's ability to provide services. Then we will explore how these services relate to human well-being for different stakeholders; how social, economic, legal, and cultural attributes of societies modify decision-making processes; and what the consequences are of such management decisions on services provision at the present time and in the future. Finally, we will discuss alternatives to ensure the long-term delivery of ecosystems services provided by SDTF.

Ecological and Historical Determinants of Human-SDTF Interaction and Ecosystem Services

Ecosystem services provided by tropical dry forests are deeply shaped by some of the forests' key ecological characteristics and various key social characteristics of the people that inhabit them.

Ecological Characteristics

Water scarcity is the major limiting factor that shapes SDTF functioning and SDTF-human interactions. Water availability varies dramatically throughout the year, and most ecosystem processes are tightly coupled to changes in water availability (chap. 9), as are most human activities and the delivery of most ecosystem services.

SDTFs are also subject to high interannual rainfall variability (chap. 9). Such unpredictability imposes a great difficulty on human decisions about ways to manipulate the ecosystem and the kinds of services that may be expected. At Chamela-Cuixmala, cattle-grazing activities are in jeopardy in very dry years and when a series of dry years occur.

Water availability also changes dramatically along space, at various scales. In general, lower sections of catchments and riparian areas have more available water over longer periods through the year (Galicia et al.

1999; Zarco 2001). These lower areas are more productive, have higher nutrient availability, and harbor taller and larger trees (chap. 7). Regionally, climatic patterns also vary among SDTFs.

Human-SDTF interactions and resulting ecosystem services are also affected by the disturbance—its magnitude, frequency, and intensity—characteristic of socioecological system dynamics. Major disturbances within SDTF-dominated regions include cyclones (García-Oliva et al. 1991), severe droughts (Segura et al. 2003), and natural fires (as in the Brazilian Cerrado; Klink and Machado 2005) or anthropogenic fires (as in Chamela, Mexico; Maass et al. 2005).

Within SDTFs there is a mosaic of geological and edaphic characteristics at various spatial scales. Shallow rocky soils are found in the Yucatán peninsula in Mexico and used for the cultivation of *Agave* spp. (Colunga-García Marín et al. 1993), while deep soils are found in the Cerrado in Brazil, recently converted for soybean production (Klink and Machado 2005).

SDTFs harbor a very large biodiversity (chap. 2, 5). Such biodiversity can play important roles in the provision of most ecosystem services, as will be further discussed.

Historical Features of the Human-SDTF Interaction

The development of Latin American civilizations choosing to live in SDTF areas relied on water provision. Mixtec pre-Hispanic terraces and water canals have been found at Hierve el Agua, Oaxaca, Mexico (Hewitt 1994).

Some SDTFs have a long and complex history of land use. In Mexico, this ecosystem provided very important crops such as maize, beans, and squashes from pre-Columbian times (Challenger 1998); SDTFs were later transformed for agriculture and cattle raising within large private proprieties called haciendas by the second half of the nineteenth century (Castillo et al. 2005). Strong impacts on Mexican SDTFs started around 1943 (Ortega 1995) and were further fostered during the 1960s and 1970s (Castillo et al. 2005) as a result of governmental programs.

Present Characteristics of the Resulting Socioecological System

There is a complex mosaic of socioecosystems derived from SDTFs that can be organized along a gradient of transformation intensity. The gradient ranges from conserved SDTF, through SDTF with sustainable extraction of *Bursera glabrifolia* in Oaxaca, Mexico (chap. 14), devastated areas of Cerrado in Brazil due to overextraction of fuelwoods for charcoal production (Ratter et al. 1997), and large fractions of Neotropical SDTF converted to pastures (Miles et al. 2006), to SDTF completely transformed into urban areas in Jamaica (Tole 2001). Along this gradient, the ability to provide different types of ecosystem services changes (Bennet and Balvanera 2007).

Cumulative transformation through management history is also relevant and can drive regime shifts, with irreversible changes in conditions (Bennet and Balvanera 2007). For instance, soil phosphorus availability changes as the number of agriculture-fallow-slash-burn cycles increases in Mexican and Brazil SDTF; it decreases because of loss in ash as a result of slash and burn in Chamela (chap. 10) and because of increasing irreversible conversion to unavailable phosphorus forms in Brazil (Lawrence and Schlesinger 2001).

The Services

SDTFs provide a variety of ecosystem services (table 15-1). These services can be classified into three major categories (MA 2003): (1) tangible resources that can be consumed or appropriated, also called provisioning services; (2) complex interactions among biotic and abiotic components of ecosystems that regulate the conditions where human beings live and work, also called regulating services; and (3) tangible or intangible benefits strongly dependent on individual or collective cultural background, also called cultural services.

Provisioning Services

Provisioning services include all well-known products obtained from SDTFs or from their transformation. Food from cattle is the most important service; transformation of SDTFs to cattle pastures has been the most important driver of land use change for this ecosystem. In the Chaco region, 85 percent of the forests have been converted to pastures in 30 years at an annual rate of 2.2 percent (Zak et al. 2004).

The second most important service obtained also from SDTF transformation is food from agricultural crops. The recent conversion of 300,000 hectares of Chaco SDTF to soybean fields is only the most recent and most dramatic example of this tendency (Grau et al. 2005).

TABLE 15-1. Benefits to stakeholders at different spatial scales from services provided by tropical dry forests

Type of service	*Ecosystem service*	*Scale of stakeholders benefited*					
		Owner	*Community*	*Municipality*	*State*	*Country*	*Planet*
Provisioning	Food (from agriculture and cattle ranching)	Pr					
	Timber	Pr					
	Nontimber forest products	Pr	Pr/Pu				
	Biofuels	Pr	Pr/Pu	Pu			
	Germplasm	Pr	Pu	Pu	Pu	Pu	
	Future options	Pr		Pu	Pu	Pu	
Regulating	Erosion regulation	Pr	Pu	Pu	Pu		Pu
	Fertility regulation	Pr					
	Plant support	Pr	Pu	Pu	Pu		
	Infiltration/ runoff regulation		Pu	Pu	Pu		
	Water temporality regulation		Pu	Pu	Pu		
	Water quality		Pu	Pu	Pu		
	Carbon storage	Pr	Pu	Pu	Pu	Pu	
	Regulation of carbon emissions		Pu	Pu	Pu	Pu	
	Regulation of albedo		Pu	Pu	Pu		
	Regulation of air temperature		Pu	Pu	Pu	Pu	
	Pollination		Pu	Pu			
	Pest regulation		Pu	Pu			
	Vector disease regulation		Pu	Pu	Pu		
	Invasion resistance		Pu	Pu			
	Seed dispersal		Pu	Pu			

(table continues)

TABLE 15-1. *(continued)*

Type of service	*Ecosystem service*	*Scale of stakeholders benefited*					
		Owner	*Community*	*Municipality*	*State*	*Country*	*Planet*
Regulating (continued)	Regulation of reliability of services	Pr	Pr/Pu	Pu	Pu	Pu	
	Regulation of vulnerability to extreme hydro-meteorological events		Pu	Pu	Pu	Pu	Pu
Cultural	Ecotourism and tourism	Pr	Pr/Pu	Pu	Pu	Pu	Pu
	Aesthetic fulfilment	Pu	Pu	Pu	Pu	Pu	Pu
	Spiritual fulfilment	Pu	Pu	Pu	Pu		
	Opportunities for work	Pr					
	Opportunities for human settlement	Pr					

Benefits are divided into private (Pr), which benefit only the owner or the manager of the plot, and public (Pu), which benefit society at large independently of ownership or capital investment.

Woody biomass, used for timber or biofuel, is extracted from intact and managed SDTF. Timber is extracted from SDTF areas with elevated water availability and primary productivity within old stands; slow-growing species that produce dense woods, such as *Tabebuia*, *Enterolobium,* and *Pirhanea*, are highly appreciated (Quesada and Stoner 2004). Selective extraction of tree individuals without any further management is common in Mexican SDTF (Maass et al. 2005); instead, liana removal and selective logging are management practices commonly used in Bolivian SDTF (Pariona et al. 2003).

The large biodiversity of SDTF makes available a wide range of non-timber forest products that are used as food, construction materials, medicine, pets, and ornaments and have many other uses (chap. 14). SDTFs also harbor important genetic diversity of staple crops and numerous species with potential future uses. SDTFs in Mexico are the habitat for *Zea diplo-*

perenis, the wild relative of maize (Lorente-Adame and Sánchez-Velásquez 1996), and many *Cucurbita* species closely related to squash (Montes-Hernández and Eguiarte 2002). Brazilian cerrado is the center of diversity of wild *Manihot* (Klink and Machado 2005).

SDTFs are at present dominated by species that have adapted to low and unpredictable water availability conditions (Holbrook et al. 1995), and secondary SDTFs are dominated by species that can grow in highly degraded environments (chap. 13). Given future predictions of increased areas with unpredictable water availability conditions in the face of climate change (Villers-Ruiz and Trejo-Vázquez 1997), and given the elevated rates at which SDTFs are being transformed and degraded (Miles et al. 2006), the germplasm found within SDTFs might become particularly important for reforestation, afforestation, and restoration efforts (Maass et al. 2005).

Regulating Services

Regulating services can be organized with respect to which major processes they are associated with. We have distinguished soil, water dynamics, climate and air quality, and biodiversity-related services.

Soil-Related Services

Regulation of soil nutrient content and availability, seen as soil fertility from the human perspective, depends on tight nutrient-recycling mechanisms. These include resorption of leaf nutrients previous to abscission, microbial immobilization of nutrients, the presence of stable soil aggregates, and dense leaf litter layers (Maass et al. 2005; chap. 7).

Soils support plant development, and their ability to do so depends on depth, texture, and density of soils and varies widely among SDTFs. As an indicator of this service, root biomass can range from 17 megagrams per hectare in Chamela-Cuixmala to 66.8 megagrams per hectare in Venezuela (chap. 7).

The delivery of soil-related ecosystem services by SDTF is greatly affected by management. For example, compared with conserving forests, slashing and burning of forest in Chamela-Cuixmala causes reduction in plant cover, destruction of the protective leaf litter layer, destabilization of soil aggregates, and 100 times more erosion in the croplands established after slash and burn (Maass et al. 1988).

Water Dynamics–Related Services

Little water is available for human populations in SDTF-dominated areas, where evapotranspirative demand exceeds precipitation during a large part of the year. In the Chamela-Cuixmala region, nearly 90 percent of the precipitation returns to the atmosphere as evapotranspiration (chap. 9). Soil water availability is further modified by (1) soil characteristics—water is available for longer periods in soils with higher clay and loam contents—and (2) slope and aspect—more water is available in north-facing slopes that are subject to lower evapotranspiration demand (Galicia et al. 1999; chap. 9).

Water quality is also related to biophysical characteristics of SDTF soil and vegetation, as well as those of freshwater and riparian neighboring systems. Increased infiltration-to-runoff ratio, reduced soil erosion, and elevated freshwater biodiversity contributed to higher water quality in Indonesia (Anbumozhi et al. 2005).

SDTF management changes the provision of these water dynamic–related services. When SDTF is converted to pasture, or when dominance of woody elements decreases as a result of frequent fires, the amount of total above- and belowground biomass is reduced, root depth diminishes, and evapotranspirative rates decrease. In the case of cerrado, grasses evapotranspire 17 to 18 percent less than woody vegetation, leading to reductions in deep water uptake and increasing deep soil water storage (Oliveira et al. 2005).

Downstream surface water and groundwater are modified when water infiltration is reduced by the partial or total removal of woody elements of SDTF. For example, in the case of Tocantins River, in the Brazilian Cerrado, river discharge increased 24 percent from 1949 to 1998; during that period, cerrado area was reduced by 3.5 million hectares because of conversion to pasture (Filoso et al. 2006).

Stream, wetland, and coastal water quality decrease when surrounding SDTFs are transformed into pastures or croplands. In the case of the SDTF at Chamela-Cuixmala, water quality along the Cuixmala River was poor in the middle part of the watershed as a result of human settlements and agricultural activities; yet, quality was best at the lower section, where SDTF was well conserved, most certainly due to infiltration, inputs of clean water, and elevated freshwater biodiversity (López-Tapia 2008).

Changes in coastal water quality are extremely important for the tourism industry developing along SDTF-dominated coastal areas (see Cultural Services below); decreases in water quality and the scenic beauty of the crystal clear coastal beaches and coral reef areas, as well as changes in sediment content and beach dynamics, are very likely to be observed.

Climate and Air Quality Regulating Services

Changes in the chemical composition of the atmosphere and the physical properties of the land's surface, which in turn affect air quality and climate regulation, have been observed when SDTFs have been removed.

SDTFs are important above- and below-ground carbon reservoirs, and their transformation results in important carbon emissions to the atmosphere and reductions in carbon stocks; the aboveground live biomass in SDTF ranges from 35 to 140 megagrams per hectare (chap. 7). Cattle ranching within SDTF is causing reductions in soil organic matter and labile carbon content, and it increases carbon dioxide flux to the atmosphere (chap. 10).

Changes in relative humidity and consequent changes in precipitation frequency are expected when SDTFs are transformed into pastures. In the case of cerrado, dry spells may become more frequent as a result of such transformations (Hoffmann and Jackson 2000). The proportion of solar energy used by vegetation in photosynthetic activity rather than radiated back to the atmosphere (albedo) changes when SDTFs are removed. Surface albedo increased from 12–13 percent to 16–19 percent when Bolivian SDTFs were replaced by pastures (Steininger et al. 2001). Changes in albedo coupled with changes in relative humidity can contribute to changes in temperature. In the case of cerrado, an increase in the mean air temperature of 0.5 degree Celsius is expected, due to conversion of SDTF to grassland (Hoffmann and Jackson 2000).

No data are available on the role played by SDTF in the regulation of air quality or with respect to soil particle content in the atmosphere. Nevertheless, an increased frequency of sandstorms is likely during the dry season in areas where SDTF's above- and belowground biomass decreases, air moisture decreases, and air temperature increases, in particular where small loam or clay soil particles are common.

Biodiversity-Related Regulating Services

The large biodiversity harbored by SDTF and the complex interactions among its components contribute to the provision of many ecosystem services; nevertheless, little information is available at present on this topic. Many ecosystem services, including soil fertility and erosion control regulation, flood control, and waste recycling, are largely due to interactions between bacteria, mycorrhizae, and micro- and macroinvertebrates (chap. 4). Regulation of pollination and pest control also depend on the maintenance of biodiversity (Maass et al. 2005; chap. 11).

Species composition, richness, and relative abundance, as well as changes in disturbance regimes, play important roles in decreasing SDTF resistance to invasive species and the functioning of the ecosystem. In the case of cerrado, invasive African grasses introduced as pasture plants, such as Molasa grass (*Melinus minutiflora*), promote fire frequency, which in turn promotes this species (Klink and Machado 2005).

Overall changes in SDTF diversity—that is, changes in the number of species and their relative abundances, composition, functional attributes, spatial distribution, and trophic diversity—will very likely contribute to changes in the magnitude and reliability of all the provisioning and regulating services mentioned above (Díaz et al. 2006). Yet information particular to SDTF on these topics is still lacking.

Changes in biodiversity together with changes in the biophysical characteristics of SDTF due to management contribute to increased vulnerability to extreme hydrometeorological events. Most Latin American SDTFs are found within 100 kilometers of the Pacific, Atlantic, or Caribbean coast and are thus subject to the incidence of cyclones. Such incidence is very likely to increase as a result of climate change (IPCC 2001). Yet the presence of coastal or near-coastal forests has been shown to play a critical role in regulating impacts of such extreme events (Kerr and Baird 2007). Thus, increased economic and social losses are to be expected as a result of extensive SDTF transformation.

Cultural Services

SDTFs bring aesthetic quality to people and are a source of enjoyment to people living in them as well as to those who visit them. People living in SDTF-dominated areas emphasize aesthetic appreciation of the shade provided by SDTF, and in particular by large trees (Chivaura-Mususa et al. 2000; Castillo et al. 2005). Inhabitants of the Chamela-Cuixmala area also associate SDTF with feelings of peace, quietness, and joy (Castillo et al. 2005).

SDTFs provide opportunities for tourism and ecotourism. The scenic combination of bright blue seas and either the gray of SDTF during the dry season or its lush green during the rainy season has drawn the attention of a growing tourism industry along the Pacific and Caribbean coasts, as in Huatulco in western Mexico, Guanacaste in Costa Rica, and Santiago de Chiquitos in Bolivia (FCBC 2003), to name a few examples.

SDTFs are important for spiritual fulfillment in many cultures, contributing to their livelihoods, cultural identities, and worldviews. For instance, *Sabal* spp. palms are particularly significant for SDTF inhabitants: their leaves are used for roof building, and they are promoted within traditional home gardens in the Yucatán peninsula (Martínez-Ballesté et al. 2006). Indigenous people often establish sacred connections with the ecosystem they live in, and this has been observed for SDTFs in Mexico (Toledo 2001) and in Bolivia (Arambiza and Painter 2006).

SDTFs provide an important opportunity for human inhabitants to work and make a living from them, as well as to live in them. In the area of Chamela-Cuixmala, local peasants perceive SDTF mainly as land that should be transformed and used for agriculture or cattle ranching, although it is known to provide services such as firewood, animals for hunting, and shade and fresh air (Castillo et al. 2005).

Ecosystem Services and Human Societies

All the above services contribute to the well-being of human societies. The nature of such benefits and the amounts provided can be assessed; benefits obtained by specific stakeholders within SDTF-based societies can be identified. The cultural, policy, institutional, and economic drivers behind SDTF management decisions that foster particular services can be analyzed (MA 2003), and the negative consequences of such management decisions on other services can also be explored.

Who Benefits from Ecosystem Services and How to Account for Such Benefits

Different stakeholders benefit in different ways from SDTF ecosystem services (table 15-1). Such benefits can be relevant for various aspects of human life, ranging from satisfaction of basic needs to security or health (MA 2003). Benefits can be accounted for in economic terms (amount of money earned) or in social terms (number of inhabitants benefited). Benefits can be received by very specific stakeholders, those who own the land or who own the infrastructure (e.g., machinery, commercialization networks) needed for obtaining benefits, and in that case can be considered private. Benefits can also be provided to societies at large, at different spatial scales, independently of ownership or capital investment, in which case they can be considered public. In some cases, combinations of private and public benefits result from private and community-established rules for access to resources.

The most tangible benefits associated with provisioning services are agricultural and pastoral goods derived from SDTF transformation. Cerrado transformation to soybean plantation is worth US$813 million per year in northern Mato Grosso (Fearnside 2001). Such benefits are private, being most relevant for those who own the lands where they are produced. Indirect public benefits derived from cash flow into the local community, municipality, state, or country could also be accounted for in the case of these and most provisioning services, but they will not be considered here. Benefits obtained from timber extraction are private, being commonly obtained by landowners, by the company in charge of forest exploitation, or in some cases by a communitarian organization that manages the forest (in which case the benefit can be considered collective). Benefits derived from extraction of biofuels can be accounted for in economic terms when associated with commercial extraction; in the case of charcoal extraction in the Cerrado area in Brazil, benefits incomes of up to US$330 million were obtained by specific private stakeholders in 2003 (Coelho et al. 2006). Yet the most tangible benefits are to the large numbers of families that depend on biofuels as their sole or main source of fuel. In Mexico, for example, 25 million people cook with them (Masera et al. 2006), but we do not know how many of these live in SDTF-dominated areas. Families may use their own lands or the surrounding communal lands to extract such resources, and benefits can be considered public in that case. Nontimber forest products found in SDTF provide a range of tangible benefits. For instance, in the Cerrado in Brazil SDTF, the collection of *Caryocar brasiliense* fruits is an important source of income for the rural population in Brazil. In Minas Gerais, the fruit collections during the summer involve 50 percent of the rural population, and the derived income represents 54.7 percent of the annual budget of the rural worker (Lopes et al. 2003). Harvesting this resource only benefits the owners of the lands where the plant is found and the individuals who perform the extraction; thus, benefits are considered private. Future potential benefits derived from germplasm and future options contained today by SDTF cannot be accounted for yet; they could provide public well-being to many areas and social actors beyond their present distribution area and SDTF owners.

Benefits derived from regulating and cultural services cannot be assessed as clearly as those from provisioning services. Gains obtained by landowners through ensuring the long-term maintenance of their land's productive capacity, or those obtained by downstream inhabitants through avoidance of sediment deposition on urban or productive areas, have not been accounted for, either in economic or social terms. Services provided by pollinators, dependent on the SDTF for their maintenance, contributed to the US$12 million harvest in 2000 of pollination-dependent crops in the Chamela-Cuixmala region (Maass et al. 2005). Yet despite the lack of concrete information on the nature of such SDTF services, we can argue that most regulating and cultural services provide public benefits, as opposed to the private benefits obtained from provisioning services (table 15-1). This is probably not as true for cultural services derived from the scenic beauty or inspirational value of places that generate private benefits to the tourism industry. Also, public benefits derived from regulation and cultural SDTF services most often encompass areas that go beyond a particular plot, ranging from the municipality to the whole globe, depending on the nature of the service.

Cultural, Policy, Institutional, and Economic Drivers

Different stakeholders interested in fostering the provision of particular services make decisions on how to manipulate SDTF. Such decisions depend on the cultural, policy, institutional, and economic context they are found in. The drivers behind SDTF transformation have been analyzed elsewhere in this volume (chap. 3), and we will only highlight some that are directly relevant to service delivery.

The generalized perception that SDTFs are not useful, along with other factors, has driven the development of policies that foster SDTF conversion in agricultural, pastoral, or urban lands. In Bolivia, macroeconomic policies have driven the conversion of SDTF into agricultural and pastoral lands (Steininger et al. 2001).

Strong institutions—systems of rules that regulate access to resources—are more commonly found when users depend importantly upon an ecosystem, when decisions are taken collectively, and when a shared vision exists about how the system works and how relevant is their sustainable use for people's livelihoods (Ostrom et al. 1999). Decisions for the conservation and sustainable management of SDTF in Yucatán have been made collectively by local communities with shared Mayan background (Dalle et al. 2006). In contrast, deforestation in Bolivian SDTF can be largely attributed to a heterogeneous group of landholders who have responded individually to market prices (Steininger et al. 2001).

Markets for SDTF products, as well as for those derived from SDTF transformation, have been driving resource extractions and land use change. Cultivated soybean area in the Brazilian Cerrado doubled in the 1990s to 21 million hectares in response to growing demand from China (Fearnside 2001).

Growing population size and growing demand for agricultural and pastoral products have contributed to SDTF transformation. Rising population growth coupled with a deep economic crisis have contributed to an annual loss of 4 percent of Jamaican SDTF (Tole 2001).

Trade-offs among Services and Stakeholders

Given that different services benefit different stakeholders and that different drivers foster the provision of different services, trade-offs arise among services and among stakeholders. At present, provisioning SDTF services have been favored over regulating services. Agricultural and pastoral services that depend on SDTF removal have been mostly favored (fig. 15-1A), as have been provisioning services that depend on the maintenance of managed SDTF (fig. 15-1B); tourism-related cultural services, that also depend on partial SDTF transformation, have also been fostered.

Yet the public and widely encompassing regulating services have seldom been taken into account; they depend on SDTF maintenance for their provision, and little is known about the amount of SDTF transformation that is compatible with such maintenance (fig. 15-1). Research is urgently needed to understand how sensitive these services are to different management regimes, whether there are thresholds beyond which loss is irreversible, and what are the relevant factors that determine the presence of such thresholds.

If values for regulating ecosystem services were weighed against those for provisioning services, clear disadvantages of SDTF transformation could be shown. Such a comparison was performed for the Del Plata Basin in South America (Viglizzo and Frank 2006), where the annual gross values of agricultural production in the Cerrado and the Chaco are between US$27 and US$31 per hectare per year, but ecosystem services provided by such ecosystems account for US$550 to US$1050 per hectare per year.

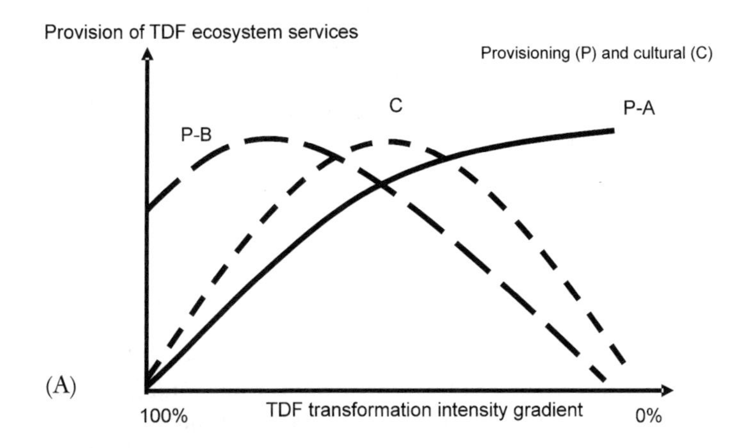

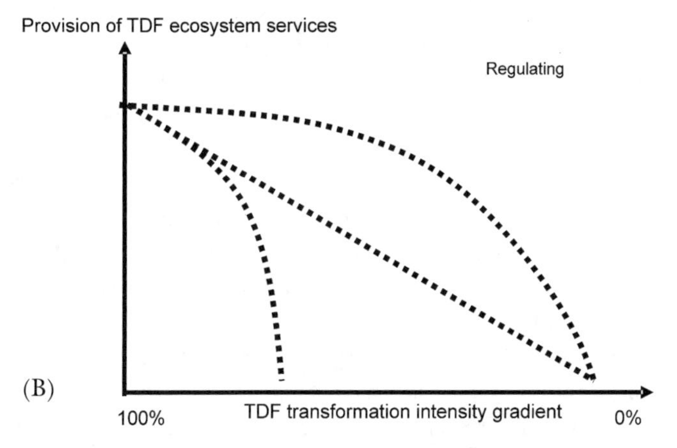

FIGURE 15-1. Trade-offs among services provided by the tropical dry forest (TDF). (A) Forest transformation is driven by the search for maximization of some provisioning services and has negative consequences on cultural and regulating services. The delivery of provisioning services such as agricultural or pastoral goods (P-A) depends on forest transformation and is thus maximal when it is largely transformed; the delivery of other provisioning services such as timber, biofuels, nontimber forest products, germplasm, and future options (P-B) is enhanced by some forest transformation but requires the maintenance of some forest area and attributes; the delivery of cultural services (C) may require some forest transformation while keeping large sections of it. (B) The delivery of most regulating services is dependent on the maintenance of forest; nevertheless, little is known on how sensitive each service is to forest transformation or whether there are thresholds beyond which delivery can be irreversibly lost.

Sustaining Future Delivery of SDTF Ecosystem Services

Present socioeconomic drivers and SDTF management strategies do not ensure the long-term maintenance of this forest, nor of the delivery of its services. Alternatives need to be found in order to manage SDTF in ways that will ensure their maintenance.

Technical Interventions

SDTFs need urgently to be conserved. Large, medium-sized, or small reserves from different land tenure schemes and containing SDTF in different successional stages need to be established and interconnected. From the ecosystem services perspective, we should be looking for SDTF with some of the following features: (1) highly diverse areas containing a large genetic and species diversity of present and potential resources; (2) forests with complex vertical structures and dense woody cover to reduce albedo; enhance infiltration; prevent soil erosion, landslides, and floods; and contribute to higher water quality; (3) species or functional groups adapted to the local water availability conditions and in accordance with local aboveground and belowground carbon stocks, to contribute to climate regulation; (4) areas that host a wide variety of pollinators as well as natural enemies of pests and disease vectors; and (5) coastal areas with elevated species and landscape diversity that can help buffer the impacts of extreme meteorological events.

Sustaining SDTFs while managing them and obtaining from them a variety of ecosystem services is still a challenge. Approaches tending to harmonize biodiversity maintenance as well as livelihoods of local inhabitants have been suggested for the conservation of a large area in the Balsas basin of Michoacán, Mexico, where surprisingly high levels of endemism have been found despite the intense SDTF management and transformation (Comisión Nacional de áreas Naturales Protegidas 2006).

At the individual plot level, agro-silvopastorial systems that provide multiple provisioning services can also be applied to the case of SDTF. For example, in the Balsas area rural communities are growing organic *Hibiscus* crops (used to make tea and juice) while conserving patches of natural vegetation to allow for the provision of additional ecosystem services (A. Burgos et al., unpublished data).

SDTF restoration and that of its associated services is urgently needed.

In some cases, restoration of basic soil support services or the retention of highly erodible sediments will be most pressing; in others, regional reductions of albedo or introduction of species with particular evapotranspirative demand or carbon storage characteristics may be relevant for climate regulation; and yet in other cases, maintenance of pollinators or the regulation of invasive species may be most urgently needed.

Economic and Financial Interventions

The creation and development of markets for sustainably obtained SDTF products and those for services SDTFs provide can contribute to their maintenance. In Bolivia, close to 1 million hectares have been certified as sustainably managed SDTF in the Santa Cruz area (Cámara Forestal de Bolivia 2006).

The development of markets and direct payments for ecosystem services have been suggested as the most suitable intervention to ensure their provision (Wunder 2007). SDTFs are not eligible for markets aimed at protecting water recharge areas, since they are found where evapotranspiration exceeds precipitation most of the year. Nevertheless, they do provide a wide variety of services for which awareness is only beginning and for which markets have not been created. In the case of carbon, Mexican SDTFs harbor about 141 tons of carbon per hectare, and avoiding deforestation could prevent the liberation of 708 million tons of carbon, which is 20 percent more than what would be prevented by avoiding deforestation of Mexican evergreen forests (Jaramillo et al. 2003). Programs aimed at ensuring high water quality through erosion control would benefit the tourism industry, while programs aimed at mitigating the impact of extreme meteorological events should be considered by local, regional, and state governments so that they might avoid associated economic and social costs. Further programs with longer-term visions should aim at the conservation of crucial germplasm and future options.

Also, major challenges include the development of adequate indicators for the delivery of the services that can be monitored to assess program success. Technical data about the way services are provided, in order to develop corresponding operational rules, as well as the quantification of the corresponding baseline values are urgently needed (Wunder 2007; Guariguata and Balvanera, in press).

Educational and Institutional Interventions: Empowerment, Collective Learning, and Governance

Sustaining functional ecosystems, provision of services, and human well-being is of paramount importance. Fostering linkages among science, policy, and local decision makers is critical to attain such goals (Castillo and Toledo 2000).

The ecosystem service perspective should also be integrated into current and future environmental education programs and activities. This is particularly true in the case of Latin American tropical countries with large areas covered by SDTF, where environmental education is expected not only to raise environmental awareness but also to promote social participation in environmental problem solving and political action (SEMARNAT 2006).

At the rural community level, where people directly make a living from the ecosystem's services, educational interventions must also address the strengthening of local capacities (Blauert and Zadeck 1999). Local institutions that have systems of norms, rules for accessing resources, and gradual sanctions for rule breaking are most needed, to teach the value of investing in the maintenance of resources (Ostrom et al. 1999) and the range of regulating and cultural services provided.

Conclusions

SDTFs provide a wide variety of ecosystem services that are crucial for human well-being, yet their provision is jeopardized by the rate of transformation of such forests. Contrary to popular wisdom, areas covered by SDTF are not useless lands, and they ensure economic and social benefits to those who own them, as well as to the surrounding communities, states, countries, and the global community. While the provision of agricultural and pastoral goods obtained by removing SDTF is the most commonly recognized service, SDTFs also contribute to human well-being through a wide variety of services, including regulation of soil fertility, erosion, the water cycle, climate, pollinators, pests, disease vectors, invasive species, and impacts of extreme meteorological events, as well as many cultural services. Until now, agricultural production has been favored at the cost of losing, probably irreversibly, most of the services SDTFs provide. Alternative management schemes, the creation of markets for the ecosystem services SDTFs provide, the empowerment of the communities that manage them, and increased awareness of the services they provide may contribute to the maintenance of a fraction of these fragile systems and the crucial ecosystem services they provide. Quantitative information is urgently needed about the provision of the different ecosystem services under different management regimes, the spatial and temporal scales at which the associated ecosystem processes operate, the economic and social benefits derived from them, and the relative success of the interventions suggested to ensure their maintenance.

Acknowledgments

Many of the ideas and perspectives addressed in this chapter were inspired by the collaborative research we have been conducting around the Chamela-Cuixmala Biosphere Reserve; we would like to thank all the participants in the Cuencas, Cuitzmala, and Mabotro projects for their insights and support, as well as CONACYT for its support through the grant SEP-CONACYT 50955. We also thank G. Daily for her input to earlier versions of this work.

Chapter 16

Climatic Change and Seasonally Dry Tropical Forests

PATRICK MEIR AND R. TOBY PENNINGTON

Recent emphasis in research on Latin American tropical forests dominates our understanding of how seasonally dry tropical forests (SDTFs) may respond to climatic change and also of how changes in land use and fire incidence may interact with this response. As a consequence, our analysis focuses mainly on how Latin American SDTF may interact with climate over the twenty-first century and the wider context within which this interaction may occur, especially with respect to rain forest. In Latin America, SDTFs generally occur where rainfall is less than 1600 millimeters per year and where the dry season is substantial, lasting at least 4 to 6 months during which precipitation is generally less than 100 millimeters per month (Gentry 1995). Apart from rainfall, SDTFs are also associated with specific edaphic factors, notably nutrient-rich soils (Ratter et al. 1973; Furley 1992; Vargas et al. 2008). In contrast, Neotropical savannas, while often occurring under identical climatic conditions, are found on nutrient-poor soils that are usually high in aluminum and sometimes seasonally flooded (fig. 16-1). Savannas and SDTFs are further distinguished ecologically by deciduousness, structure, and fire resistance. Savanna trees are frequently evergreen, whereas most SDTF species are deciduous or semideciduous. Savannas are open, grass-rich vegetation, whereas SDTFs have a closed canopy with few understory grasses. Without human intervention, although SDTF trees may occasionally possess fire adaptations (e.g., thick corky bark), they lack the widespread fire-resistant features above and below ground that are characteristic of the woody flora of Neotropical savannas, which frequently burn during the dry season because of the presence of flammable C4 grasses (Taiz and Zeiger 2006). While SDTF, savanna, and rain forest have complex relationships and perhaps represent points on a continuum of vegetation types, there are in general clear ecological differences between savanna and SDTF that must be taken into account when considering the effects of climate on tropical forests.

Although the climatic parameters under which Neotropical SDTF grows are found elsewhere in the Paleotropics, it is far from clear whether vegetation such as "monsoon forest" in Asia is equivalent, because it lacks key attributes of Neotropical SDTF, including the abundant succulent element of the flora, which is especially characteristic of the drier formations (Pennington et al. 2009). Elsewhere, the distinction between SDTF and related vegetation types may occur at lower rainfall thresholds, as has been

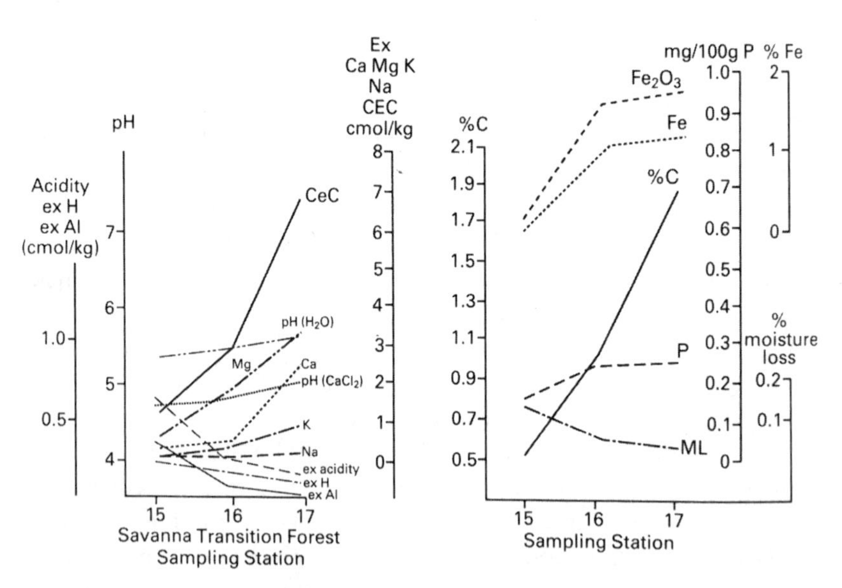

FIGURE 16-1. Soil properties across a savanna-forest boundary on Maracá Island, Brazil. A visible and sharp boundary in vegetation occurs between sampling stations 15 and 16 over a distance of a few meters. There is a large difference between forest and savanna in levels of cation exchange (ex) capacity, acidity, aluminum (Al), magnesium (Mg), calcium (Ca), and potassium (K). The data illustrate the determining influence of soil properties on natural vegetation cover under an identical climate. Figure reproduced with permission from Furley 1992.

reported for savanna-forest boundaries in Africa (Lloyd et al. 2008), but the distinction among forest biome types in all regions is also moderated by soil fertility and/or fire incidence (Sarmiento 1992; Mooney et al. 1995).

SDTFs occupy more than 1.05 million square kilometers, about 40 percent of the tropical forest landmass (Miles et al. 2006), and consequently their effect on the interactions between the tropical land surface and the atmosphere may be substantial. However, because of their smaller stature and occurrence on more fertile soils, often near human settlements, they have experienced historically high rates of deforestation, with less than 10 percent of mature forest remaining in some areas (e.g., Janzen 1988c; Trejo and Dirzo 2000). The primary uses of these forests have been for selective logging, cultivation, or conversion to grazing (Vargas et al. 2009), and the proximity of these human activities increases the risk of fire for remaining areas of SDTF.

Smaller in stature than rain forests, SDTFs tend to have closed or nearly closed canopies, with a biomass ranging between 40 and 150 tons of carbon per hectare and a leaf area index of 3 to 5 square meters of leaf area per square meter of ground area (Murphy and Lugo 1986a). Net primary productivity is lower for SDTF than for rain forest, partly because of the smaller leaf area index but mainly because of the dry season decline in assimilation rates. However, drought tolerance is often much greater than found in rain forests; notably, widely ranging adaptations to moisture stress coexist despite the lower tree species diversity, and these adaptations frame the likely responses by SDTF to climatic change. Access to soil moisture resources is a further key determinant of primary productivity by SDTF, with site- and species-based differences in soil profile depth, soil physical properties, and rooting properties strongly influencing functional variation among different forest formations.

Given the diversity of SDTF types, the detailed impacts of climatic change on them are also likely to be diverse, but they can be considered at three interconnected temporal scales: responses operating from seconds to seasons that are predominantly physical and physiological; responses occurring on a multiyear to decadal basis that are dominated by changes in structure, mortality, and reproduction; and responses occurring over centuries or millennia. This chapter focuses mainly on the first two, reviewing process-level understanding of the responses by SDTF to changes in climate that may be seasonal, episodic such as the impact of climatic perturbations of the El Niño/Southern Oscillation, or secular such as twenty-first-century climatic warming. Process-level vegetation models are required to predict terrestrial ecosystem-climate responses, and although fine-scale site-specific forest growth models have been available for some years, the main tool now used to examine ecosystem-climate interactions is the dynamic global vegetation model (DGVM) because this can be coupled dynamically to the principal type of model used for climate prediction, the general circulation model (GCM).

Climatic Drivers

There is intermodel variance in GCM predictions of twenty-first-century climate, and as a consequence most climate change scenarios are now reported as multimodel analyses (IPCC 2007). This approach is particularly necessary at the regional scale, as large differences can emerge in simple two-model comparison studies. This is especially true for regions where SDTFs are found, since the spatial resolution required to pinpoint effects on existing SDTFs is usually higher than the resolution of GCM analytical results, which may be reported at a grid scale of 200 by 200 kilometers (DGVM outputs can also be generated at this scale, although they have been reported at finer resolutions too, e.g., 50 by 50 kilometers; Sitch et al. 2003). Notwithstanding these concerns, there is substantial consistency in temperature predictions in the recent 23-model IPCC analysis, with estimated warming rates for regions containing SDTF consistent with or slightly above the global predicted means of approximately 2 to 4 degrees Celsius warming by 2100 (Christensen et al. 2007). However, scenarios for regional alterations to rainfall are more uncertain, with no clear global pattern emerging (Christensen et al. 2007).

In the absence of secondary factors, warming on the land surface without increased rainfall results in a reduction in soil moisture availability, or drought. The most recent IPCC examination of the twenty-first-century hydrological cycle (Bates et al. 2008) used a 15-model ensemble analysis to indicate that mixed effects on soil moisture availability were likely in the African, Asian, and Australian tropics but that a 5 to 15 percent reduction in soil moisture availability was expected by the late twenty-first century across much of tropical Latin America. Regional analysis for Amazonia has further confirmed that drought may become a key climatic driver during the twenty-first century because of a possible increase in El Niño/Southern Oscillation frequency (Timmerman 1999) and because of the regional impacts on precipitation of warming of the northern tropical Atlantic Ocean relative to the south (Cox et al. 2008). Further, focusing the IPCC multimodel analysis on Amazonia, Malhi et al. (2008) demonstrated a 20 to

70 percent agreement among GCMs in the prediction of substantial dry season reductions in precipitation across Amazonia, with the greatest certainty of drought in the eastern portion of the region. Other alterations to key climate variables, such as changes in radiation flux caused by differences in cloud cover, are thought to impact primary production (Mercado et al. 2009) and may also influence tree reproduction (Wright et al. 1999). However, there remains significant uncertainty in GCM predictions of cloud cover over this century, and hence the majority of our analysis is restricted to the likely effects of temperature and moisture on SDTF, together with a consideration of the impacts of increased carbon dioxide concentration.

Land use change impacts are likely to exacerbate the effects of climatic warming, and this may be especially true for SDTFs, given that they are mostly already highly disturbed and fragmented (Murphy and Lugo 1986a; Kauffman et al. 2003). Although the effects of small-scale land clearance are thought to be mixed and may sometimes lead to localized increases in precipitation (Werth and Avissar 2002), widespread forest conversion to pasture and agriculture is expected to reduce rainfall through differential effects on latent and sensible heat transfer (Nobre et al. 1991; Werth and Avissar 2002; Costa et al. 2007). Increased regional atmospheric aerosol loading resulting from fire and land use change could also cause widespread reductions in precipitation (IPCC 2007). Overall, these results imply an increased frequency of extreme events at seasonal and longer timescales.

Modeling Tropical Woody Vegetation Changes in Response to Climatic Change

The ecological differences between SDTF and savanna are relevant for understanding both twenty-first-century land surface–climate interaction scenarios and paleoecological evidence for vegetation change. A variety of models have been used to examine the impact of climatic warming and drying on tropical forests. A relatively straightforward approach is to consider near-equilibrium ecosystem-climate relationships, whereby the natural environmental envelope experienced by existing forests is used to inform how forest distribution may change under climatic forcing. Driving an equilibrium vegetation model with a variety of climate scenarios, Sampaio, Nobre et al. (2007) predicted a range of rain forest–to–savanna switches in Amazonia following warming and drying. A recent bioclimatic analysis of the distribution of current natural Amazonian vegetation and likely twenty-first-century regional climate change also indicated a shift from rain forest (fig. 16-2; Malhi et al. 2009). While there is substantial variation in the climate anomaly trajectories predicted by the different GCMs in figure 16-2, the majority imply a drought-driven transition from rain forest to some form of vegetation comprising species better adapted to seasonal drought. However, the nature of this drought-adapted vegetation will depend strongly upon edaphic factors and the frequency of fire. We suggest that if fires are frequent and, critically, if soils are not sufficiently nutrient rich (Furley 1992), the tree species more likely to thrive either will be those found in the often narrow (3 to 4 kilometers wide; Ratter 1992) transitional ecotone between the cerrado savanna and Amazon rain forest or they will be true savanna species. Transitional, marginal Amazonian forest such as this is found on poor soils and is confusingly termed *mata seca*, or "dry

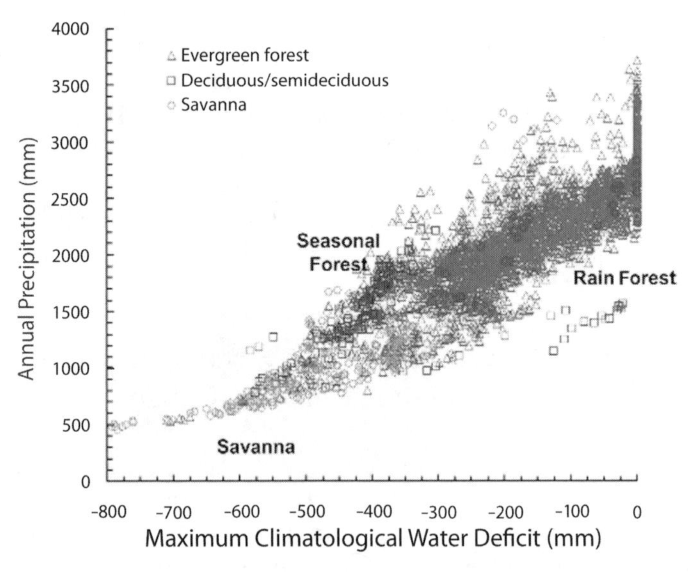

FIGURE 16-2. (A) The relationship between vegetation type and rainfall regime for tropical South America. Rainfall data are derived from the Tropical Rainfall Measuring Mission for the period 1998–2006; vegetation data are from Eva et al. 2004. The mean climatological water deficit is defined as the most negative value of the climatological water deficit, attained over a year, where the monthly change in water deficit is calculated from precipitation minus evapotranspiration (both in millimeter/month). The zone suggested for "seasonal forest" (sensu Malhi et al. 2009, but likely to represent SDTF, transition forest or savanna transition forest) sits above the shaded area, where annual precipitation is more than 1500 mm and mean climatological water deficit is between –200 and –300. Figure from Malhi et al. 2009.

forest" (Pires 1974; Pires and Prance 1985), but it is unrelated to SDTF as defined here. It is mostly evergreen and consists mainly of geographically widespread rain forest species with broad ecological tolerance (Ratter 1992). Under stronger drying and more frequent fire, a further transition to more open savannas might occur from this vegetation type. Because large areas of rain forest, including many in the drought-threatened eastern Amazon, are underlain by poor soils (Quesada, Lloyd et al. 2009), a transition to forest dominated by SDTF species is unlikely, as they favor nutrient-rich soils.

Both present-day and paleoecological evidence indicate that savanna-rain forest and SDTF-rain forest boundaries can change rapidly under a climatic trend of increasing rainfall but that both savanna and SDTF are much more resistant to change during drought. Ratter, Richards et al. (1973), Ratter, Askew et al. (1978), and Marimon et al. (2006) have observed the spread of Amazonian rain forest into the cerrados of Mato Grosso in Brazil by means of surveys carried out over 40 years since the late 1960s. In contrast to the nutrient-poor soils in this area, the soils are more fertile in the Chiquitano region of Bolivia, where Mayle et al. (2004) have used fossil pollen to show that rain forest spread relatively recently into areas that were covered previously by SDTF. In both cases the changes were probably driven by increasing rainfall. Under drought, by contrast, because of gener-

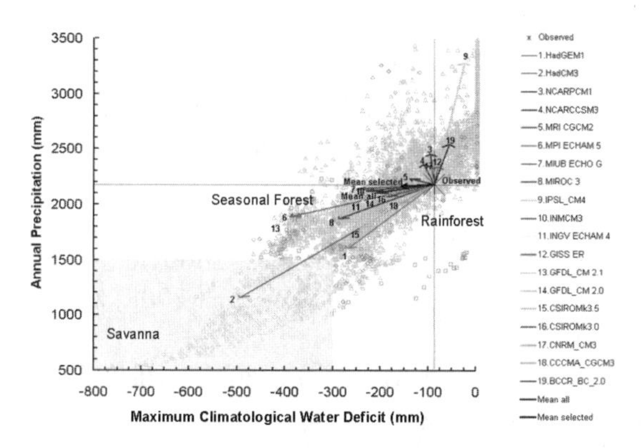

FIGURE 16-2. (B) An evaluation of 19 GCM simulations of the change in rainfall regime in eastern Amazonia for the twenty-first century. The arrows represent the trajectories of changes in rainfall regime for each GCM, when recalculated as relative changes forced to start from the observed climatology (New et al. 2000). The tip of each arrow indicates the predicted late-twenty-first-century rainfall. The zone suggested for non-rain forest vegetation ("seasonal forest" and "savana" sensu Malhi et al. 2009), as in A, sits above the shaded zone. Figure from Malhi et al. 2009.

ally poor soils, one would expect to see rain forest–to–savanna transitions along the southern fringe of Amazonia, and this is corroborated by fossil pollen evidence that suggests expansion of savannas into the southern edge of Amazonia during the cool dry periods of the last glacial maximum (Van der Hammen 1972). SDTFs are, like savanna, strongly drought tolerant. Dendrochronological evidence from the SDTF of the Pacific slope of northern Peru suggests that most tree growth is concentrated in years of unusually high precipitation coincident with El Niño/Southern Oscillation events (P. Zavallos-Pollito, pers. comm.), indicating a physiology that can tolerate several years of suboptimal growth conditions. Furthermore, molecular phylogenetic studies of species and populations of woody species in the SDTF of inter-Andean valleys in Peru and Bolivia indicate remarkable population stability over timescales of millions of years (up to about 10 million) in small geographic areas (Pennington et al. 2009). Given the large climatic fluctuations over these geological timescales, especially the drier climates of glacial periods of the Pleistocene, these data further support the expectation of drought resistance in SDTF.

Accurate prediction of the responses by tropical forests to drought requires greater model resolution than has hitherto been achieved, especially in the representation of ecological processes. Although equilibrium vegetation models suggest switches from one vegetation type to another, the actual response by vegetation to climate will be defined by multiple ecological and physical processes, leading to either gradual or punctuated changes in vegetation over several stages of development. Modeling these changes is desirable, as it enables a time-series quantification of the interaction between vegetation and the atmosphere, and in principle it provides a better tool by which natural resources may be managed at large scales. Process-based dynamic vegetation models (e.g., DGVMs) have the structure to capture the relevant bioclimatic interactions, although they require observation-based parameterization (Meir et al. 2008) and can be computationally expensive.

A widely discussed scenario of Amazonian dieback in response to twenty-first-century warming and drought emerged from two early DGVM modeling analyses (A. White et al. 1999; Cox et al. 2000). DGVMs necessarily contain simplified representations of canopy structure, soil processes, and functional diversity among species, and these first climate-coupled model analyses contained much uncertainty (Friedlingstein et al. 2006; Meir et al. 2006) as they were based on very limited field observations, especially in relation to the distinctions between rain forest and other tropical vegetation types, such as SDTF. As a consequence, there is now a focus on using field-based measurements and experimentation to reduce uncertainty in DGVM predictions for tropical forests (Keller et al. 2009). Model advances are needed in the representation of the supply of moisture and nutrients (especially phosphorus) to plants and of key physiological processes, including respiration, transpiration, and photosynthesis, together with their influences on mortality and reproduction (e.g., Fisher et al. 2007; Fisher et al. 2008; Baker et al. 2008; Valdespino et al. 2009). Further, an expansion of the functional ("species") diversity embedded in DGVMs is probably needed to take account of the important ecological differences among SDTF, savanna, and rain forest (Mayle et al. 2004; Fyllas et al. 2009). However, devising the most computationally efficient way to do this and also incorporating the impacts of land use change remain a substantial challenge (Meir et al. 2006; Golding and Betts 2008).

Sensitivity in the Carbon Cycle to Climate and Atmospheric Composition

Understanding the cycling of carbon and water lies at the heart of quantifying forest-atmosphere interactions. Net carbon gain by plants is also used in DGVMs as a proxy to determine whether one vegetation type dominates or is replaced by another (Cramer et al. 2001). Hence, carbon has been a key currency used to formulate vegetation-climate change scenarios (Friedlingstein et al. 2006; IPCC 2007). In the absence of abiotic oxidation events such as fire, the balance of respiration and photosynthesis is the principal determinant of the net ecosystem productivity (NEP). This balance depends on the environmental responses by respiration in soil and plants, and by photosynthesis.

Soil Respiration

The emission of carbon dioxide from soil, or "soil respiration" (R_s), is the second largest flux in the terrestrial carbon cycle and comprises both heterotrophic respiration of soil organic matter by microbes and autotrophic respiration by plant roots (Trumbore 2006). However, despite its importance to the carbon cycle, only a few measurements of R_s are available for SDTF. R_s varies in SDTF between 0.3 and 4.0 micromole per square meter per second and annually between 6 and 13 megagrams carbon per hectare per year (Raich and Schlesinger 1992; Cuevas 1995; Vargas and Allen 2008). Although the lower end of the range of instantaneous flux rates is lower than for rain forests, presumably because of the impact of the extended dry season in SDTFs, the annual values fit within the range observed for tropical upland and lowland forests (3 to 20 megagrams carbon per hectare per year; Schwendenmann and Veldkamp 2006; Zimmermann et al. 2009a).

On a diurnal basis, variation in R_s appears to be larger than observed in lowland tropical rain forests (up to 1 micromole per square meter per second), and this variation is unlikely to be related principally to diurnal changes in soil moisture. Vargas and Allen (2008) calculated that without allowing for diurnal variability, the annual R_s sum might be in error by up to 10 percent for a Mexican SDTF. The mechanistic basis of the day-night difference was unclear in this study, as soil temperature was not correlated in a simple way to R_s, although alterations in R_s might have been related to changes in plant root or microbial metabolic activity (Tang and Baldocchi 2005). Few reports exist of diurnal patterns in tropical forests, but Zimmermann et al. (2009b), working in a montane tropical forest in Peru, highlighted diurnal flux variation of up to 3 micromoles per square meter per second and were able to attribute the majority of this variation to differences in litter temperature rather than soil temperature. As with Vargas and Allen (2008), Zimmermann et al. (2009b) demonstrated that R_s calculated on the basis of daytime measurements alone would result in a large error in the annual sum (of up to 60 percent), underlining the need for future studies to account carefully for the full diurnal cycle in R_s before scaling flux rates to longer timescales.

Over seasonal and annual timescales, however, there is good evidence that as with other tropical forests, variation in soil moisture is the main climatic determinant of R_s (Davidson et al. 2000; Schwendenmann et al. 2003; Meir et al. 2008). The response by R_s to soil moisture is best described currently as a function of soil water potential, rather than the simpler measure of soil moisture content by volume, and measurements have usually indicated either a linear response (Davidson et al. 2000; but see Davidson et al. 2008) or a nonlinear, peaked response surface (Schwendenmann et al. 2003; Sotta et al. 2007; fig. 16-3). If dry season soil moisture constraints are strong, R_s for tropical rain forest can decline by more than 20 percent with respect to wet season values (e.g., Sotta et al. 2007), and thus better quantification of this response for SDTF should be a priority. A need for improved understanding of the moisture response by R_s has also been noted for other tropical forests (Meir et al. 2008), as the parameterization in many DGVMs and other biogeochemical models may be inaccurate, potentially leading to large errors when R_s models are driven using climate perturbation data for events such as the El Niño/Southern Oscillation (Keller et al. 2009). Clearly, modeling of R_s for tropical forests such as SDTFs requires substantial attention to new field data. The representation of these processes is likely to remain partially empirical for the foreseeable future (Davidson and Janssens 2006), but incorporation of them into a single process-based framework is a realistic aim, particularly through improvement of the mechanistic basis of moisture controls on R_s.

Photosynthesis and Plant Respiration

Canopy-scale gas exchange by SDTF results from the combined activity of a variety of different life-forms, different photosynthetic pathways, and different patterns in dry-wet season phenology. Evergreen, deciduous, and succulent woody plants coexist in some types of tropical dry forest (Sarmiento 1972). Evergreen sclerophyllous woody plants exchange leaves during the wet season without losing canopy cover (Marin and Medina 1981), but in deciduous species, canopy leaf-out is strongly related to dry season timing, usually occurring just before or just after the beginning of the rains, with differences partly dependent on variation in soil moisture supply and stem water storage (Lieberman and Lieberman 1984; Reich and Borchert 1984; Medina 1995).

Drought tolerance mechanisms are varied among dry forest tree species, and this is partly reflected in the spectrum of responses to drought in terms of the extent (or occurrence) of leaf abscission during the dry season. Variation in minimum leaf water potential (cf. Fisher et al. 2006), stem and leaf hydraulic conductance (e.g., Brodribb and Holbrook 2003), stomatal conductance (e.g., Choat et al. 2006), access to and tolerance of soil water reserves (Engelbrecht et al. 2007; Fisher et al. 2008; Goldstein et al. 2008), osmotic adjustment (Mulkey et al. 1991; Medina 1995), and stem and root water storage (Borchert 1994b) have all been identified as physiological response axes among which variation may confer more or less resistance to seasonal drought in woody species using the C3 photosynthetic mechanism. CAM-based photosynthesis is also adopted by some dry forest species, including succulents, epiphytes, and terrestrial rosettes. This is of note where seasonal drought is severe, because the CAM mechanism confers high water-use efficiency through the accumulation of carbon dioxide at night (as vacuolar malic acid), thus enabling the fixation of carbon dioxide without substantive daytime water loss through stomata

(A)

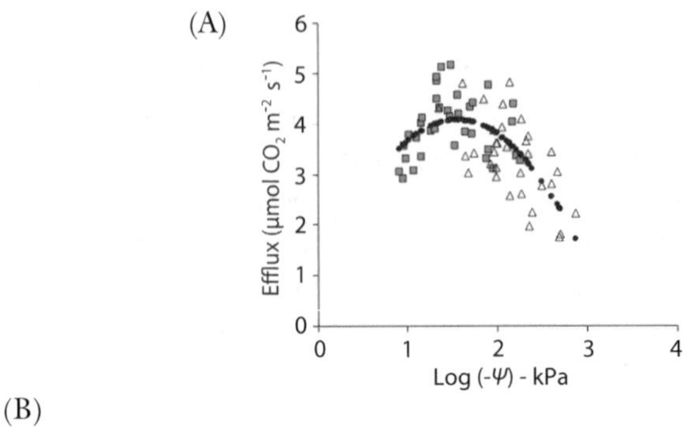

(B)

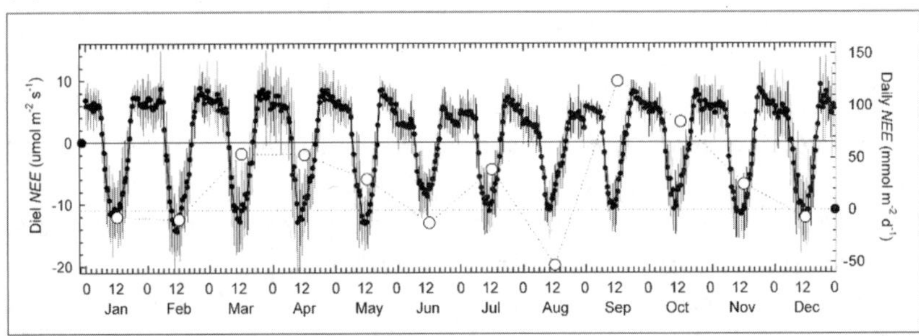

FIGURE 16-3. Moisture constraints on assimilation and respiration. (A) Chamber-based measurements of the relationship between soil water potential and soil respiration (R_s, micromoles per square meter per second; $R^2 = 0.43$, $P < 0.001$). The data are from a soil moisture reduction experiment in rain forest. Figure reproduced with permission from Sotta et al. 2007. (B) Eddy covariance measurement of net ecosystem exchange (NEE) in a seasonally dry transition forest in Sinop, Mato Grosso, Brazil. During the 4-month dry season (May–August), ecosystem respiration is more sensitive to moisture stress than is maximum photosynthesis (a decline of 28 percent in ecosystem respiration vs. 5 percent in maximum photosynthesis; Vourlitis et al. 2005). Data show changes in diel and daily NEE of CO_2. Positive values indicate CO_2 release to the atmosphere (mean NEE plus or minus SD) of 30-minute data (spots) and average daily NEE by month (circles); the horizontal solid line is diel NEE = 0, and the horizontal dashed line is daily NEE = 0. Figure reproduced with permission from Vourlitis et al. 2005.

(e.g., Lüttge 1997). However, the total contribution to canopy-scale gas exchange by the CAM pathway is only significant where CAM species such as columnar and shrubby cacti dominate. Lianas, on the other hand, use the dominant C3 photosynthetic pathway but are surprisingly abundant in drier forests, reaching up to 34 percent of all species and 16 percent of all individuals (Gentry and Dodson 1987). Water use by lianas is incompletely understood, but their maximum photosynthetic capacities can be higher than neighboring C3 trees (Domingues et al. 2007), and as they have the capacity to dominate the upper canopy of some stands, their overall contribution to canopy-scale gas exchange may be substantial (van der Heijden and Phillips 2008).

In addition to affecting growth through variation in leaf fall or leaf flush in deciduous species, differences in water transport properties strongly affect photosynthetic performance in many SDTF trees. Choat et al. (2006), comparing deciduous and evergreen trees, demonstrated that leaf abscission in deciduous species was strongly sensitive to changes in soil moisture status and closely related to sharp reductions in stomatal conductance, but not reductions in leaf water potential. Those leaves that were not shed under drought retained a relatively high leaf water potential, and overall dry season canopy gas exchange declined instead through reductions in leaf area and stomatal conductance. Evergreen species, on the other hand, responded to strong moisture constraints through reductions in both leaf gas exchange capacity and leaf water potential. These differences in the mode of response to drought may also reflect differences in the depth of soil moisture access (Sobrado 1993) and are probably further influenced by plasticity in the hydraulic conductance of the root-atmosphere continuum. Hydraulic capacity has been shown to correlate strongly with photosynthetic capacity during the dry season (fig. 16-4; Brodribb et al. 2002). Thus, when soil moisture is limiting, variability in the hydraulic conductivity of both xylem and leaves can constrain leaf and canopy assimilation rates and may directly lead to the loss of photosynthetic capacity before senescence and abscission (Brodribb and Holbrook 2003).

The coordination of photosynthesis with respiration that often occurs in tropical trees (Meir et al. 2001) implies that respiration should decline with photosynthetic capacity, thus theoretically impacting net leaf carbon gain minimally. However, in some tropical forests this relationship has been found to break down under drought because of moisture-stress-related increases in respiration. Although laboratory evidence on the response to drought by respiration in leaves is mixed (Flexas et al. 2006; Atkin and Macherel 2008), experimental drought of rain forest has resulted in increased respiration rates in leaves (Meir et al. 2008) and roots (Metcalfe et al. 2008). This effect has also been observed in leaves during a natural dry season in Mato Grosso, Brazil, in the transitional Amazonian forest vegetation between rain forest and savanna (Miranda et al. 2005). The physiological basis for these responses to drought remains unclear, although increased dark respiration rates have been observed elsewhere in the leaves of dry-adapted species (Wright et al.

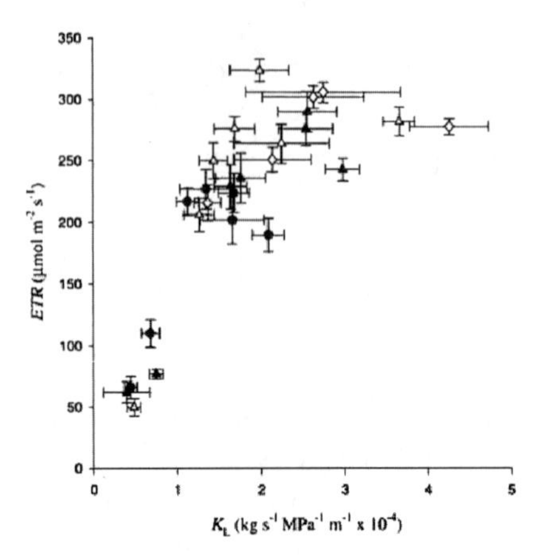

FIGURE 16-4. Mean leaf-specific hydraulic conductivity plus or minus SE (K_L) versus mean photosynthetic capacity (electron transfer rate, ETR) plus or minus SE for branch tips of 12 species investigated in a SDTF in Costa Rica. Nondeciduous species were measured in the mid (closed triangles) and late (spots) dry season and the wet season in July (open triangles); deciduous species were measured in July (open diamonds). A strong initial response of ETR to K_L appears to saturate when K_L is in excess of approximately 3 kilograms per second per megapascal meters. Figure reproduced with permission from Sotta et al. 2007 and Vourlitis et al. 2005.

2006) and the effect is consistent with a need to maintain ion gradients (cf. Mulkey et al. 1991) and increase protein repair rates; it is possible that the same physiological drought responses are found in SDTF trees.

Quantifying the effects in SDTF of the mosaic of different physiological responses to drought at the scale of the canopy potentially requires substantial modeling detail in terms of process representation and input data from individual species. Alternatively, above-canopy measurements using eddy covariance to monitor the biophysical activity of several square kilometers of forest can be used to quantify seasonal and interannual transitions in NEP, water use, and energy exchange by vegetation and to test larger-scale models of these processes. We are unaware of any eddy covariance measurements made over strictly defined SDTF, but Vourlitis et al. (2004) and Vourlitis et al. (2005) have used this technique to document seasonality in the exchange of carbon dioxide, water vapor, and energy by a transitional Amazonian forest near Sinop, Mato Grosso, Brazil. This forest occupies the ecotone between rain forest and savanna and experiences strong dry seasons and more substantial seasonal leaf fall than Amazonian rain forest, even though few trees are truly deciduous. Despite likely variation at the levels of species and individuals, these canopy-scale data have demonstrated that the gross environmental response characteristics of the Sinop transitional forest can be represented by relatively simple empirical models driven by soil moisture availability, temperature, radiation intensity, and atmospheric water vapor pressure deficit (Vourlitis et al. 2005). Initial measurements suggested that, overall, this rain forest–savanna transition forest is in carbon balance on an annual basis and during normal rainfall (Vourlitis et al. 2004), although the forest was considered at risk of losing carbon (i.e., a negative NEP) under a warming scenario. Understanding the mechanisms determining the net carbon balance is critical to reliable predictions of forest-atmosphere interactions, and an interesting finding in this work was that the determinants of seasonal variation in NEP varied strongly and also differed from the assumptions underlying recent modeling approaches. Under seasonal drought at this site, the maximum photosynthetic rate was resistant to drought, declining by only 5 percent, while ecosystem respiration was strongly drought sensitive, declining by up to 28 percent (fig. 16-4), partly because of large reductions in R_s. Thus, variation in NEP during the year was strongly dependent on variation in respiration processes in soil and plants, not just photosynthesis. Although recent attempts to model the impacts of drought on rain forest gas exchange have been successful (Fisher et al. 2007; Baker et al. 2008), the role of respiration in NEP has not been fully accounted for in tropical forests (Meir et al. 2008). Modeling these effects accurately and over longer time periods remains a challenge, but clearly a closer focus on the effects of climate on the interplay between photosynthesis and respiration in both soil and plants is needed to fully capture the response to climatic change by SDTF.

Mortality, Reproduction, and Resilience

Short-term physiological responses to drought have impacts over the long term, most notably through negative effects on growth and mortality. The direct impact of mortality on photosynthesis is immediate, largely through the loss of functional leaf area. The effects of tree mortality on R_s are strongly constrained by soil moisture availability, but over longer time periods, mortality will also affect the allocation of newly acquired carbon to belowground components (Tang and Baldocchi 2005; Brando et al. 2008). Hence, lagged reductions in soil carbon stock and R_s can be expected in the decades following substantial tree mortality events under climatic forcing.

Understanding tree death is central to modeling the impacts of climatic change on NEP and to modeling vegetation change in response to climate. Although the proximate reasons for the death of individual trees are varied

(Martini et al. 2008), in the absence of human intervention, the underlying biological reasons for mortality are physiological or pathogenic; otherwise, death results from substantial physical disturbance events, such as fire (e.g., Fensham et al. 2003) or hurricanes (e.g., Vargas et al. 2008). As SDTFs are subjected annually to strong seasonality in rainfall, understanding the process of drought-related mortality is important. McDowell et al. (2008) review proposed plant physiological responses to moisture stress (Tardieu and Simonneau 1998), and they distinguish between species that use strong stomatal control to maintain leaf water potential above a minimum value ("isohydry") and those that rely instead on maintaining a water supply from the soil even at low soil-moisture availability ("anisohydry"). By exerting minimal stomatal control under moisture limitation, anisohydric plants risk mortality by hydraulic failure under severe drought (Tyree and Sperry 1988; West et al. 2008). In contrast, isohydric plants reduce (although they do not remove) the risk of hydraulic failure by exerting strong stomatal control under moisture stress. However, by doing so, they risk death by "carbon starvation" because of the negative impact on photosynthesis of reduced stomatal conductance and high leaf temperatures and/or through increased susceptibility to pathogenic attack (Ayres and Lombardero 2000; Meir et al. 2006; McDowell et al. 2008).

There is evidence that some Amazonian rain forest trees are isohydric (Fisher et al. 2007), implying that the dominant mode of tree death during extreme drought in these ecosystems is carbon starvation, or physical and biological damage resulting from reduced carbon acquisition. Reduced carbon supply has also been implicated in observed declines in flower and fruit production in tropical forests (Brando et al. 2006; Zimmerman et al. 2007), and so carbon starvation may both increase mortality and reduce fecundity among species not resistant to drought. In contrast, experimental studies on tree seedlings that naturally experience a wide range of seasonal drought intensities in Panama have indicated that hydraulic failure is the principal reason for death under strong moisture stress (Kursar et al. 2009). Thus, studies available from tropical forests suggest that more than one physiological mechanism causing death in SDTF trees is likely and that, in any case, stomatal adaptations to drought are likely to be complemented in many species by adaptive leaf abscission, with consequent impacts on the supply of carbon to support metabolism. Quantifying the combined importance of isohydric species, anisohydric species, and moisture-stress-related leaf abscission limits may be a useful way to advance the mechanistic representation of drought impacts on mortality in SDTF.

Improved modeling of the physiological basis of changes in tree mortality and reproduction is clearly needed to represent properly the change in SDTF in response to warming and drought, but as important is the response under drought by SDTF to fire. While climate may force vegetation change gradually, even in the face of species management (Maranz 2009), fire is often the switch that converts a drought-stressed-but-resistant forest to a more open woody or grass-dominated vegetation type (Vourlitis et al. 2004; Hutyra et al. 2005). Fire reduces forest biomass (Kauffman et al. 2003) and soil carbon stocks (Castellanos et al. 2001; Powers 2004), and smoke from forest fires can affect the surface energy balance (Wang and Christopher 2006) and local to regional rainfall patterns (Nepstad et al. 2001; Werth and Avissar 2002). The interplay of drought and fire maintains the open nature of some savannas (Fensham et al. 2003), and this process is beginning to be incorporated into vegetation models, together with the physical and atmospheric effects of fire.

The role of fire in maintaining successional processes in SDTF has been discussed by Vieira and Scariot (2006). Where species are fire resistant, recovery may be rapid, but given an adequate natural seed source, reestablishment and relatively fast regrowth after complete forest loss are also characteristic of SDTF because of the high soil fertility (Murphy and Lugo 1986a). The effect is to impart substantial resilience to, and relatively rapid recovery from, disturbance and, in some cases, disturbance by fire. In view of this, Ewel (1980) suggested that recovery by SDTF from disturbance was rapid in comparison to rain forest, especially as the floristic complexity of the mature vegetation was lower. Lebrija-Trejos et al. (2008) tested this idea in southern Mexico and found that resilience was not intrinsically higher in SDTF following extensive disturbance by agriculture. However, studying recovery of SDTF from severe fires in the northeastern Yucatán, Mexico, Vargas et al. (2008) were able to demonstrate that recovery rates for carbon stocks were relatively rapid, probably the result of rapid reestablishment and growth of new trees, rather than regrowth of existing damaged trees. In this latter study, belowground carbon stocks apparently returned to prefire values within 20 years, and aboveground carbon stocks returned to prefire values within 80 years.

Future Impacts of Land Use and Climate on SDTF

The impacts of land use and climate can be divided into an analysis of the ecological response by forest to climate and of the integrated effects on SDTF of changes in climate, fire incidence, and land use.

Ecological Responses to Climatic Warming and Drying

Short-term fluxes of gross primary production, respiration, or evapotranspiration (e.g., Vourlitis et al. 2005; Vargas et al. 2008) have been measured and relatively well parameterized for some sites. This information provides a key test bed for vegetation-climate models (Fisher et al. 2007; Baker et al. 2008), but because of a lack of experimental and long-term observational evidence, mechanistic modeling of decade- or century-scale vegetation-atmosphere interactions for SDTF remains reliant principally on physiological principles. For example, in a recent review, Lloyd and Farquhar (2008) argued that the positive effect of increased atmospheric carbon dioxide concentration on photosynthesis in tropical forests is likely to outweigh any negative impacts of warming through concurrent temperature increases, and the overall effect will probably balance in favor of a positive impact on NEP. The wide range of adaptations to drought in SDTF trees is therefore likely to be enhanced at higher atmospheric carbon dioxide concentrations because plants will be able to reduce water loss through stomatal closure without reducing photosynthesis (Kirschbaum et al. 1996). However, future patterns of transpiration remain unclear, as under drought and/or warming, a drier atmosphere will impose a bigger atmospheric demand on evaporation, potentially leading instead to higher rates of evapotranspiration (Salazar et al. 2007) irrespective of reductions in stomatal conductance. Plant respiration has previously been considered to be strongly sensitive to temperature, but more recent analyses favor the hypothesis that plant tissues may acclimate to higher temperatures (Atkin and Tjoelker 2003), and this would also confer resistance to climatic drying and warming through reduced use of plant carbon reserves. Additional uncertainty remains as to whether the response in stomatal conductance to increased atmospheric carbon dioxide concentration over the long term will be similar to the declines reported from short-term measurements, and the impact of this response may in any case be further modified by changes in leaf area index and the response to drought by plant respiration, as discussed earlier (Atkin and Macherel 2008; Meir et al. 2008).

Linking these physiological responses to the longer-term ecological processes that determine vegetation change, such as mortality and reproduction, remains a modeling frontier (McDowell et al. 2008), and currently most DGVMs only contain crude representations of the fundamental connections among these processes (Moorcroft 2006). However, although little process-based vegetation modeling specific to SDTF has been reported, new dynamic vegetation models (Moorcroft 2006) and new multidecadal observational data sets from lowland rain forests (e.g., Philips et al. 2009) offer the near-term prospect of rapid improvements in the ecological representation of the response to twenty-first-century climate change by tropical forests (Meir et al. 2006).

Fire and Land Use

The role of fire in maintaining both open and closed tropical woody vegetation types has been discussed elsewhere (Maranz 2009; Durigan 2006). However, the impact of fire on many tropical forest ecosystems has increased markedly in intensity in recent decades because of human land use (Cochrane and Laurance 2002). The rate of conversion of SDTF to grazing or agriculture is particularly high because of the relative ease of access and the relatively high soil fertility of these forests (Kauffman et al. 2003; Vargas et al. 2009). This makes SDTF more vulnerable to fire, under a scenario of climatic warming and drying, than many other tropical forests. A key concern is of high-intensity fires leading to substantial loss of vegetation cover; if local seed sources are not available, this will be followed by restricted regrowth of woody SDTF vegetation.

The networks of positive feedbacks among climatic warming and drying, land conversion, forest fragmentation, and fire have been described in detail for lowland rain forest (Nepstad et al. 1999; Nepstad et al. 2001; Soares-Filho et al. 2006). But for no tropical region have coupled dynamic models been developed to synthesize interactions of climate, vegetation, deforestation, and fire. In a recent attempt to address this issue for South America, including many areas of SDTF, Golding and Betts (2008) superimposed deforestation scenarios (Soares-Filho et al. 2006; van Vuuren et al. 2007) and a simple climate-driven fire model (Noble et al. 1980; Hoffmann et al. 2003) upon an ensemble climate model analysis using GCM output that also incorporated a simplified representation of the response to climate by vegetation (HadCM3; Collins et al. 2001; Harris et al. 2008). This first integrated analysis demonstrated a substantially increased fire risk across lowland Amazonia by 2020 and a "high" risk of fire across 50 percent of the region by 2080. Some of the areas of high fire risk overlap with current areas occupied by SDTF and underline the need to quantify the role of fire in near-term and decadal-scale analyses of the response by SDTF to climatic warming and drying in the twenty-first century.

Conclusions

The responses by tropical forests to drought and fire have become touchstone issues in environmental science and governance in recent years, especially in Latin America. Consequently, increased attention has focused on predicting the effects of climate on tropical forests, of which SDTFs comprise a significant land area.

SDTFs represent a heterogeneous but very widely occurring vegetation type. They are floristically distinct from rain forest and savanna and exhibit discrete ecological characteristics and histories. Although the climate response characteristics of the wet-season carbon and water cycles of SDTFs are unlikely to differ widely from those of rain forest, the range of physiological and phenological adaptations to drought that are exhibited by SDTF tree species shows that, like many savanna species, they are much more resistant to seasonal and multiyear soil moisture deficits than rain forest species. By contrast, while fire resistance in SDTF species is less than in many savanna trees, some adaptation to fire is evident, perhaps placing them at an advantage over rain forest species. In the absence of land use change, we argue that it is these ecological differences, together with the key tendency for SDTFs to require fertile soils, that will set apart the role played by SDTFs in tropical forest–climate interactions in the twenty-first century.

Climate and vegetation modeling studies have focused on the potential for rain forest to degrade into lower-biomass forest types, but no process-based modeling study has yet distinguished between rain forest–SDTF and rain forest–savanna transitions. Modern and paleoecological evidence suggest that while savanna or SDTF may aggrade into rain forest in a climatic wetting scenario, drought-impacted rain forest will only degrade into SDTF where soils are fertile and seed sources are sufficient. Many Amazonian rain forest soils are too infertile to support SDTF, making more likely a drought-driven transition to vegetation dominated by species such as those currently found in the transitional forest between Amazon rain forest and the cerrado savannas. Only where soils are more fertile will drought favor SDTF species. These simple but important ecological rules need to be incorporated rapidly into current modeling analyses if realistic scenarios of vegetation transitions across Amazonia are to be simulated.

The relatively new modeling frameworks of DGVMs provide a way of dynamically and mechanistically representing the processes governing the responses to drought by rain forest, SDTF, and savanna and of coupling them with both climate and land use models. Integrated modeling of this sort strengthens environmental policy and governance and can substantially improve the prospects for reducing the rate of tropical forest loss (Nepstad et al. 2001; Mitchell et al. 2008). The first such integrated steps in tropical land-atmosphere modeling are beginning to be taken, but future work will need to pay close attention to both the differences in the ecology of tropical forest types and the land use risks to which they may be exposed.

Acknowledgments

We thank the Royal Society of Edinburgh and the University of Edinburgh for research fellowship support to PM, and the Royal Botanic Garden Edinburgh for research support to RTP. We also thank J. Ratter and F.E. Mayle for expert and useful reviews of an earlier version of this manuscript.

Chapter 17

Synthesis and Promising Lines of Research on Seasonally Dry Tropical Forests

HAROLD A. MOONEY

In the mid 1990s the first global synthesis of our knowledge of the biology of seasonally dry tropical forests (SDTFs) was published (Bullock et al. 1995). The motivation for that synthesis was the fact that vast areas of tropical dry forests of the world were poorly studied and yet they represent one of the most threatened ecosystems of the world. These systems provide a vast treasure of biological information on the adaptive modes of organisms to an environment that is not thermally limited but where water availability varies greatly within and between seasons. It was hoped that revealing the knowledge that had accrued would be a stimulus to further research on these unique ecosystems as well as new efforts for their conservation.

A comparison of the 1995 volume with this book shows that much new information has accumulated on the biology of these systems and their conservation. Also, a reading of both volumes gives us a glimpse of the increasing attention toward not only studying what we have left but also learning about restoring what we have lost. This volume differs from the earlier volume in that it concentrates solely on the Neotropics, whereas the former treatment was somewhat more global in coverage.

The seasonally dry tropical forest, as an ecosystem type, has a widespread distribution in the Neotropics, extending from Mexico to

Argentina. A detailed floristic analysis by Linares-Palomino et al. indicates that there are strong floristic affinities among these systems within the so-called Pleistocenic Arc of now-isolated pockets of SDTF stretching from Argentina to Paraguay to Brazil. The more northerly representatives of the SDTF, in the Caribbean region and Central America, are, however, not closely related to those of the south. The few widespread species that occur in both the north and the south likely represent long-distance dispersal events. Caetano and Naciri, in a more geographically limited study but utilizing genetic data from widespread species, support the Pleistocenic Arc hypothesis.

Chapter 3, by Sánchez-Azofeifa and Portillo-Quintero, underscores the large losses that have occurred over time to the SDTFs. Their analysis indicates that in tropical regions, SDTFs have been the first systems to fall to agricultural development and continue to represent prime conversion targets for pasturage, as well as intensive and subsistence agriculture. Remote sensing of the status of these forests is being utilized to a greater degree to document these trends, but there are still issues in the classification of the remotely sensed images due to the complexity of these forests in terms of tree density and even leaf duration. Further, the variety of image scales utilized in the analysis of the remotely sensed images has led to variability in the estimates of their extent. However, there is little question about the continuing conversion trends of the SDTF and the large losses that have occurred in all of the Neotropical countries where conversion occurs.

The inventory of the biodiversity richness of the SDTF indicates that this is a work still much in progress. Wall et al. make the case that our knowledge of the diversity of soil fauna is still at a rudimentary stage, even more so than in most of the other ecosystems of the world. Given the central role that decomposers play in ecosystem functioning, and the adaptive challenges of the seasonally dry system, soil biology should be a lure for future research effort that will be very fruitful.

Looking more generally at insect diversity, Hanson notes that although, in general, insect species richness is greater in wet forests than in dry forests, as is the case for plant species, there are some exceptions—beetle species, bee genera, and nocturnal moths reversing this trend. Interestingly, it appears that a number of species previously restricted to the dry forest are now invading areas of wet tropical forest that have been converted to agricultural use.

Our knowledge about SDTF mammals is certainly incomplete, as Stoner and Timm illustrate. There are indications that mammal diversity in this system is as great as that in the tropical wet forests. There is a surprising lack of information on the physiological ecology of SDTF mammals, which needs attention since the seasonally dry and warm conditions of this system put enormous stress on mammals, and no doubt greater study will reveal new pathways for surviving this metabolically challenging environment. However, there is a limited pool of scientists specializing in this research area. This is indicated by the fact that the overwhelming number of studies of mammals in the SDTF are centered in only four countries—Mexico, Costa Rica, Brazil, and Panama. Large mammals characteristic of this system are among the most threatened in tropical areas because of the enormous loss and fragmentation of this system, although some small-bodied species have actually benefited.

Three chapters in this volume focus on ecosystem processes in the SDTF—Jaramillo et al., Maass and Burgos, and García-Oliva and Jaramillo. The first two of these studies provide comprehensive and long-term data sets for understanding the dynamics of ecosystems that occur in highly variable environments. The normal "snapshot" studies of community and ecosystem dynamics can lead to misleading interpretations of the events controlling system processes through time. At the Chamela long-term research site, the annual growth period can be as short as 3 months and as long as nearly 8 months. There is a legacy effect of these episodic events that may play out over many years in the structuring and functioning of this ecosystem type and on those societies that are dependent on the services they provide, as Maass and Burgos clearly illustrate. Knowing the combinations of rainfall years that precede any given new year gives early warning on how to prepare for the year ahead.

Giraldo and Holbrook discuss the search for ecosystem function predictability at the level of the responses of individual trees. They reveal that some of the triggers that have been implicated for leaf-out time and leaf abscission are mediated through changes in the "plumbing" of the trees relating to changes in the hydraulic conductivity. These findings open up a new area for research on the determinants of the duration of the canopy—certainly a key element in controlling ecosystem resource processing.

The carbon and nutrient dynamics of these systems have certainly been elaborated to a greater extent since the 1990s, although for some parameters, such as net primary productivity, there were no new published data in the interim. The values we do have indicate a net primary productivity of between 11 and 13.5 megagrams per hectare per year with 40 percent of this allocated belowground. Fine-root production is mainly restricted to the upper soil profile, with more than 40 percent occurring in the upper 10 centimeters. Virtually all of the incoming rain is taken up by the vegetation, with very little runoff feeding the stream systems. No doubt there is considerable variation in these values from site to site and from region to region, but we cannot capture this from the scant data available.

A detailed comparison of breeding systems, pollination, and fruit dispersal modes in wet and dry tropical forests by Quesada et al. indicates that both systems are highly dependent on insects for pollination success but with a greater number of obligate outcrossing species in the SDTF. In contrast, plants of the SDTF are less dependent on animal dispersal of their fruits. Those species of the SDTF are particularly vulnerable to forest fragmentation if dependent on pollination by insects of low range extent. There can be pollinator substitutes in disrupted habitats, such as feral honey bees, but these bees are preferentially attracted to those species with mass flowering.

What has been the impact of forest conversion to agricultural landscapes on biological diversity? With such a small fraction of the original SDTF remaining, this is certainly a key question. Chazdon and coworkers explicitly examine such conversions, first in general and then in a specific locality characterized by small plot conversions that utilized limited equipment and pesticide use. In the general case, they recount the findings of countryside biogeography that there is more native diversity remaining than originally predicted in farming systems but that this depends on agricultural practices and time since conversion, concluding that it is important to work with farmers and their land use constraints, legal or otherwise, to optimize the maintenance of biodiversity. Surely this is a crucial adjunct to conservation approaches through reserve formation and further gives the possibility of building resilience to the changing climate, since management is the prevailing practice. In the case of "light" conversion to agriculture, it appears that post-abandonment succession returns the native biodiversity in less than a decade through natural succession. This focus of optimizing "working landscapes" for biodiversity preservation, plus more attention to restoration possibilities, certainly represents a shift from the thinking embodied in the 1995 volume on the SDTF. However, there are also concerns about the loss of genetic diversity due to impacts on species distributions and reproductive systems. Then the dramatic loss of nutrient capital in conversion certainly would have some impact on the time of full recovery of these systems after abandonment—but nonetheless, there is no question that we will need to think more deeply about conservation of the components of the SDTF during the continuing onslaught of population growth, system conversion, and climate change. Detailed studies on the restoration possibilities for various elements of the SDTF are revealing. The oak species studied by Cavender-Bares and colleagues is capable of being restored if a crucial element is provided: aided dispersal of the acorns.

One of the remarkable things about the SDTF is the enormous number of species that have value for society, be it as food, fiber, waxes, medicines, beverages, building and handicraft material, and so forth. Peters in his chapter describes how community-based management of the valuable resource of these forests has resulted in their long-term sustainability. The application of this model depends on the land tenure system and a full development of the economic and social potential of the resources available. Another powerful approach to evaluating the resources from the SDTF that are of benefit to society is the evaluation of their ecosystem services. This work has just begun, as Balvanera and colleagues note. They make the case that the value of the services provided by these systems can exceed that provided by conversion of these forests for many agricultural uses. The challenge is to provide information about all the items in the ecosystem resource base, such as carbon sequestration, and the values of those services to society in the current marketplace, and to foster the development of new markets for system products.

The concluding chapter of the volume looks to the future. All of our knowledge about ecosystem structure and function is built on our knowledge of how they respond to the local climate through time. The SDTF, of course, is remarkable in its adaptability to extreme variation in the availability (timing and amount) of precipitation. Predictions about the future status of the SDTF is uncertain for a number of reasons; for one, the climate models differ considerably in their predictions about rainfall change, the main controller of this system, and further, the vegetation response modeling capacity available at present cannot fully capture even the information that we do have about the physiological responses of the many constituent species of these systems. We are still at the point, however, that land use conversion remains the major threat to the viability of the SDTF.

This volume shows that significant progress has been achieved in our understanding of the ecology and conservation biology of SDTFs since the last synthesis of knowledge. However, it is clear that more needs to be learned about this critical tropical ecosystem to help stem the forces of disruption that are destroying their capacity to continue to deliver ecosystem services to society. The volume identifies some of the most pressing lacunae that need to be filled to ensure the conservation of the remaining areas of SDTF and the restoration of degraded areas. It of course will take more than better scientific knowledge alone to accomplish conservation and restoration of the unique resources provided by the SDTF. It will take the involvement of the many stakeholders who live in these unique landscapes and who are dependent on the sustainability of these systems.

REFERENCES

Abrams, M.D. 1990. Adaptations and responses to drought in *Quercus* species of North America. *Tree Physiology* 7: 227–38.

Achard, F., H.D. Eva, H.J. Stibig, P. Mayaux, J. Gallergo, T. Richards, and J.P. Malingreau. 2002. Determination of deforestation rates of the world's humid tropical forests. *Science* 297: 999–1002.

Acosta, G.J., C. Illsley, and A. Flores. 1998. Producción foliar en *Brahea dulcis* (HBK) Mart., Arecaceae, en Topiltepec, Guerrero. Memorias 7th Congreso Latinoamericano de Bótanica y XIV Congreso de Bótanica, México, D.F.

Adams, D.C., J. Gurevitch, and M.S. Rosenberg. 1997. Resampling tests for meta-analysis of ecological data. *Ecology* 78: 1277–83.

Adis, J. 1988. On the abundance and density of terrestrial arthropods in Central Amazonian dryland forests. *Journal of Tropical Ecology* 4: 19–24.

Adis, J., E. Riberio, J. de-Morias, and E.T.S. Cavalcante. 1989. Vertical distribution and abundance of arthropods from white wand soil of a Neotropical Campinarana forest during the dry season. *Studies on Neotropical Fauna and Environment* 24: 201–11.

Aguilar, J., J. Acoasta, C. Illsley, T. Gómez, J. García, and E. Quintanar. 1997. *La Palma y el Monte. Hacia un Major Manejo Comunitario*. Cuaderno de trabajo. Grupo de Estudios Ambientales-SSS Sanzekan Tinemi. México, D.F.

Aguilar, R., L. Ashworth, L. Galetto, and M.A. Aizen. 2006. Plant reproductive susceptibility to habitat fragmentation: Review and synthesis through a meta-analysis. *Ecology Letters* 9: 968–80.

Aguilar, R., M. Quesada, L. Ashworth, Y. Herrerias-Diego, and J. Lobo. 2008. Genetic consequences of habitat fragmentation in plant populations: Susceptible signals in plant traits and methodological approaches. *Molecular Ecology* 17: 5177–88.

Aide, T.M. 1993. Patterns of leaf development and herbivory in a tropical understory community. *Ecology* 74: 455–66.

Aizen, M.A., L. Ashworth, and L. Galetto. 2002. Reproductive success in fragmented habitats: Do compatibility systems and pollination specialization matter? *Journal of Vegetation Science* 13: 885–92.

Aizen, M.A., and P. Feinsinger. 1994. Habitat fragmentation, native insect pollinators, and feral honey bees in Argentine chaco serrano. *Ecological Applications* 4: 378–92.

——. 2003. Bees not to be? Responses of insect pollinator faunas and flower pollination to habitat fragmentation. In *Disruptions and Variability: The Dynamics of Climate, Human Disturbance and Ecosystems in the Americas*, ed. G.A. Bradshaw, P.A. Marquet, and H.A. Mooney, 111–29. Berlin: Springer-Verlag.

Alatorre Frenk, G. 2000. *La Construcción de una Cultural Gerencial Democrática en las Empresas Forestales Communitarias*. Mexico City: Casa Juan Pablos, Procuraduría Agraria.

Albuquerque, U.P., and L.H.C. Andrade. 2002. Conhecimento botanico tradicional e conservação em uma area de caatinga no estado de Pernambuco, nordeste do Brasil. *Acta Botanica Brasilica* 16: 273–85.

Albuquerque, U.P., L.H.C. Andrade, and A.C.O. Silva. 2005. Use of plant resources in a seasonal dry forest (northeastern Brazil). *Acta Botanica Brasilica* 19: 27–38.

Aldhous, R. 1993. Tropical deforestation: Not just a problem in Amazonia. *Science* 259: 1390.

Aldrich, P.R., and J.L. Hamrick. 1998. Reproductive dominance of pasture trees in a fragmented tropical forest mosaic. *Science* 281: 103–5.

Aldrich, P.R., J.L. Hamrick, P. Chavarriaga, and G. Kochert. 1998. Microsatellite analysis of demographic genetic structure in fragmented populations of the tropical tree *Symphonia globulifera*. *Molecular Ecology* 7: 933–44.

Allen, A.C. 1991. Estudo da biologia reprodutiva de duas espécies florestais (aroeira e gonçalo alves) da região do cerrado. In *Boletim de Pesquisa em Andamento, Cenargen* 2: 1–5.

Allen, E.B., M.E. Allen, L. Egerton-Warburton, L. Corkidi, and A. Gómez-Pompa. 2003. Impacts of early- and late-seral mycorrhizae during restoration in seasonal tropical forest, Mexico. *Ecological Applications* 13: 1701–17.

Allen, E.B., E. Rincón, M.F. Allen, A. Pérez-Jiménez, and P. Huante. 1998. Disturbance and seasonal dynamics of mycorrhizae in a tropical deciduous forest in Mexico. *Biotropica* 30: 261–74.

Almeida, C., and U.P. Albuquerque. 2002. Uso e conservação de plantas e animais medicinais no estado de Pernambuco (Nordeste do Brasil); um estudo de caso. *Interciencia* 27: 276–85.

Alvarez-Santiago, S.A. 2002. Efecto de la perturbación en la interacción micorrízica vesículo-arbuscular en un ecosistema tropical estacional. MS diss., Universidad Nacional Autónoma de México.

Álvarez-Yépiz, J., A. Martínez-Yrízar, A. Búrquez, and C. Lindquist. 2008. Variation in vegetation structure and soil properties related to land-use history of old-growth and secondary tropical dry forests in northwestern Mexico. *Forest Ecology and Management* 256: 355–66.

Alves-Costa, C.P., G.A.B. da Fonseca, and C. Christófaro. 2004. Variation in the diet of the brown-nosed coati (*Nasua nasua*) in southeastern Brazil. *Journal of Mammalogy* 85: 478–82.

Alvim, P.T. 1960. Moisture as a requirement for flowering of coffee. *Science* 132: 354.

Anaya, C.A., F. García-Oliva, and V.J. Jaramillo. 2007. Rainfall and labile carbon availability control litter nitrogen dynamics in a tropical dry forest. *Oecologia* 150: 602–10.

Anbumozhi, V., J. Radhakrishnan, and E. Yamaji. 2005. Impact of riparian buffer zones on water quality and associated management considerations. *Ecological Engineering* 24: 517–23.

Anderson, O.R. 2002. Laboratory and field-based studies of abundances, small-scale patchiness and diversity of gymnamoebae in soils of varying porosity and organic content: Evidence of microbiocoenoses. *Journal of Eukaryotic Microbiology* 49: 17–23.

APG (Angiosperm Phylogeny Group). 2003. An update of the Angiosperm Phylogeny Group classification for the orders and families of flowering plants: APG II. *Botanical Journal of the Linnean Society* 141: 399–436.

Appanah, S. 1985. General flowering in the climax rain forests of south-east Asia. *Journal of Tropical Ecology* 1: 225–40.

Arambiza, E., and M. Painter. 2006. Biodiversity conservation and the quality of life of indigenous people in the Bolivian Chaco. *Human Organization* 65: 20–34.

Argueta, V.A., A. Cano, and M. Rodarte. 1994. *Atlas de las Plantas de la Medicina Tradicional Mexicana*. México: Instituto Nacional Indigenista.

Arias-Cóyotl, E., K.E. Stoner, and A. Casas. 2006. Effectiveness of bats as pollinators of *Stenocereus stellatus* (Cactaceae) in wild, managed in situ, and cultivated populations in La Mixteca Baja, central Mexico. *American Journal of Botany* 93: 1675–83.

Arizmendi, M.C., L. Márquez-Valdelamar, and J.F. Ornelas. 2002. Avifauna de la región de Chamela, Jalisco. In *Historia Natural de Chamela*, ed. F.A. Noguera, J.H. Vera-Rivera, A.N. García-Aldrete, and M. Quesada Avendaño, 297–329. Mexico City: Instituto de Biología, Universidad Nacional Autónoma de México.

Armbruster, W.S., C.S. Keller, M. Matsuki, and T.P. Clausen. 1989. Pollination of *Dalechampia magnoliifolia* (Euphorbiaceae) by male euglossine bees (Apidae: Euglossini). *American Journal of Botany* 76: 1279–85.

Arroyo-Mora, J.P., G.A. Sánchez-Azofeifa, B. Rivard, J.C. Calvo, and D.H. Janzen. 2005. Dynamics in landscape structure and composition for the Chorotega region, Costa Rica from 1960 to 2000. *Agriculture, Ecosystems and Environment* 106: 27–39.

Aschoff, J. 1982. The circadian rhythm of body temperature as a function of body size. In *A Companion to Animal Physiology*, ed. C.R. Taylor, K. Johansen, and L. Bolis, 173–88. New York: Cambridge University Press.

Ashton, P.S., T.J. Givnish, and S. Appanah. 1988. Staggered flowering in the Dipterocarpaceae: New insights into floral induction and the evolution of mast fruiting in the aseasonal tropics. *American Naturalist* 231: 44–66.

Ashworth, L., R. Aguilar, L. Galetto, and M.A. Aizen. 2004. Why do pollination generalist and specialist plant species show similar reproductive susceptibility to habitat fragmentation? *Journal of Ecology* 92: 717–19.

Asquith, N.M., and M. Mejía-Chang. 2005. Mammals, edge effects, and the loss of tropical forest diversity. *Ecology* 86: 379–90.

Atkin, O.K., and D. Macherel. 2008. The crucial role of plant mitochondria in orchestrating drought tolerance. *Annals of Botany* 103: 581–97.

Atkin, O.K., and M.G. Tjoelker. 2003. Thermal acclimation and the dynamic response of plant respiration to temperature. *Trends in Plant Science* 8: 343–51.

ATTA. 2004. Sistema de información de INBio, Animalia. http://www.inbio.ac.cr/bims/k02/p05/c029.htm.

Attum, O. 2007. Can landscape use be among the factors that potentially make some ungulates species more difficult to conserve? *Journal of Arid Environments* 69: 410–19.

Audet, D., and D.W. Thomas. 1997. Facultative hypothermia as a thermoregulatory strategy in the phyllostomid bats, *Carollia perspicillata* and *Sturnira lilium*. *Journal of Comparative Physiology, B: Biochemical, Systemic, and Environmental Physiology* 167: 146–52.

Austin, A.T., L. Yahdjian, J.M. Stark, J. Belnap, A. Porporato, U. Norton, D.A. Ravetta, and S.M. Schaeffer. 2004. Water pulses and biogeochemical cycles in arid and semiarid ecosystems. *Oecologia* 141: 221–35.

Ávila Cabadilla, L.D., K.E. Stoner, and M. Henry. 2009. Composition, structure and diversity of a phyllostomid bat community in different successional stages of a tropical dry forest. *Forest Ecology and Management* 258: 986–96.

Avise, J.C. 1994. *Molecular Markers, Natural History and Evolution*. London: Chapman and Hall.

Ayres, M.P., and M.J. Lombardero. 2000. Assessing the consequences of global change for forest disturbances for herbivores and pathogens. *Science of the Total Environment* 262: 263–86.

Bagchi, S., S.P. Goyal, and K. Sankar. 2003. Niche relationships of an ungulate assemblage in a dry tropical forest. *Journal of Mammalogy* 84: 981–88.

Baker, I.T., L. Prihodko, A.S. Denning, M. Goulden, S. Miller, and H.R. Da Rocha. 2008. Seasonal drought stress in the Amazon: Reconciling models and observations. *Journal of Geophysical Research* 113: G00B01, doi: 10.1029/2007JG000644.

Baker, J.R. 1938. The evolution of breeding seasons. In *Evolution: Essays on Aspects of Evolutionary Biology*, ed. G.R. de Beer, 161–77. Oxford: Oxford University Press.

Baker, J.R., and Z. Baker. 1936. The seasons in a tropical rain-forest (New Hebrides). Part 3. Fruit-bats (Pteropidae). *Journal of the Linnean Society of London Zoology* 40: 123–41.

Baker, T.R., E.N.H. Coronado, O.L. Phillips, J. Martin, M.F. van der Heijden, M. Garcia, and J. Silva-Espejo. 2007. Low stocks of coarse woody debris in a southwest Amazonian forest. *Oecologia* 152: 495–504.

Balvanera, P., E. Lott, G. Segura, C. Seibe, and A. Islas. 2002. Patterns of beta diversity in a Mexican topical dry forest. *Journal of Vegetation Science* 13: 145–58.

Bandelt, H.J., P. Forster, and A. Röhl. 1999. Median-joining networks for inferring intraspecific phylogenies. *Molecular Biology and Evolution* 16: 37–48.

Barberena-Arias, M.F. 2000. A comparison of litter macroarthropod diversity among three habitats in Mona Island. *Acta Cientifica* 14: 51–59.

———. 2008. Single tree species effects on temperature, nutrients and arthropod diversity in litter and humus in the Guánica dry forest. PhD diss., University of Puerto Rico.

Bardgett, R.D., G.W. Yeates, and J.M. Anderson. 2005. Patterns and determinants of soil biological diversity. In *Biological Diversity and Function in Soils*, ed. R.D. Bardgett, M.B. Usher, and D.W. Hopkins, 100–118. Cambridge: Cambridge University Press.

Barrance, A.J., L. Flores, E. Padilla, J.E. Gordon, and K. Schreckenberg. 2003. Trees and farming in the Honduran dry zone of southern Honduras. 1. Campesino tree husbandry practices. *Agroforestry Systems* 59: 97–106.

Barrios, E. 2007. Soil biota, ecosystem services and land productivity. *Ecological Economics* 64: 269–85.

Bartholome, E., A.S. Belward, F. Achard, S. Bartalev, C. Carmona-Moreno, H. Eva, S. Fritz, J. Gre'goire, P. Mayaux, and H.J. Stibig. 2002. *GLC 2000—Global Land Cover Mapping for the year 2000—Project Status November 2002, EUR 20524 EN*. Luxembourg: Publication of the European Commission.

Basile, A.C., J.A. Sertie, P.C.D. Freitas, and A.C. Zanini. 1988. Anti-inflammatory activity of oleoresin from Brazilian copaiba. *Journal of Ethnopharmacology* 22: 101–9.

Bates, B.C., Z.W. Kundzewicz, S. Wu, and J.P. Palutikof, ed. 2008. *Climate Change and Water*. Technical Paper of the Intergovernmental Panel on Climate Change. Geneva: IPCC Secretariat.

Bawa, K.S. 1974. Breeding system of tree species of a lowland tropical community. *Evolution* 28: 85–92.

———. 1990. Plant-pollinator interactions in tropical rainforests. *Annual Review of Ecology and Systematics* 21: 399–422.

Bawa, K.S., S.H. Bullock, D.R. Perry, R.E. Coville, and M.H. Grayum. 1985. Reproductive biology of tropical lowland rain forest trees. 2. Pollination systems. *American Journal of Botany* 72: 346–56.

Bawa, K.S., and P.A. Opler. 1975. Dioecism in tropical forest trees. *Evolution* 29: 167–79.

Bayart, F., and B. Simmen. 2005. Demography, range use, and behavior in black lemurs (*Eulemur macaco macaco*) at Ampasikely, northwest Madagascar. *American Journal of Primatology* 67: 299–312.

Bayma, C. 1958. *Carnaúba*. Rio de Janeiro: Ministerio da Agriculura, Serviço de Informação Agrícola.

Beard, J.S. 1955. The classification of tropical American vegetation types. *Ecology* 36: 89–100.

Bengtsson, J., A. Bergman, M. Olsson, and J. Örberg. 2003. Reserves, resilience and dynamic landscapes. *Ambio* 32: 389–96.

Bennet, E.M., and P. Balvanera. 2007. The future of production systems in a globalized world: Challenges and opportunities in the Americas. *Frontiers in Ecology and the Environment* 5: 191–98.

Bennett, A.F., J.Q. Radford, and A. Haslem. 2006. Properties of land mosaics: Implications for nature conservation in agricultural environments. *Biological Conservation* 133: 250–64.

Bermingham, E., and C. Moritz. 1998. Comparative phylogeography: Concepts and applications. *Molecular Ecology* 7: 367–69.

Bernard, R.T.F., and G.S Cumming. 1997. African bats: Evolution of reproductive patterns and delays. *Quarterly Review of Biology* 72: 253–74.

Betts, R., and N. Golding. 2008. Fire risk in Amazonia due to climate change in the HaD.C.M3 climate model: Potential interactions with deforestation. *Global Biogeochemical Cycles* 22: GB4007, doi: 10.1029/2007GB003166.

Bibby, C.J., N.D. Burgess, D.A. Hill, and S.H. Mustoe. 2000. *Bird Census Techniques*. 2nd ed. London: Academic Press.

Bicca-Marques, J.C., and D.F. Gomes. 2005. Birth seasonality of *Cebus apella* (Platyrrhini, Cebidae) in Brazilian zoos along a latitudinal gradient. *American Journal of Primatology* 65: 141–47.

Bierregaard, R.O., T.E. Lovejoy, V. Kapos, A.A. Dossantos, and R.W. Hutchings. 1992. The biological dynamics of tropical rain-forest fragments. *BioScience* 42: 859–66.

Bignell, D.E., J. Tondoh, L. Dibog, S.P. Huang, F. Moreira, D. Nwaga, B. Pashanasi, E. Guiaraees-Pereira, F.X. Susilo, and M. Swift. 2005. Below-ground biodiversity assessment: Developing a key functional group approach in best-bet alternatives to slash and burn. In *Slash-and-Burn Agriculture: The Search for Alternatives*, ed. C.A. Palm, S.A. Vosti, P.A. Sánchez, and P.J. Ericksen, 119–42. New York: Columbia University Press.

Billings, R.F., and P.J. Schmidtke. 2002. *Central America southern pine beetle/fire management assessment*. U.S. Agency for International Development, Guatemala–Central America Program. USDA Foreign Agricultural Service/International Cooperation and Development. http://www.ccad.ws/.

Birkhead, T.R., and A.P Møller. 1993. Sexual selection and the temporal separation of reproductive events: Sperm storage data from reptiles, birds and mammals. *Biological Journal of the Linnean Society* 50: 295–311.

Blakesley, D., S. Elliot, C. Kuarak, P. Navakitbumrung, S. Zangkum, and V. Anusarnsunthorn. 2002. Propagating framework tree species to restore seasonally dry tropical forest: Implications of seasonal seed dispersal and dormancy. *Forest Ecology and Management* 164: 31–38.

Blauert, J., and S. Zadeck. 1999. *Mediación para la Sustentabilidad, Construyendo Políticas desde las Bases*. Mexico: Plaza y Valdés, British Council, IDS Sussex, CIESAS.

Boag, B., and G.W. Yeates. 1998. Soil nematode biodiversity in terrestrial ecosystems. *Biodiversity and Conservation* 7: 617–30.

Boege, K., and R. Dirzo. 2004. Intraspecific variation in growth, defense and herbivory in *Dialium guianense* (Caesalpiniaceae) mediated by edaphic heterogeneity. *Plant Ecology* 175: 59–69.

Bonaccorso, E., I. Koch, and A.T. Peterson. 2006. Pleistocene fragmentation of Amazon species' ranges. *Diversity and Distributions* 12: 157–64.

Bonaccorso, F.J., and B.K. McNab. 1997. Plasticity of energetics in blossom bats (Pteropodidae): Impact on distribution. *Journal of Mammalogy* 78: 1073–88.

Boonman, A., E. Prinsen, F. Gilmer, U. Schurr, A.J.M. Peeters, L. Voesenek, and T.L. Pons. 2007. Cytokinin import rate as a signal for photosynthetic acclimation to canopy light gradients. *Plant Physiology* 143: 1841–52.

Booth, M.S., J.M. Stark, and E. Rastetter. 2005. Controls on nitrogen cycling in terrestrial ecosystems: A synthetic analysis of literature data. *Ecology* 75: 139–57.

Borchert, R. 1994a. Induction of rehydration and bud break by irrigation or rain in deciduous trees of a tropical dry forest in Costa Rica. *Trees—Structure and Function* 8: 198–204.

——. 1994b. Soil and stem water storage determine phenology and distribution of tropical dry forest trees. *Ecology* 75: 1437–49.

Borchert, R., and G. Rivera. 2001. Photoperiodic control of seasonal development and dormancy in tropical stem-succulent trees. *Tree Physiology* 21: 213–21.

Borchert, R., G. Rivera, and W. Hagnauer. 2002. Modification of vegetative phenology in a tropical semi-deciduous forest by abnormal drought and rain. *Biotropica* 34: 27–39.

Boshier, D.H., J.E. Gordon, and A.J. Barrance. 2004. Prospects for *circa situm* tree conservation in Mesoamerican dry-forest agro-ecosystems. In *Biodiversity Conservation in Costa Rica*, ed. G.W. Frankie, A. Mata, and S.B. Vinson, 210–26. Berkeley: University of California Press.

Boucher, D.H. 1981. Seed predation by mammals and forest dominance by *Quercus oleoides*, a tropical lowland oak. *Oecologia* 49: 409–14.

——. 1983. *Quercus oleoides* (Roble Encino, Oak). In *Costa Rican Natural History*, ed. D.H. Janzen, 319–22. Chicago: University of Chicago Press.

Brando, P., D. Ray, D. Nepstad, G. Cardinot, L. Curran, and R. Oliviera. 2006. Effects of partial throughfall exclusion on the phenology of *Coussarea racemosa* (Rubiaceae) in an east-central Amazon rainforest. *Oecologia* 150: 181–89.

Brando, P.M., D.C. Nepstad, E.A. Davidson, S.E. Trumbore, D. Ray, and P. Camargo. 2008. Drought effects on litterfall, wood production and belowground carbon cycling in an Amazon forest: Results of a throughfall reduction experiment. *Philosophical Transactions of the Royal Society B* 363: 1839–48.

Bray, D.B., L. Merino-Pérez, and D. Barry. 2005. *The Community Forests of Mexico: Managing for a Sustainable Landscape*. Austin: University of Texas Press.

Bray, D.B., L. Merino-Pérez, P. Negreros-Castillo, G. Segura-Warnholtz, J.M. Torres-Rojo, and H.F.M. Vester. 2003. Mexico's community-managed forests as a global model for sustainable landscapes. *Conservation Biology* 17: 672–77.

Bridgewater, S., J.A. Ratter, and J.F. Ribeiro. 2004. Biogeographic patterns, beta-diversity and dominance in the cerrado biome of Brazil. *Biodiversity and Conservation* 13: 2295–318.

Brodribb, T.J., and N.M. Holbrook. 2003. Changes in leaf hydraulic conductance during leaf shedding in seasonally dry tropical forest. *New Phytologist* 158: 295–303.

——. 2005. Leaf physiology does not predict leaf habit: Examples from tropical dry forest. *Trees—Structure and Function* 19: 290–95.

Brodribb, T.J., N.M. Holbrook, E.J. Edwards, and M.V. Gutiérrez. 2003. Relations between stomatal closure, leaf turgor and xylem vulnerability in eight tropical dry forest trees. *Plant Cell and Environment* 26: 443–50.

Brodribb, T.J., N.M. Holbrook, and M.V. Gutiérrez. 2002. Hydraulic and photosynthetic co-ordination in seasonally dry tropical forest trees. *Plant, Cell and Environment* 25: 1435–44.

Brown, A.D., and G.E. Zunino. 1990. Dietary variability in *Cebus apella* in extreme habitats: Evidence for adaptability. *Folia Primatologica* 54: 187–95.

Bruijnzeel, L.A. 1990. *Hydrology of Moist Tropical Forest and Effects of Conversion: A State of Knowledge Review*. Paris: UNESCO-IMP, Humid Tropics Program.

——. 1991. Nutrient input-output budgets of tropical forest ecosystems: A review. *Journal of Tropical Ecology* 7: 1–24.

Bruna, E.M., and W.J. Kress. 2002. Habitat fragmentation and the demographic structure of an Amazonian understory herb (*Heliconia acuminata*). *Conservation Biology* 16: 1256–66.

Bucci, S.J., F.G. Scholz, G. Goldstein, F.C. Meinzer, and L.D.L. Sternberg. 2003. Dynamic changes in hydraulic conductivity in petioles of two savanna tree species: Factors and mechanisms contributing to the refilling of embolized vessels. *Plant, Cell and Environment* 26: 1633–45.

Buchanan-Wollaston, V., S. Earl, E. Harrison, E. Mathas, S. Navabpour, T. Page, and D. Pink. 2003. The molecular analysis of leaf senescence: A genomics approach. *Plant Biotechnology Journal* 1: 3–22.

Buchmann, S.L., and G.P. Nabhan. 1996. *The Forgotten Pollinators*. Washington, DC: Island Press.

Budowski, G. 1987. Living fences in tropical America, a wide-spread agroforestry practice. In *Agroforestry: Realities, Possibilities, and Potentials*, ed. H.L. Gholz, 169–78. Dordrecht, Netherlands: Martinus Nijhoff Publishers.

Bueno, A.A., and J.C. Motta-Junior. 2006. Small mammal selection and functional response in the diet of the maned wolf, *Chrysocyon brachyurus* (Mammalia: Canidae), in southeast Brazil. *Mastozoología Neotropical* 13: 11–19.

Bullock, S.H. 1985. Breeding systems in the flora of a tropical deciduous forest in México. *Biotropica* 17: 287–301.

——. 1995. Plant reproduction in Neotropical dry forests. In *Seasonally Dry Tropical Forests*, ed. S.H. Bullock, H.A. Mooney, and E. Medina, 277–303. Cambridge: Cambridge University Press.

——. 1997. Effects of seasonal rainfall on radial growth in two tropical tree species. *International Journal of Biometeorology* 41: 13–16.

——. 2002. *Cordia elaeagnoides* D.C. (Boraginaceae) Barcino. In *Historia Natural de Chamela*, ed. F.A. Noguera, J.H. Vera-Rivera, A.N. García-Aldrete, and M. Quesada Avendaño, 151–53. Mexico City: Instituto de Biología, Universidad Nacional Autónoma de México.

Bullock, S.H., H.A. Mooney, and E. Medina. 1995. *Seasonally Dry Tropical Forests*. Cambridge: Cambridge University Press.

Bullock, S.H., and J.A. Solís-Magallanes. 1990. Phenology of canopy trees of a tropical deciduous forest in Mexico. *Biotropica* 22: 22–35.

Bumrungsri, S., W. Bumrungsri, and P.A. Racey. 2007. Reproduction in the short-nosed fruit bat in relation to environmental factors. *Journal of Zoology* (London) 272: 73–81.

Burda, H., R.L. Honeycutt, S. Begall, O. Locker-Grutjen, and A. Scharff. 2000. Are naked and common mole-rats eusocial and if so, why? *Behavioral Ecology and Sociobiology* 47: 293–303.

Burgos, A. 1999. Dinámica hidrológica del bosque tropical seco en Chamela, Jalisco, México. MS diss., Universidad Nacional Autónoma de México.

Burgos, A., and M. Maass. 2004. Vegetation change associated with land-use in tropical dry forest areas of western Mexico. *Agriculture, Ecosystems and Environment* 104: 475–81.

Burnham, R.J., and N.L. Carranco. 2004. Miocene winged fruits of *Loxopterygium* (Anacardiaceae) from the Ecuadorian Andes. *American Journal of Botany* 91: 1767–73.

Burns, J.M., and D.H. Janzen. 2001. Biodiversity of pyrrhopygine skipper butterflies (Hesperiidae) in the Area de Conservación Guanacaste, Costa Rica. *Journal of the Lepidopterists' Society* 55: 15–43.

Buschbacher, R., C. Uhl, and E. Serrao. 1988. Abandoned pastures in eastern Amazonia. 2. Nutrient stocks in soil and vegetation. *Journal of Ecology* 76: 682–99.

Bye, R. 1981. Quelites: Ethnoecology of edible greens—Past, present, and future. *Journal of Ethnobiology* 1: 109–23.

——. 1989. Plantas útiles del bosque tropical caducifolio de Chihuahua, México. In *Programa y Resúmenes, Reunión Etnobotánica Ecológica Regional de Selvas Bajas Caducifolias (Bosque tropical caducifolia) y vegetación asociada en México*, ed. L. Cervantes and R. Bye, 19–20. Mexico City: Instituto de Biología, Universidad Nacional Autónoma de México.

——. 1995. Ethnobotany of the Mexican tropical dry forests. In *Seasonally Dry Tropical Forests*, ed. S.H. Bullock, H.A. Mooney, and E. Medina, 423–38. Cambridge: Cambridge University Press.

Cabin, R.J., S.J. Weller, C.H. Lorence, T.W. Flynn, A.K. Sakai, D. Sandquist, and L.J. Hadway. 2002. Effects of long-term ungulate exclusion and recent alien species control on the preservation and restoration of a Hawaiian tropical dry forest. *Conservation Biology* 14: 439–53.

Cabrera, A.L., and A. Willink. 1980. *Biogeografía de América Latina*. Washington, DC: Organización de Estados Americanos.

Caetano, S., L. Nusbaumer, and Y. Naciri. 2008. Chloroplast and microsatellite markers in *Astronium urundeuva* and close species (Anacardiaceae): Toward the definition of a species complex? *Candollea* 63: 115–30.

Caetano, S., D. Prado, R.T. Pennington, S. Beck, A. Oliveira-Filho, R. Spichiger, and Y. Naciri. 2008. The history of seasonally dry tropical forest in eastern South America: Inferences from the genetic structure of the tree *Astronium urundeuva* (Anacardiaceae). *Molecular Ecology* 17: 3147–59.

Caetano, S., P. Silveira, R. Spichiger, and Y. Naciri-Graven. 2005. Identification of microsatellite markers in a Neotropical seasonally dry forest tree, *Astronium urundeuva* (Anacardiaceae). *Molecular Ecology Notes* 5: 21–23.

Cairns, M.A., P.K. Haggerty, R. Alvarez, B.H.J. De Jong, and I. Olmsted. 2000. Tropical Mexico's recent land-use change: A region's contribution to the global carbon cycle. *Ecological Applications* 10: 1426–41.

Cairns, M.A., I. Olmsted, J. Granados, and J. Argaez. 2003. Composition and aboveground tree biomass of a dry semi-evergreen forest of Mexico's Yucatan peninsula. *Forest Ecology and Management* 186: 125–32.

Caldwell, M.M., T.E. Dawson, and J.H. Richards. 1998. Hydraulic lift: Consequences of water efflux from the roots of plants. *Oecologia* 113: 151–61.

Calvo-Alvarado, J., B. McLennan, A. Sánchez-Azofeifa, and T. Garvin. 2009. Deforestation and forest restoration in Guanacaste, Costa Rica: Putting conservation policies in context. *Forest Ecology and Management* 258: 931–40.

Cámara Forestal de Bolivia. 2006. Certificación Forestal. *Sector Forestal*. Santa Cruz, Bolivia.

Campbell, D.G., D.C. Daly, G.T. Prance, and U.B. Maciel. 1986. Quantitative ecological inventory of terra firma and várzea tropical forest on the Rio Xingu, Brazilian Amazon. *Brittonia* 38: 369–93.

Campo, J., V.J. Jaramillo, and J.M. Maass. 1998. Pulses of soil P availability in a tropical dry forest: Effects of seasonality and level of wetting. *Oecologia* 115: 167–72.

Campo, J., J.M. Maass, V.J. Jaramillo, and A. Martínez-Yrízar. 2000. Calcium, potassium, and magnesium cycling in a Mexican tropical dry forest ecosystem. *Biogeochemistry* 49: 21–36.

Campo, J., J.M. Maass, V.J. Jaramillo, A. Martínez-Yrízar, and J. Sarukhán. 2001. Phosphorus cycling in a Mexican tropical dry forest ecosystem. *Biogeochemistry* 53: 161–79.

Campo, J., and C. Vázquez-Yanes. 2004. Effects of nutrient limitation on aboveground carbon dynamics during tropical dry forest regeneration in Yucatán, México. *Ecosystems* 7: 311–19.

Canadell, J., R.B. Jackson, J.B. Ehleringer, H.A. Mooney, O. Sala, and E.D. Schulze. 1996. Maximum rooting depth of vegetation types at the global scale. *Oecologia* 108: 583–95.

Cantú-Salazar, L., M.G. Hidalgo-Mihart, C.A. López-González, and A. González-Romero. 2005. Diet and food resource use by the pygmy skunk (*Spilogale pygmaea*) in the tropical dry forest of Chamela, Mexico. *Journal of Zoology* (London) 267: 283–89.

Cárdenas, I., and J. Campo. 2007. Foliar nitrogen and phosphorus resorption and decomposition in the nitrogen-fixing tree *Lysiloma microphyllum* in primary and secondary seasonally tropical dry forests in Mexico. *Journal of Tropical Ecology* 23: 107–13.

Cárdenas, G., A. Cardozo, G. Castro, E. Comiskey, F. Estela, R. Greenberg, M. Ibrahim, E.J. Molina, E. Murgueitio, and L.G. Naranjo. 2000. *Recovering Paradise: Making Pasturelands Productive for People and Biodiversity*. Proceedings of the First International Workshop on Bird Conservation in Livestock Production Systems, Airlie Conference Center, Virginia, April 13, 2000.

Caron, H., S. Dumas, G. Marque, C. Messier, E. Bandou, R.J. Petit, and A. Kremer. 2000. Spatial and temporal distribution of chloroplast DNA polymorphism in a tropical tree species. *Molecular Ecology* 9: 1089–98.

Casas, A., and J. Caballero. 1996. Traditional management and morphological variation in *Leucaena esculenta* (Moc et Sessé ex A. DC.) Benth. (Leguminosae: Mimosoideae) in the Mixtec region of Guerrero, Mexico. *Economic Botany* 50: 167–81.

Casas, A., B. Pickersgill, J. Caballero, and A. Valient-Banuet. 1997. Ethnobotany and the process of domestication of the xoconochtli, *Stenocereus stellatus* (Cactaeae), in the Tehuacán Valley and La Mixteca Baja, Mexico. *Economic Botany* 51: 279–92.

Casas, A., M.C. Vázquez, J.L. Viveros, and J. Caballero. 2006. Plant management among the Nahua and the Mixtec in the Balsas River Basin: An ethnobotanical approach to the study of plant domestication. *Human Ecology* 24: 455–78.

Cascante, A., M. Quesada, J.A. Lobo, and E.J. Fuchs. 2002. Effects of dry tropical forest fragmentation on the reproductive success and genetic structure of the tree *Samanea saman*. *Conservation Biology* 16:137–47.

Castellanos, J., V.J. Jaramillo, R.L Sanford Jr., and J.B. Kauffman. 2001. Slash-and-burn effects on fine root biomass and productivity in a tropical dry forest ecosystems in México. *Forest Ecology and Management* 148: 41–51.

Castellanos, J., J. M. Maass, and J. Kummerow. 1991. Root biomass of a dry deciduous tropical forest in Mexico. *Plant and Soil* 131: 225–28.

Castillo, A., A. Magaña, A. Pujadas, L. Martínez, and C. Godínez. 2005. Undertanding the interaction of rural people with ecosystems: A case study in a tropical dry forest of Mexico. *Ecosystems* 8: 630–43.

Castillo, A., and V.M. Toledo. 2000. Applying ecology in the third world: The case of Mexico. *BioScience* 50: 66–76.

Castillo, G. 1993. Contribución al conocimiento sobre *Brahea dulcis* (HBK) Mart. en la Region Mixteca de Cárdenas, Oaxaca. PhD diss., Universidad Autónoma de Chapingo, México.

Cavender-Bares, J. 2007. Inter- and intraspecific variation in PSII sensitivity to chilling and freezing stress in the American live oak species complex corresponds to latitude. *Photosynthesis Research*, doi: 10.1007/s11120-007-9215-8.

Cavender-Bares, J., and F.A. Bazzaz. 2000. Changes in drought response strategies with ontogeny in *Quercus rubra*: Implications for scaling from seedlings to mature trees. *Oecologia* 124: 8–18.

Cavender-Bares, J., and N.M. Holbrook. 2001. Hydraulic properties and freezing-induced cavitation in sympatric evergreen and deciduous oaks with contrasting habitats. *Plant, Cell and Environment* 24: 1243–56.

Cavender-Bares, J., K. Kitajima, and F.A. Bazzaz. 2004. Multiple trait associations in relation to habitat differentiation among 17 Florida oak species. *Ecological Monographs* 74: 635–62.

Cavender-Bares, J., L. Sack, and J. Savage. 2007. Atmospheric and soil drought reduce nocturnal conductance in live oaks. *Tree Physiology* 27: 611–20.

Cavers, S., C. Navarro, and A.J. Lowe. 2003. Chloroplast DNA phylogeography reveals colonization history of a Neotropical tree, *Cedrela odorata* L., in Mesoamerica. *Molecular Ecology* 12: 1451–60.

CCAD. 2002. *Nature, People and Well Being: Mesoamerica Fact Book*. Paris: CCAD–World Bank.

Ceballos, G. 1989. Population and community ecology of small mammals from tropical deciduous and arroyo forests in western México. PhD diss., University of Arizona.

——. 1990. Comparative natural history of small mammals from tropical forests in western Mexico. *Journal of Mammalogy* 71: 263–66.

——. 1995. Vertebrate diversity, ecology, and conservation in Neotropical dry forests. In *Seasonally Dry Tropical Forests*, ed. S.H. Bullock, H.A. Mooney, and E. Medina, 195–220. Cambridge: Cambridge University Press.

Ceballos, G., and A. García. 1995. Conserving Neotropical biodiversity: The role of dry forests in western Mexico. *Conservation Biology* 9: 1349–53.

Ceballos, G., A. García, L. Martínez, E. Espinosa, J. Bezaury, and R. Dirzo. 2010. *Diversidad, amenazas y áreas prioritarias para la conservación de las selvas secas del oeste de México*. CONABIO—UNAM, Mexico City.

Ceballos, G., and A. Miranda. 1986. *Los mamíferos de Chamela, Jalisco: Manual de campo*. Mexico City: Instituto de Biología, Universidad Nacional Autónoma de México.

——. 2000. *Guía de Campo de los Mamíferos de la Costa de Jalisco*. México, D.F.: Fundación Ecológica de Cuixmala A.C.

Ceballos, G., and D. Navarro L. 1991. Diversity and conservation of Mexican mammals. In *Latin American Mammalogy: History, Biodiversity, and Conservation*, ed. M.A. Mares and D.J. Schmidly, 167–98. Norman: University of Oklahoma Press.

Ceballos, G., and G. Oliva. 2005. *Los Mamíferos Silvestres de México*. México, D.F.: Comisión Nacional para el Conocimiento y Uso de la Biodiversidad and Fondo de Cultura Económica.

Cerri, C.C., B. Volkhoff, and F. Andreaux. 1991. Nature and behavior of organic matter in soil under natural forest, and after deforestation burning and cultivation, near Manaus. *Forest Ecology and Management* 38: 247–57.

Cervantes, L. 1988. Intercepción de lluvia por el dosel en una comunidad tropical. Ingeniería Hidráulica en México. *Segunda Época* 2: 38–43.

Chacón, M., and C.A. Harvey. 2006. Live fences and landscape connectivity in a Neotropical agricultural landscape. *Agroforestry Systems* 68: 15–26.

Challenger, A. 1998. *Utilización y Conservación de los Ecosistemas Terrestres de México. Pasado, Presente y Futuro*. México, D.F.: Comisión Nacional para el Conocimiento y Uso de la Biodiversidad.

Chamberlain, J.R., C.E. Hughes, and N.W. Galwey. 1996. Patterns of variation in the *Leucaena shannonii* alliance (Leguminosae: Mimosoideae). *Silvae Genetica* 45: 1–7.

Chapman, C.A. 1987. Flexibility in diets of three species of Costa Rican primates. *Folia Primatologica* 49: 90–105.

Chapotin, S.M., N.M. Holbrook, S.R. Morse, and M.V. Gutiérrez. 2003. Water relations of tropical dry forest flowers: Pathways for water entry and the role of extracellular polysaccharides. *Plant Cell and Environment* 26: 623–30.

Chapotin, S.M., J.H. Razanameharizaka, and N.M. Holbrook. 2006. Baobab trees (*Adansonia*) in Madagascar use stored water to flush new leaves but not to support stomatal opening before the rainy season. *New Phytologist* 169: 549–59.

Chase, M.R., C. Moller, R. Kesseli, and K.S. Bawa. 1996. Distant gene flow in tropical trees. *Science* 383: 398–99.

Chauvel, A., M. Grimaldi, E. Barrios, E. Blanchart, T. Desjardins, M. Sarrazin, and P. Lavelle. 1999. Pasture damage by an Amazonian earthworm. *Nature* 398: 32–33.

Chazdon, R.L., S. Careaga, C. Webb, and O. Vargas. 2003. Community and phylogenetic structure of reproductive traits of woody species in wet tropical forests. *Ecological Monographs* 73: 331–48.

Chazdon, R.L., C.A. Harvey, O. Komar, D.M. Griffith, B.G. Ferguson, M. Martínez-Ramos, H. Morales, R. Nigh, L. Soto-Pinto, M. van Breugel, and S.M. Philpott. 2009. Beyond reserves: A research agenda for conserving biodiversity in human-modified tropical landscapes. *Biotropica* 41: 142–53.

Chazdon, R.L., S.G. Letcher, M. van Breugel, M. Martínez-Ramos, F. Bongers, and B. Finegan. 2007. Rates of change in tree communities of secondary Neotropical forests following major disturbances. *Philosophical Transactions of the Royal Society B* 362: 273–89.

Chazdon, R.L., C.A. Peres, D. Dent, D. Sheil, A.E. Lugo, D. Lamb, N.E. Stork, and S. Miller. 2009. The potential for species conservation in tropical secondary forests. *Conservation Biology* 23(6): 1406–17.

Chen, J., and J.M. Stark. 2000. Plant species effects and carbon and nitrogen cycling in a sagebrush-crested wheatgrass soil. *Soil Biology and Biogeochemistry* 32: 47–57.

Chiappy-Jhones, C., V. Rico-Gray, L. Gama, and L. Giddings. 2001. Floristic affinities between the Yucatan peninsula and some karstic areas of Cuba. *Journal of Biogeography* 28: 535–42.

Chiarello, A.G. 1999. Effects of fragmentation of the Atlantic forest on mammal communities in south-eastern Brazil. *Biological Conservation* 89: 71–82.

——. 2003. Primates of the Brazilian Atlantic Forest: The influence of forest fragmentation on survival. In *Primates in Fragments: Ecology and Conservation*, ed. L.K. Marsh, 99–121. New York: Kluwer Academic/Plenum Publishers.

Chibnick, M. 2003. *Crafting Tradition: The Making and Marketing of Oaxacan Wood Carvings*. Austin: University of Texas Press.

Chivaura-Mususa, C., B. Campbell, and W. Kenyon. 2000. The value of mature trees in arable fields in the smallholder sector, Zimbabwe. *Ecological Economics* 33: 395–400.

Choat, B., M.C. Ball, J.G. Luly, C.F. Donnelly, and J.A.M. Holtum. 2006. Seasonal patterns of leaf gas exchange and water relations in dry rain forest trees of contrasting leaf phenology. *Tree Physiology* 26: 657–64.

Choat, B., M.C. Ball, J.G. Luly, and J.A.M. Holtum. 2005. Hydraulic architecture of deciduous and evergreen dry rainforest tree species from north-eastern Australia. *Trees—Structure and Function* 19: 305–11.

Chomentowski, M., B. Salas, and D.S. Skole. 1994. Landsat Pathfinder project advances deforestation mapping. *GIS World* 7: 34–38.

Christensen, J.H., B. Hewitson, A. Busuioc, A. Chen, X. Gao, I. Held, R. Jones, R.K. Kolli, W.T. Kwon, R. Laprise, et al. 2007. Regional climate projections. In *Climate Change 2007: The Physical Science Basis*. Contribution of Working Group 1 to the 4th Assessment Report of the Intergovernmental Panel on Climate Change, ed. D. Solomon et al., 847–940. Cambridge: Cambridge University Press.

Cintrón, B.B., and A.E. Lugo. 1990. Litterfall in a subtropical dry forest: Guánica Puerto Rico. *Acta Científica* (San Juan) 4: 37–49.

Clark, D.A., S. Brown, D.W. Kicklighter, J.Q. Chambers, J.R. Thomlinson, J. Ni, and E. Holland. 2001. Net primary production in tropical forests: An evaluation and synthesis of existing field data. *Ecological Applications* 11: 371–84.

Cleland, E.E., I. Chuine, A. Menzel, H.A. Mooney, and M.D. Schwartz. 2007. Shifting plant phenology in response to global change. *Trends in Ecology and Evolution* 22: 357–65.

Clutton-Brock, T.H., S.D. Albon, and F.E. Guinness. 1989. Fitness costs of gestation and lactation in wild mammals. *Nature* 337: 260–62.

Coburn, D.K., and F. Geiser. 1998. Seasonal changes in energetics and torpor patterns in the subtropical blossom-bat *Syconycteris australis* (Megachiroptera). *Oecologia* 113: 467–73.

Cochrane, M.A., and W.F. Laurance. 2002. Fire as a large-scale edge effect in Amazonian forests. *Journal of Tropical Ecology* 18: 311–25.

Cockrum, E.L. 1969. Migration in the guano bat, *Tadarida brasiliensis*. *Miscellaneous Publication, University of Kansas Museum of Natural History* 51: 303–36.

——. 1991. Seasonal distribution of northwestern populations of the long-nosed bats, *Leptonycteris sanborni* family Phyllostomidae. *Anales del Instituto de Biología de la Universidad Nacional Autónoma de México, Serie Zoológica* 62: 181–202.

Codex Mendoza. 1978. Reproduction from the manuscript in the Bodleian Library. Miller Graphics.

Coelho, L.M., J.L. Pereira, N. Calegario, and M. Lopes. 2006. Análise Longitudinal dos Preços do Carvão Vegetal, no Estado de Minas Gerais. *árvore* 30: 429–38.

Cohen, A.C., and J.D. Pinto. 1977. An evaluation of xeric adaptiveness of several species of blister beetles (Meloidae). *Annals of the Entomological Society of America* 70: 741–49.

Coleman, D.C., D.A. Crossley Jr., and P.F. Hendrix. 2004. *Fundamentals of Soil Ecology*. 2nd ed. San Diego: Elsevier Academic Press.

Collevatti, R.G., D. Gratapaglia, and J.D. Hay. 2001. Population genetic structure of the endangered tropical tree species *Caryocar brasiliense* based on variability at microsatellite loci. *Molecular Ecology* 7:1275–81.

——. 2003. Evidences for multiple maternal lineages of *Caryocar brasiliense* populations in the Brazilian Cerrado based on the analysis of chloroplast DNA sequences and microsatellite haplotype variation. *Molecular Ecology* 12: 105–15.

Collins, M., S.F.B. Tett, and C. Cooper. 2001. The internal climate variability of a HaD.C.M3, a version of the Hadley Centre coupled model without flux adjustments. *Climate Dynamics* 17: 61–81.

Colón, S.M., and A.E. Lugo. 2006. Recovery of a subtropical dry forest after abandonment of different land uses. *Biotropica* 38: 354–64.

Colunga-García Marín, P., J. Coello-Coello, L. Espejo-Peniche, et al. 1993. *Agave* studies in Yucatan, Mexico. I. Past and present germplasm diversity and uses. *Economic Botany* 47: 328–340.

Comisión Nacional de Áreas Naturales Protegidas. 2006. Estudio previo justificativo para el establecimiento del Área Natural Protegida Reserva de la Biosfera "Zicuirán-Infiernillo." México D.F: Comisión Nacional de Áreas Naturales Protegidas.

Comps, B., D. Gömöry, J. Letouzey, B. Thiébaut, and R.J. Petit. 2001. Diverging trends between heterozygosity and allelic richness during postglacial colonization in the European beech. *Genetics* 157: 389–97.

Connor, E.F. 1986. The role of Pleistocene forest refugia in the evolution and biogeography of tropical biotas. *Trends in Ecology and Evolution* 1: 165–68.

Cordeiro, J.N., and H.F. Howe. 2001. Low recruitment of trees dispersed by animals in African forest fragments. *Conservation Biology* 15: 1733–41.

Costa, M.H., S.N.M. Yanagi, P. Souza, A. Ribeiro, and E.J.P. Rocha. 2007. Climate change in Amazonia caused by soybean cropland expansion, as compared to change caused by pastureland expansion. *Geophysical Research Letters* 34: L07706, doi: 10.1029/2007GL029271.

Cox, P.M., R.A. Betts, C.D. Jones, S.A. Spall, and I.J. Totterdell. 2000. Acceleration of global warming due to carbon-cycle feedbacks in a coupled climate model. *Nature* 408: 184–87.

Cox, P.M., P.P. Harris, C. Huntingford, R.A. Betts, M. Collins, C.D. Jones, T.E. Jupp, J.A. Marengo, and C.A. Nobre. 2008. Increasing risk of Amazonian drought due to decreasing aerosol pollution. *Nature* 453: 212–15.

Cramer, W., A. Bondeau, F.I. Woodward, I.C. Prentice, R. Betts, V. Brovkin, P.M. Cox, V. Fischer, J.A. Foley, A.D. Friend, C. Kucharik, M.R. Lomas, N. Ramankutty, S. Sitch, B. Smith, A. White, and C. Young-Molling. 2001. Global responses of terrestrial ecosystem structure and function to CO_2 and climate change: Results from six dynamic global vegetation models. *Global Change Biology* 7: 357–73.

Creighton, G.K., and A.L. Gardner. 2007. Genus *Thylamys* Gray, 1843. In *Mammals of South America*. Vol. 1. *Marsupials, Xenarthrans, Shrews, and Bats*, ed. A.L. Gardner, 107–17. Chicago: University of Chicago Press.

Croat, T.B. 1979. The sexuality of the Barro Colorado Island flora (Panama). *Phytologia* 42: 319–48.

Cruz-Neto, A.P., and A.S. Abe. 1997. Taxa metabólica e termoregulacão no morçego nectarívoro, *Glossophaga soricina* (Chiroptera, Phyllostomidae). *Revista Brasileira de Biologia* 57: 203–9.

Cuartas-Hernández, S., and J. Núñez-Farfán. 2006. The genetic structure of tropical understory herb *Dieffenbachia seguine* L. before and after forest fragmentation. *Evolutionary Ecology Research* 8: 1061–75.

Cuevas, E. 1995. Biology of the belowground system of tropical dry forests. In *Seasonally Dry Tropical Forests*, ed. S.H. Bullock, H.A. Mooney, and E. Medina, 362–78. Cambridge: Cambridge University Press.

Culik, M., and D. Zeppelini-Filho. 2003. Diversity and distribution of Collembola (Arthropoda: Hexapoda) of Brazil. *Biodiversity and Conservation* 12: 1119–43.

Cullen, L., Jr., R.E. Bodmer, and C.V. Valladares Pádua. 2000. Effects of hunting in habitat fragments of the Atlantic forests, Brazil. *Biological Conservation* 95: 49–56.

Daily, G.C., G. Ceballos, J. Pacheco, G. Suzán, and A. Sánchez-Azofeifa. 2003. Countryside biogeography of Neotropical mammals: Conservation opportunities in agricultural landscapes of Costa Rica. *Conservation Biology* 17: 1814–26.

Daily, G.C., P.R. Ehrlich, and G.A. Sánchez-Azofeifa. 2001. Countryside biogeography: Use of human-dominated habitats by the avifauna of southern Costa Rica. *Ecological Applications* 11: 1–13.

Daily, G.C., P.A. Matson, and P.M. Vitousek. 1997. Ecosystem services supplied by soil. In *Nature's Services: Societal Dependence on Natural Ecosystems*, ed. G.C. Daily, 113–32. Washington, DC: Island Press.

Dalal, R.C., and C. Mayer. 1987. Long-term trends in fertility of soils under continuous cultivation and cereal cropping in Southern Queensland. 3. Distribution and kinetics of soil organic matter in particle-size and density fraction. *Australian Journal of Soil Research* 25: 83–93.

Dalecky, A., S. Chauvet, S. Ringuet, O. Classens, J. Judas, M. Larue, and J.F. Cosson. 2002. Large mammals on small islands: Short term effects of forest fragmentation on the large mammal fauna in French Guiana. *Revue d'Ecologie la Terre et la Vie* Supplement 8: 145–64.

Dall, S.R.X., and I.L. Boyd. 2004. Evolution of mammals: Lactation helps mothers to cope with unreliable food supplies. *Proceedings of the Royal Society B* 271: 2049–57.

Dalle, S.P., S. De Blois, J. Caballero, and T. Johns. 2006. Integrating analyses of local land-use regulations, cultural perceptions and land-use/land cover data for assessing the success of community-based conservation. *Forest Ecology and Management* 222: 370–83.

Dalling, J.W., H.C. Muller-Landau, S.J. Wright, and S.P. Hubbell. 2002. Role of dispersal in the recruitment limitation of Neotropical pioneer species. *Journal of Ecology* 90: 714–27.

da Silva, A.P., Jr., and A.R. Mendes Ponetes. 2008. The effect of a mega-fragmentation process on large mammal assemblages in the highly-threatened Pernambuco Endemism Centre, north-eastern Brazil. *Biodiversity and Conservation* 17: 1455–64.

Daubenmire, R. 1972a. Ecology of *Hyparrhenia rufa* (Nees) in derived savanna in northwestern Costa Rica. *Journal of Applied Ecology* 9: 11–23.

——. 1972b. Phenology and other characteristics of tropical semi-deciduous forest in northeastern Costa Rica. *Journal of Ecology* 60: 147–70.

Dausmann, K.H., J. Glos, J.U. Ganzhorn, and G. Heldmaier. 2004. Hibernation in a tropical primate. *Nature* 429: 825–26.

——. 2005. Hibernation in the tropics: Lessons from a primate. *Journal of Comparative Physiology, B: Biochemical, Systemic, and Environmental Physiology* 175: 147–55.

Davidson, E.A., C.J.R. de Carvalho, I.C.G. Vieira, R.D. Figueiredo, P. Moutinho, F.Y. Ishida, M.T.P. dos Santos, J.B. Guerrero, K. Kalif, and R.T. Saba. 2004. Nitrogen and phosphorus limitation of biomass growth in a tropical secondary forest. *Ecological Applications* 14: 150–63.

Davidson, E.A., and I.A. Janssens. 2006. Temperature sensitivity of soil carbon decomposition and feedbacks to climate change. *Nature* 440: 165–73.

Davidson, E.A., and W. Kingerlee. 1997. A global inventory of nitric oxide emissions from soils. *Nutrient Cycling Agroecosystem* 48: 37–50.

Davidson, E.A., P.A. Matson, P.M. Vitousek, R. Riley, K. Dunkin, G. García-Méndez, and J.M. Maass. 1993. Processes regulating soil emissions of NO and N^2O in a seasonally dry tropical forest. *Ecology* 74: 130–39.

Davidson, E.A., D.C. Nepstad, F.Y. Ishida, and P.M. Brando. 2008. Effects of an experimental drought and recovery on soil emissions of carbon dioxide, methane, nitrous oxide, and nitric oxide in a moist tropical forest. *Global Change Biology* 14: 2582–90.

Davidson, E.A., L.V. Verchot, J.H. Cattanio, I.L. Ackerman, and J.E.M. Carvalho. 2000. Effects of soil water content on soil respiration in forests and cattle pastures of eastern Amazonia. *Biogeochemistry* 48: 53–69.

Davis, M.B. 1983. Quaternary history of deciduous forests of eastern North America and Europe. *Annals of the Missouri Botanical Garden* 70: 550–63.

Dawson, W.R. 1955. The relation of oxygen consumption to temperature in desert rodents. *Journal of Mammalogy* 36: 543–53.

Dayanandan, S., J. Dole, K. Bawa, and R. Kessel. 1999. Population structure delineated with microsatellite markers in fragmented populations of a tropical tree, *Carapa guianensis* (Meliaceae). *Molecular Ecology* 8:1585–92.

Decaëns, T., J.J. Jiménez, E. Barros, A. Chauvel, E. Blanchart, C. Fragoso, and P. Lavelle. 2004. Soil macrofaunal composition in permanent pastures derived from tropical forest or savanna. *Agriculture, Ecosystem and Environment* 103: 2004.

DeClerck, F., R. Chazdon, K. Holl, J. Milder, B. Finegan, A. Martinez-Salinas, P. Imbach, L. Canet, and Z. Ramos. 2010. Biodiversity conservation in human-modified landscapes of Mesoamerica: Past, present and future. *Biological Conservation* doi: 10.1016/j.biocon.2010.03.026.

Delaney, M., S. Brown, A.E. Lugo, A. Torres-Lezama, and N. Bello-Quintero. 1997. The distribution of organic carbon in major components of forests located in five life zones of Venezuela. *Journal of Tropical Ecology* 13: 697–708.

——. 1998. The quantity and turnover of dead wood in permanent forest plots in five life zones of Venezuela. *Biotropica* 30: 2–11.

Delgado-Salinas, A., R. Bibler, and M. Lavín. 2006. Phylogeny of the genus *Phaseolus* (Leguminosae): A recent diversification in an ancient landscape. *Systematic Botany* 31: 779–91.

Delorme, M., and D.W. Thomas. 1996. Nitrogen and energy requirements of the short-tailed fruit bat (*Carollia perspicillata*): Fruit bats are not nitrogen constrained. *Journal of Comparative Physiology, B: Biochemical, Systematic, and Environmental Physiology* 166: 427–34.

Demesure, B., B. Comps, and R.J. Petit. 1996. Chloroplast DNA phylogeography of the common beech (*Fagus sylvatica* L.) in Europe. *Evolution* 50: 2515–20.

de Nettancourt, D. 2001. *Incompatibility and Incongruity in Wild and Cultivated Plants*. Berlin: Springer-Verlag.

Denslow, J.S., A.L. Uowolo, and R.F. Hughes. 2006. Limitations to seedling establishment in a mesic Hawaiian forest. *Oecologia* 148: 118–28.

DeVries, P.J. 1987. *The Butterflies of Costa Rica and Their Natural History: Papilionidae, Pieridae, Nymphalidae*. Princeton, NJ: Princeton University Press.

———. 1997. *The Butterflies of Costa Rica and Their Natural History*. Vol. 2. *Riodinidae*. Princeton, NJ: Princeton University Press.

Díaz, S., J. Fargione, F.S. Chapin III, and D. Tilman. 2006. Biodiversity loss threatens human well-being. *PLoS Biology* 4: e277, doi: 10.1371/journal.pbio.0040277.

Díaz, S.V. 1997. Dinámica de nitrógeno y de fósforo en la hojarasca de una selva baja caducifolia en Chamela, Jalisco, México. BS diss., Universidad Nacional Autónoma de México.

Di Bella, C.M., E.G. Jobbaggy, J.M. Paruelo, and S. Pinnock. 2006. Continental fire density patterns in South America. *Global Ecology and Biogeography* 15:192–99.

Di Bitetti, M.S., and C.H. Janson. 2000. When will the stork arrive? Patterns of birth seasonality in Neotropical primates. *American Journal of Primatology* 50: 109–30.

Dick, C.W. 2001. Genetic rescue of remnant tropical trees by an alien pollinator. *Proceedings of the Royal Society of London B* 268: 2391–96.

Dick, C.W., K. Abdul-Salim, and E. Bermingham. 2003. Molecular systematic analysis reveals cryptic tertiary diversification of a widespread tropical rain forest tree. *American Naturalist* 162: 691–703.

Dickie, I.A., and P.B. Reich. 2005. Ectomycorrhizal fungal communities at forest edges. *Journal of Ecology* 93: 244–55.

Dickie, I.A., S.A. Schnitzer, P.B. Reich, and S.E. Hobbie. 2007. Is oak establishment in old-fields and savanna openings context dependent? *Journal of Ecology* 95: 309–20.

Dirzo, R., and K. Boege. 2009. Patterns of herbivory and defense in tropical dry and rain forests. In *Tropical Forest Community Ecology*, ed. W. Carson and S. Schnitzer, 63–78. Chichester, UK: Wiley-Blackwell.

Dirzo, R., and C. Domínguez. 1995. Plant-herbivore interactions in Mesoamerican tropical dry forest. In *Seasonally Dry Tropical Forests*, ed. S. H. Bullock, H.A. Mooney, and E. Medina, 304–25. Cambridge: Cambridge University Press.

Dirzo, R., and A. Miranda. 1991. Altered patterns of herbivory and diversity in the forest understory: A case study of the possible consequences of contemporary defaunation. In *Plant-Animal Interactions: Evolutionary Ecology in Tropical and Temperate Regions*, ed. P.W. Price, T.M. Lewinsohn, G.W. Fernandes, and W.W. Benson, 273–87. New York: Wiley and Sons.

Dirzo, R, and P.H. Raven. 2003. Global state of biodiversity and loss. *Annual Review of Environment and Resources* 28: 137–67.

Domec, J.C., A. Noormets, J.S. King, G. Sun, S.G. McNulty, M.J. Gavazzi, J.L. Boggs, and E.A. Treasure. 2009. Decoupling the influence of leaf and root hydraulic conductances on stomatal conductance and its sensitivity to vapour pressure deficit as soil dries in a drained loblolly pine plantation. *Plant Cell and Environment* doi: 10.1111/j.1365-3040.2009.01981.x.

Domec, J.C., F.G. Scholz, S.J. Bucci, F.C. Meinzer, G. Goldstein, and R. Villalobos-Vega. 2006. Diurnal and seasonal variation in root xylem embolism in Neotropical savanna woody species: Impact on stomatal control of plant water status. *Plant Cell and Environment* 29: 26–35.

Domingues, T.F., L.A. Martinelli, and J.R. Ehleringer. 2007. Ecophysiological traits of plant functional groups in forest and pasture ecosystems from eastern Amazonia, Brazil. *Plant Ecology* 193: 101–12.

Donald, P.F. 2004. Biodiversity impacts of some agricultural commodity production systems. *Conservation Biology* 18: 17–38.

Dosch, J.J., C.J. Peterson, and B.L. Haines. 2007. Seed rain during initial colonization of abandoned pastures in the premontane wet forest zone of southern Costa Rica. *Journal of Tropical Ecology* 23: 151–59.

Duncan, F.D., B. Drasnov, and M. McMaster. 2002. Metabolic rate and respiratory gas-exchange patterns in tenebrionid beetles from the Negev Highlands, Israel. *Journal of Experimental Biology* 205: 791–98.

Dunn, P., L. DeBano, and G. Eberlein. 1979. Effects of burning on chaparral soils. 2. Soil microbes and nitrogen mineralization. *Soil Science Society of America Journal* 43: 509–14.

Dunn, R.R. 2004. Recovery of faunal communities during tropical forest regeneration. *Conservation Biology* 18: 302–9.

Durigan, G. 2006. Observations on the southern Cerrados and their relationship with the core area. In *Neotropical Savannas and Seasonally Dry Forests: Plant Diversity, Biogeography, and Conservation*, ed. R.T. Pennington, G.P. Lewis, and J.A. Ratter, 67–78. Boca Raton, FL: CRC Press.

Dutech, C., L. Maggia, C. Tardy, H.I. Joly, and P. Jarne. 2003. Tracking a genetic signal of extinction-recolonization events in a Neotropical tree species: *Vouacapoua americana* Aublet in French Guiana. *Evolution* 57: 2753–64.

Dwyer, J.D. 1951. The Central America, West Indian, and South American species of *Copaifera* (Caesalpiniaceae). *Brittonnia* 7: 143–72.

Eaton, J.M., and D. Lawrence. 2006. Woody debris stocks and fluxes during succession in a dry tropical forest. *Forest Ecology and Management* 232: 46–55.

Ehrlich, D., E.F. Lambin, and J.P. Malingreau. 1997. Biomass burning and broad scale land-cover changes in western Africa. *Remote Sensing of the Environment* 61: 201–9.

Ellingson, L.J., J.B. Kauffman, D.L. Cummings, R.L. Sanford Jr., and V.J. Jaramillo. 2000. Soil N dynamics associated with deforestation, biomass burning, and pasture conversion in a Mexican tropical dry forest. *Forest Ecology and Management* 137: 41–51.

Elliott, S., P. Navakitbumrung, C. Kuarak, S. Zangkum, V. Anusarnsunthorn, and D. Blakesley. 2003. Selecting framework trees for restoring seasonally dry tropical forests in northern Thailand based on field performance. *Forest Ecology and Management* 184: 177–91.

Engelbrecht, B.J., L.S. Comita, R. Condit, T.A. Kursar, M.T. Tyree, B.L. Turner, and S.P. Hubbell. 2007. Drought sensitivity shapes species distribution patterns in tropical forests. *Nature* 447: 80–82.

Ennos, R.A., W.T. Sinclair, X.S. Hu, and A. Langdon. 1999. Using organelle markers to elucidate the history, ecology and evolution of plant populations. In *Molecular*

Systematics and Plant Evolution, ed. P.M. Hollingsworth, R.M.Bateman, and R.J. Gornall, 1–19. London: Taylor and Francis.

Enquist, B.J., and J. Leffler. 2001. Long term tree ring chronologies from sympatric tropical dry-forest trees: Individualistic responses to climatic variation. *Journal of Tropical Ecology* 17: 41–60.

Esquivel, H., M. Ibrahim, C.A. Harvey, C. Villanueva, T. Benjamin, and F.L. Sinclair. 2003. Árboles dispersos en potreros de fincas ganaderas en un ecosistema seco de Costa Rica. *Agroforestería en las Américas* 10: 24–29.

Esquivel, M.J., C.A. Harvey, B. Finegan, F. Casanoves, and C. Skarpe. 2008. Effects of pasture management on the natural regeneration of Neotropical trees. *Journal of Applied Ecology* 45: 371–80.

Estrada, A. 2008. Fragmentación de la selva y agrosistemas como reservoirios de conservación de la fauna silvestre en Los Tuxtlas, México. In *Evaluacion y Conservación de Biodiversidad en Paisajes Fragmentados de Mesoamérica*, ed. C.A. Harvey and J.C. Saénz, 327–48. Heredia, Costa Rica: INBio.

Estrada, A., and R. Coates-Estrada. 1996. Tropical rain forest fragmentation and wild populations of primates at Los Tuxtlas, Mexico. *International Journal of Primatology* 17: 759–83.

——. 2001. Bat species richness in live fences and in corridors of residual rain forest vegetation at Los Tuxtlas, Mexico. *Ecography* 24: 94–102.

Estrada, A., P. Commarano, and R. Coates-Estrada. 2000. Bird species richness in vegetation fences and in strips of residual rain forest vegetation at Los Tuxtlas, Mexico. *Biodiversity and Conservation* 9: 1399–1416.

Eva, H.D., A.S. Belward, E.E. De Miranda, C.M. Di Bella, V. Gond, O. Huber, S. Jones, M. Sgrenzaroli, and S. Fritz. 2004. A land cover map of South America. *Global Change Biology* 10: 731–44.

Eva, H.D., and E.F. Lambin. 2000. Fires and land-cover change in the tropics: A remote sensing analysis at the landscape scale. *Journal of Biogeography* 27: 765–76.

Ewel, J. 1980. Tropical succession: Manifold routes to maturity. *Biotropica* 12: 2–7.

Excoffier, L., G. Laval, and S. Schneider. 2005. Arlequin, ver. 3.0: An integrated software package for population genetics data analysis. *Evolutionary Bioinformatics Online* 1: 47–50.

Excoffier, L., P.E. Smouse, and J.M. Quattro. 1992. Analysis of molecular variance inferred from metric distances among DNA haplotypes: Application to human mitochondrial DNA restriction data. *Genetics* 131: 479–91.

Fahrig, L. 2003. Effects of habitat fragmentation on biodiversity. *Annual Review of Ecology Evolution and Systematics* 34: 487–515.

Fajardo, L., V. González, J. Nassar, P. Lacabana, C.A. Portillo, F. Carrasquet, and J.P. Rodríguez. 2005. Tropical dry forests of Venezuela: Characterization and current conservation status. *Biotropica* 37: 531–46.

FAO. 2000. *Non-Wood Forest Products Study for Mexico, Cuba, and South America*. Forest Resources Assessment Working Paper 11. Rome: Food and Agriculture Organization.

——. 2003. *Non-Wood News*. No. 10. Wood and Non-Wood Products Utilization Branch, FAO Forest Products Division. Rome: Food and Agriculture Organization.

——. 2005. *Global Forest Resources Assessment 2005*. Rome: Food and Agriculture Organization.

Fautin, R.W. 1946. Biotic communities of the Northern Desert Shrub Biome in western Utah. *Ecological Monographs* 16: 251–310.

Fayenuwo, J.O., and L.B. Halstead. 1974. Breeding cycle of straw-colored fruit bat, *Eidolon helvum*, at Ile-Ife, Nigeria. *Journal of Mammalogy* 55: 453–54.

FCBC (Fundación para la Conservación del Bosque Chiquitano). 2003. Hacia un turismo sostenible. In *Programa Piloto de Educación Ambiental para la Conservación del Bosque Seco Chiquitano, Cerrado y Pantanal Boliviano*. Santa Cruz, Bolivia: FCBC.

Fearnside, P.M. 2001. Soybean cultivation as a threat to the environment in Brazil. *Environmental Conservation* 28: 23–38.

Fensham, R.J., R.J. Fairfax, D.W. Butler, and D.M.J.S. Bowman. 2003. Effects of fire and drought in a tropical eucalypt savanna colonized by rain forest. *Journal of Biogeography* 30: 1405–14.

Fietz, J., and J.U. Ganzhorn. 1999. Feeding ecology of the hibernating primate *Cheirogaleus medius*: How does it get so fat? *Oecologia* 121: 157–64.

Filoso, S., L.A. Martinelli, R.W. Howarth, E.W. Boyer, and F. Dentener. 2006. Human activities changing the nitrogen cycle in Brazil. *Biogeochemistry* 79: 61–89.

Finegan, B., and R. Nasi. 2004. The biodiversity and conservation potential of shifting cultivation landscapes. In *Agroforestry and Biodiversity Conservation in Tropical Landscapes*, ed. G. Schroth, G.A.B. da Fonseca, C.A. Harvey, C. Gason, H.L. Vasconcelos, and A.M.N. Izaac, 153–97. Washington, DC: Island Press.

Fischer, C.R., D.P. Janos, D.A. Perry, and R.G. Linderman. 1994. Mycorrhiza inoculum potentials in tropical secondary succession. *Biotropica* 26: 369–77.

Fisher, J.B., G. Angeles, F.W. Ewers, and J. López Portillo. 1997. Survey of root pressure in tropical vines and woody species. *International Journal of Plant Sciences* 158: 44–50.

Fisher, R., M. Williams, R. Lobo do Vale, A.C. Lola da Costa, and P. Meir. 2006. Evidence from Amazonian forests is consistent with isohydric control of leaf water potential. *Plant, Cell, and Environment* 29: 151–65.

Fisher, R.A., M. Williams, A.L. Costa, Y. Malhi, R.F. da Costa, S. Almeida, and P. Meir. 2007. The response of an Eastern Amazonian rain forest to drought stress: Results and modelling analyses from a throughfall exclusion experiment. *Global Change Biology* 13: 2361–78.

Fisher, R.A., M. Williams, M. de Lourdes Ruivo, A.L. Costa, and P. Meir. 2008. Evaluating climatic and soil water controls on evapotranspiration at two Amazonian rainforest sites. *Agricultural and Forest Meteorology* 148: 850–61.

Fleming, T.H. 1971a. *Artibeus jamaicensis*: Delayed embryonic development in a Neotropical bat. *Science* 171: 402–4.

——. 1971b. Population ecology of three species of Neotropical rodents. *Miscellaneous Publications, Museum of Zoology, University of Michigan* 143: 1–77.

——. 1974. The population ecology of two species of Costa Rican heteromyid rodents. *Ecology* 55: 493–510.

——. 1977. Response of two species of tropical heteromyid rodents to reduced food and water availability. *Journal of Mammalogy* 58: 102–6.

——. 1988. *The Short-tailed Fruit Bat: A Study in Plant-Animal Interactions*. Chicago: University of Chicago Press.

Fleming, T.H., E.T. Hooper, and D.E. Wilson. 1972. Three Central American bat communities: Structure, reproductive cycles, and movement patterns. *Ecology* 53: 555–69.

Fleming, T.H., R.A. Núñez, and L.S.L. Sternberg. 1993. Seasonal changes in the diets of migrant and non-migrant nectarivorous bats as revealed by carbon stable isotope analysis. *Oecologia* 94: 72–75.

Flexas, J., J. Bota, J. Galmés, H. Medrano, and M. Ribas-Carbo 2006. Keeping a positive carbon balance under adverse conditions: Responses of photosynthesis and respiration to water stress. *Physiologia Planterum* 127: 343–52.

Foerster, C.R., and C. Vaughan. 2002. Home range, habitat use, and activity of Baird's tapir in Costa Rica. *Biotropica* 34: 423–37.

Foissner, W. 1995. Tropical protozoan diversity: 80 ciliate species (Protozoa, Ciliophora) in a soil sample from a tropical dry forest of Costa Rica, with descriptions of 4 new genera and 7 new species. *Archiv Fur Protistenkunde* 145: 37–79.

Foissner, W. 1997a. Global soil climate (Protozoa: Ciliophora) diversity: A probability based approach using large sample collections from Africa, Australia and Antarctica. *Biodiversity Conservation* 6: 1627–38.

——. 1997b. Soil ciliates (Protozoa: Ciliophora) from evergreen rain forests of Australia, South America and Costa Rica: Diversity and description of new species. *Biology and Fertility of Soils* 25: 317–39.

Fragoso, C., G.G. Brown, J.C. Patron, E. Blanchart, P. Lavelle, B. Pashanasi, B. Senapati, and T. Kumar. 1997. Agricultural intensification, soil biodiversity and agroecosystem function in the tropics: The role of earthworms. *Applied Soil Ecology* 6: 17–35.

Fragoso, C., S.W. James, and S. Borges. 1995. Native earthworms of the north Neotropical region: Current status and controversies. In *Earthworm Ecology and Biogeography in North America*, ed. P.F. Hendrix, 67–116. Boca Raton, FL: Lewis Publishers.

Fragoso, J.M.V. 1998. Home range and movement patterns of white-lipped peccary (*Tayassu pecari*) herds in the northern Brazilian Amazon. *Biotropica* 30: 458–69.

Franco, A.C., M. Bustamante, L.S. Caldas, G. Goldstein, F.C. Meinzer, A.R. Kozovits, P. Rundel, and V.T.R. Coradin. 2005. Leaf functional traits of Neotropical savanna trees in relation to seasonal water deficit. *Trees—Structure and Function* 19: 326–35.

Frankie, G.W., H.G. Baker, and P.A. Opler. 1974. Comparative phenological studies of trees in tropical wet and dry forests in lowlands of Costa Rica. *Journal of Ecology* 62: 881–919.

Frankie, G.W., W.A Haber, S.B. Vinson, K.S. Bawa, P.S. Ronchi, and N. Zamora. 2004. Flowering phenology and pollination systems diversity in the seasonal dry forest. In *Biodiversity Conservation in Costa Rica: Learning the Lessons in a Seasonal Dry Forest*, ed. G.W. Frankie, A. Mata, and S.B. Vinson, 17–29. Berkeley: University of California Press.

Frankie, G.W., S.B. Vinson, M.A. Rizzardi, T.L. Griswold, S. O'Keefe, and R.R. Snelling. 1998. Diversity and abundance of bees visiting a mass flowering tree species in disturbed seasonal dry forest, Costa Rica. *Journal of the Kansas Entomological Society* 70: 281–96.

Fredericksen, T.S., and B. Mostacedo. 2000. Regeneration of timber species following selection logging in a Bolivian tropical dry forest. *Forest Ecology and Management* 131: 47–55.

Friedlingstein, P., P.M. Cox, R.A. Betts, L. Bopp, W. Von Bloh, V. Brovkin, P. Cadule, S. Doney, M. Eby, I. Fung, G. Bala, J. John, C.D. Jones, F. Joos, T. Kato, M. Kawamiya, W. Knorr, K. Lindsay, H.D. Matthews, T. Raddatz, P. Rayner, C. Reick, E. Roeckner, K.G. Schnitzler, R. Schnur, K. Strassman, A.J. Weaver, C. Yoshikawa, and N. Zeng. 2006. Climate-carbon cycle feedback analysis: Results from the C4MIP model intercomparison. *Journal of Climate* 19: 3337–53.

Fuchs, E.J., J.A. Lobo, and M. Quesada. 2003. Effects of forest fragmentation and flowering phenology on the reproductive success and mating patterns on the tropical dry forest tree, *Pachira quinata* (Bombacaceae). *Conservation Biology* 17: 149–57.

Furley, P. 1992. Edaphic changes at the forest-savanna boundary with particular reference to the Neotropics. In *Nature and Dynamics of Forest-Savanna Boundaries*, ed. P. Furley, J. Ratter, and J. Proctor, 91–117. London: Chapman and Hall.

Fyllas, N.M., S. Patiño, T.R. Baker, G. Bielefeld Nardoto, L.A. Martinelli, C.A. Quesada, R. Paiva, M. Schwarz, V. Horna, L.M. Mercado, A. Santos, L. Arroyo, E.M. Jiménez, F.J. Luizão, D.A. Neill, N. Silva, A. Prieto, A. Rudas, M. Silviera, I.C.G. Vieira, G. López-González, Y. Malhi, O.L. Phillips, and J. Lloyd. 2009. Basin-wide variations in foliar properties of Amazonian forest: Phylogeny, soils and climate. *Biogeosciences Discussions* 6: 3707–69.

Galetti, M., A. Keuroghlian, L. Hanada, and M. Inez Morato. 2001. Frugivory and seed dispersal by the lowland tapir (*Tapirus terrestris*) in southeast Brazil. *Biotropica* 33: 723–26.

Galetti, M., and F. Pedroni. 1994. Seasonal diet of capuchin monkeys (*Cebus apella*) in a semideciduous forest in South-East Brazil. *Journal of Tropical Ecology* 10: 27–39.

Galicia, L., J. López-Blanco, A.E. Zarco-Arista, V. Filip, and F. García-Oliva. 1999. The relationship between solar radiation interception and soil water content in a tropical deciduous forest in Mexico. *Catena* 36: 153–64.

Galicia, L., A.E. Zarco-Arista, K. Mendoza-Robles, J.L. Palacio-Prieto, and A. García-Romero. 2008. Land use/cover, landforms and fragmentation patterns in a tropical dry forest in the southern Pacific region of Mexico. *Singapore Journal of Tropical Geography* 29: 137–54.

Galindo-González, J., and V.J. Sosa. 2003. Frugivorous bats in isolated trees and riparian vegetation associated with human-made pastures in a fragmented tropical landscape. *Southwestern Naturalist* 48: 579–89.

Gallardo, J.F., and M.I. González. 2004. Sequestration of carbon in Spanish deciduous oak forest. *Advances in GeoEcology* 37: 341–51.

Gallopin, G. 2006. Linkages between vulnerability, resilience, and adaptive capacity. *Journal of Global Environmental Change* 16: 293–303

Gan, S.S., and R.M. Amasino. 1995. Inhibition of leaf senescence by autoregulated production of cytokinin. *Science* 270: 1986–88.

———. 1996 Cytokinins in plant senescence: From spray and pray to clone and play. *Bioessays* 18: 557–65.

García-Méndez, G., J.M. Maass, P.A. Matson, and P.M. Vitousek. 1991. Nitrogen transformations and nitrous oxide flux in a tropical deciduous forest in Mexico. *Oecologia* 88: 362–66.

García-Oliva, F., A. Camou, and J.M. Maass. 2002. El clima de la región central de la costa del pacífico mexicano. In *Historia Natural de Chamela*, ed. F.A. Noguera, J.H. Vera-Rivera, A.N. García-Aldrete, and M. Quesada Avendaño, 3–10. Mexico City: Instituto de Biología, Universidad Nacional Autónoma de México.

García-Oliva, F., I. Casar, P. Morales, and J.M. Maass. 1994. Forest-to-pasture conversion influences on soil organic carbon dynamics in a tropical deciduous forest. *Oecologia* 99: 392–96.

García-Oliva, F., E. Ezcurra, and L. Galicia. 1991. Patterns of rainfall distribution in the central Pacific coast of Mexico. *Geografiska Annaler Series A, Physical Geography* 73: 179–86.

García-Oliva, F., J.F. Gallardo, N.M. Montaño, and P. Islas. 2006. Soil carbon and nitrogen dynamics followed by a forest-to-pasture conversion in western Mexico. *Agroforestry Systems* 66: 93–100.

García-Oliva, F., G. Hernández, and J.F. Gallardo. 2006. Comparison of ecosystem C pools in three forests in Spain and Latin America. *Annals of Forest Science* 63: 519–23.

García-Oliva, F., and J.M. Maass. 1998. Efecto de la transformación de la selva a pradera sobre la dinámica de los nutrientes en un ecosistema tropical estacional en México. *Boletín de la Sociedad Botánica de México* 62: 39–48.

García-Oliva, F., J.M. Maass, and L. Galicia. 1995. Rainstorm analysis and rainfall erosivity of a seasonal tropical region with a strong cyclonic influence on the Pacific coast of Mexico. *Journal of Applied Meteorology* 34: 2491–98.

García-Oliva, F., M. Oliva, and B. Sveshtarova. 2004. Effect of soil macroaggregates crushing on C mineralization in a tropical deciduous forest ecosystem. *Plant and Soil* 259: 297–305.

García-Oliva, F., R.L. Sanford Jr., and E. Kelly. 1999a. Effect of burning of tropical deciduous forest soil in Mexico on the microbial degradation of organic matter. *Plant and Soil* 206: 29–36.

——. 1999b. Effects of slash-and-burn management on soil aggregate organic C and N in a tropical deciduous forest. *Geoderma* 88: 1–12.

García-Oliva, F., B. Sveshtarova, and M. Oliva. 2003. Seasonal effects on soil organic carbon dynamics in a tropical deciduous forest ecosystem in western Mexico. *Journal of Tropical Ecology* 19: 179–88.

Gardner, T.A., J. Barlow, R. Chazdon, R.M. Ewers, C.A. Harvey, C.A. Peres, and N.S. Sodhi. 2009. Prospects for tropical forest biodiversity in a human-modified world. *Ecology Letters* 12: 561–82.

Garwood, N.C. 1983. Seed germination in a seasonal tropical forest in Panama: A community study. *Ecological Monographs* 53: 158–81.

Gauld, I. 1997. The Ichneumonidae of Costa Rica, 2. *Memoirs of the American Entomological Institute* 57: 1–485.

——. 2000. The Ichneumonidae of Costa Rica, 3. *Memoirs of the American Entomological Institute* 63: 1–453.

Gauld, I., C. Godoy, R. Sithole, and G.J. Ugalde. 2002. The Ichneumonidae of Costa Rica, 4. *Memoirs of the American Entomological Institute* 66: 1–768.

Gauld, I.D., and D.H. Janzen. 2004. The systematics and biology of the Costa Rican species of parasitic wasps in the *Thyreodon* genus group (Hymenoptera: Ichneumonidae). *Zoological Journal of the Linnean Society* 141: 297–51.

Gauld, I.D., J.A.G. Ugalde, and P. Hanson. 1998. Guía de los Pimplinae de Costa Rica (Hymenoptera: Ichneumonidae). *Revista Biología Tropical* 46: 1–189.

Geiselman, C.K., S.A. Mori, and F. Blanchard. (2002 onwards). Database of Neotropical bat/plant interactions. http://www.nybg.org/botany/tlobova/mori/batsplants/database/dbase_frameset.htm.

Geiser, F., R.L. Drury, B.M. McAllan, and D.H. Wang. 2003. Effects of temperature acclimation on maximum heat production, thermal tolerance, and torpor in a marsupial. *Journal of Comparative Physiology, B: Biochemical, Systemic, and Environmental Physiology* 173: 437–42.

Geßler, A., H.M. Duarte, A.C. Franco, U. Lüttge, E.A. de Mattos, M. Nahm, P.J.F.P. Rodrigues, F.R. Scarano, and H. Rennenberg. 2005. Ecophysiology of selected tree species in different plant communities at the periphery of the Atlantic Forest of SE-Brazil III: Three legume trees in a semi-deciduous dry forest. *Trees* 19: 523–30.

Genet, J.A., K.S. Genet, T.M. Burton, P.G. Murphy, and A.E. Lugo. 2001. Response of termite community and wood decomposition rates to habitat fragmentation in a subtropical dry forest. *Tropical Ecology* 42: 35–49.

Gentry, A.H. 1982a. Neotropical floristic diversity: Phylogeographical connections between Central and South America, Pleistocene climatic fluctuations, or an accident of the Andean orogeny? *Annals of the Missouri Botanical Garden* 69: 557–93.

——. 1982b. Patterns of Neotropical plant species diversity. *Evolutionary Biology* 15: 1–84.

——. 1982c. Phytogeographic patterns as evidence for a Chocó refuge. In *Biological Diversification in the Tropics*, ed. G.T. Prance, 112–36. New York: Columbia University Press.

——. 1988. Tree species richness of upper Amazonian forests. *Proceedings of the U.S. National Academy of Sciences* 85: 156–59.

——. 1991. The distribution and evolution of climbing plants. In *The biology of vines*, ed. F.G.E. Putz and H.A. Mooney, 3–42. Cambridge: Cambridge University Press.

——. 1995. Diversity and floristic composition of Neotropical dry forests. In *Seasonally Dry Tropical Forests*, ed. S.H. Bullock, H.A. Mooney, and E. Medina, 147–94. Cambridge: Cambridge University Press.

Gentry, A.H., and C.H. Dodson. 1987. Diversity and biogeography of Neotropical vascular epiphytes. *Annals of the Missouri Botanical Garden* 74: 205–33.

Gentry, A.H., and J. Terborgh. 1990. Composition and dynamics of the Cocha Cashu mature floodplain forest. In *Four Neotropical Rainforests*, ed. A.H. Gentry, 542–64. New Haven: Yale University Press.

Gerhardt, K. 1993. Tree seedling development in tropical dry abandoned pasture and secondary forest in Costa Rica. *Journal of Vegetation Science* 4: 95–102.

Gessel, S.P., D.W. Cole, D. Johnson, and J. Turner. 1980. The nutrient cycles of two Costa Rican forests. In *Progress in Ecology*, ed. V.P. Agarwal and V.K. Sharma, 23–44. New Delhi: Today and Tomorrow's Printers and Publishers.

Ghazoul, J. 2005. Pollen and seed dispersal among dispersed plants. *Biological Reviews* 80: 413–43.

Ghazoul, J., K.A. Liston, and T.J.B. Boyle. 1998. Disturbance-induced density-dependent seed set in *Shorea siamensis* (Dipterocarpaceae), a tropical forest tree. *Journal of Ecology* 86: 462–73.

Ghazoul, J., and M. McLeish. 2001. Reproductive ecology of forest trees in logged and fragmented habitats in Thailand and Costa Rica. *Plant Ecology* 153: 335–45.

Ghazoul, J., and R.U. Shaanker. 2004. Sex in space: Pollination among spatially isolated plants. *Biotropica* 36: 128–30.

Giardina, C.P., R.L. Sanford Jr., and I.C. Dockersmith. 2000a. Changes in soil phosphorus and nitrogen during slash-and-burn clearing of a dry tropical forest. *Soil Science Society of America Journal* 64: 399–405.

——. 2000b. The effects of slash burning on ecosystems nutrients during the land preparation phase of shifting cultivation. *Plant and Soil* 220: 247–60.

Gigord, L., F. Picot, and J. Shycoff. 1999. Effects of habitat fragmentation on *Dombeya acutangula* (Sterculiaceae), a native tree on La Réunion (Indian Ocean). *Biological Conservation* 88: 43–51.

Giller, K.E., M.H. Beare, P. Lavelle, A.M.N. Izac, and M.J. Swift. 1997. Agricultural intensification, soil biodiversity and agroecosystem function. *Applied Soil Ecology* 6: 3–16.

Gillespie, T. 1999. Life history characteristics and rarity of woody plants in tropical dry forest fragments of Central America. *Journal of Tropical Ecology* 15: 637–49.

Gillespie, T.W., A. Grijalva, and C.N. Farris. 2000. Diversity, composition, and structure of tropical dry forests in Central America. *Plant Ecology* 147: 3–47.

Giri, C., and C. Jenkins. 2005. Land cover mapping of Greater Mesoamerica using MODIS data. *Canadian Journal of Remote Sensing* 31: 274–82.

Givnish, T.J. 2002. Adaptive significance of evergreen vs. deciduous leaves: Solving the triple paradox. *Silva Fennica* 36: 703–43.

Glander, K.E. 1978. Howling monkey feeding behavior and plant secondary compounds: A study of strategies. In *The Ecology of Arboreal Folivores*, ed. G.G. Montgomery, 561–73. Washington, DC: Smithsonian Institution Press.

——. 1981. Feeding patterns in mantled howling monkeys. In *Foraging Behavior: Ecological, Ethological and Psychological Approaches*, ed. A.C. Kamil and T.D. Sargent, 231–57. New York: Garland Press.

——. 1983. *Alouatta palliata* (Congo, howling monkey, howler monkey). In *Costa Rican Natural History*, ed. D.H. Janzen, 448–49. Chicago: University of Chicago Press.

Gliwicz, J. 1984. Population dynamics of the spiny rat *Proechimys semispinosus* on Orchid Island (Panama). *Biotropica* 16: 73–78.

Golding, N., and R. Betts. 2008. Fire risk in Amazonia due to climate change in the HaD.C.M3 climate model: Potential interactions with deforestation. *Global Biogeochemical Cycles* 22: GB4007, doi: 10.1029/2007GB003166.

Goldstein, G., F.C. Meinzer, S.J. Bucci, F.G. Scholz, A.C. Franco, and W.A. Hoffmann. 2008. Water economy of Neotropical savanna trees: Six paradigms revisited. *Tree Physiology* 28: 395–404.

González, G. 2002. Soil organisms and litter decomposition. In *Modern Trends in Applied Terrestrial Ecology*, ed. R.S. Ambasht and N.K. Ambasht, 315–30. New York: Kluwer Academic/Plenum Publishers.

González, G., R. Ley, S.K. Schmidt, X. Zou, and T.R. Seastedt. 2001. Soil ecological interactions: Comparisons between tropical and subalpine forests. *Oecologia* 128: 549–56.

González, G., and T.R. Seastedt. 2000. Comparison of the abundance and composition of litter fauna in tropical and subalpine forests. *Pedobiologia* 44: 549–56.

——. 2001. Soil fauna and plant litter decomposition in tropical and subalpine forests. *Ecology* 82: 955–64.

González-Astorga, J., and J. Núñez-Farfán. 2001. Effect of habitat fragmentation on the genetic structure of the narrow endemic *Brongniartia vazquezii*. *Evolutionary Ecology Research* 3: 861–72.

González-Iturbe, J.A., I. Olmsted, and F. Tun-Dzul. 2002. Tropical dry forest recovery after long term Henequen (sisal, Agave foucroydes Lem.) plantation in northern Yucatan, Mexico. *Forest Ecology and Management* 167: 67–82.

González-Ruiz, T. 1997. Efecto de la humedad del suelo en la biomasa microbiana de un ecosistema tropical estacional. BS diss., Universidad Nacional Autónoma de México.

González-Ruiz, T., V.J. Jaramillo, J.J. Peña-Cabriales, and A. Flores. 2008. Nodulation dynamics and nodule activity in leguminous tree species of a Mexican tropical dry forest. *Journal of Tropical Ecology* 24: 107–10.

González-Zamora, A., V. Arroyo-Rodríguez, Ó.M. Chaves, S. Sánchez-López, K.E. Stoner, and P. Riba-Hernández. 2009. Diet of spider monkeys (*Ateles geoffroyi*) in Mesoamerica: Current knowledge and future directions. *American Journal of Primatology* 71: 8–20.

Good-Avila, S.V., and A.G. Stephenson. 2002. The inheritance of modifiers conferring self-fertility in the self-incompatible perennial, *Camapnula rapunculoides* L. (Campanulaceae). *Evolution* 56: 263–272.

———. 2003. Parental effects in a partially self-incompatible herb *Campanula rapunculoides* (Campanulaceae): Influence of variation in the strength of self-incompatibility on seed set and progeny performance. *American Naturalist* 161: 615–30.

Gordon, J.E., E. Bowen-Jones, and M.A. González. 2006. What determines dry forest conservation in Mesoamerica? Opportunism and pragmatism in Mexican and Nicaraguan protected areas. In *Neotropical Savannas and Seasonally Dry Forests: Plant Diversity, Biogeography, and Conservation*, ed. R.T. Pennington, G.P. Lewis, and J.A. Ratter, 343–57. Boca Raton, FL: CRC Press.

Gordon, J.E., W.D. Hawthorne, A. Reyes-García, G. Sandoval, and A.J. Barrance. 2004. Assessing landscapes: A case study of tree and shrub diversity in the seasonally dry tropical forests of Oaxaca, Mexico and southern Honduras. *Biological Conservation* 117: 429–42.

Gorresen, P.M., and M.R. Willig. 2004. Landscape responses of bats to habitat fragmentation in Atlantic Forest of Paraguay. *Journal of Mammalogy* 85: 688–97.

Gottsberger, G., and I. Silberbauer-Gottsberger. 1983. Dispersal and distribution in the cerrado vegetation of Brazil. *Sonderbaende des Naturwissenschaftlichen Vereins in Hamburg* 7: 315–52.

Gould, W.A., G. González, and G. Carrero-Rivera. 2006. Structure and composition of vegetation along an elevational gradient in Puerto Rico. *Journal of Vegetation Science* 17: 653–64.

Goulden, M.L. 1996. Carbon assimilation and water-use efficiency by neighboring Mediterranean-climate oaks that differ in water access. *Tree Physiology* 16: 417–24.

Goulden, M.L., J.W. Munger, S.M. Fan, B.C. Daube, and S.C. Wofsy. 1996. Exchange of carbon dioxide by a deciduous forest: Response to interannual climate variability. *Science* 271: 1576–78.

Gove, A.D., J.D. Majer, and V. Rico-Gray. 2005. Methods for conservation outside of formal reserve systems: The case of ants in the seasonally dry tropics of Veracruz, Mexico. *Biological Conservation* 126: 328–38.

Grassman, L.I., M.E. Tewes, N.J. Silvy, and K. Kreetiyutanont. 2005. Spatial organization and diet of the leopard cat (*Prionailurus bengalensis*) in north-central Thailand. *Journal of Zoology* 266: 45–54.

Grau, H.R., N.I. Gasparri, and T.M. Aide. 2005. Agriculture expansion and deforestation in seasonally dry forests of north-west Argentina. *Environmental Conservation* 32: 140–48.

Greenberg, R., P. Bichier, and J. Sterling. 1997. Acacia, cattle, and migratory birds in southeastern Mexico. *Biological Conservation* 80: 235–47.

Greene, D.F., M. Quesada, and I.C. Calogeropoulus. 2008. Dispersal of seeds by the tropical breeze. *Ecology* 89: 118–25.

Griffith, D.M. 2000. Agroforestry: A refuge for tropical biodiversity after fire. *Conservation Biology* 14: 325–26.

Griscom, H.P., B.W. Griscom, and M.S. Ashton. 2009. Forest regeneration from pasture in the dry tropics of Panama: Effects of cattle, exotic grass, and forested riparia. *Restoration Ecology* 17: 117–26.

Griswold, T., F.D. Parker, and P.E. Hanson. 1995. The bees (Apidae). In *The Hymenoptera of Costa Rica*, ed. P.E. Hanson and I.D. Gauld, 650–91. Oxford: Oxford University Press.

Griz, L., and I. Machado. 2001. Fruiting phenology and seed dispersal syndromes in Caatinga, a tropical dry forest in the northeast of Brazil. *Journal of Tropical Ecology* 17: 303–21.

Grombone-Guarantini, M.T., and R. Ribeiro-Rodrigues. 2002. Seed bank and seed rain in a seasonal semi-deciduous forest in south-eastern Brazil. *Journal of Tropical Ecology* 18: 759–74.

Guariguata, M., and P. Balvanera. In press. Tropical forest service flows: Improving our understanding of the biophysical dimension of ecosystem services. *Forest Ecology and Management*.

Guo, L.B., and R.M. Gifford 2002. Soil carbon stocks and land use changes: A meta analysis. *Global Change Biology* 8: 345–60.

Guo, Y.F., and S.S. Gan. 2005. Leaf senescence: Signals, execution, and regulation. *Current Topics in Developmental Biology* 71: 83.

Gurevitch, J., and L.V. Hedges. 2001. Meta-analysis: Combining the results of independent experiments. In *Design and Analysis of Ecological Experiments*, ed. S.M. Scheiner and J. Gurevitch, 378–98. New York: Oxford University Press.

Güsewell, S. 2004. N:P ratios in terrestrial plants: Variation and functional significance. *New Phytologist* 164: 243–66.

Gutiérrez-Soto, M.V., A. Pacheco, and N.M. Holbrook. 2008. Leaf age and the timing of leaf abscission in two tropical dry forest trees. *Trees—Structure and Function* 22: 393–401.

Haase, R., and R. Hirooka. 1998. Structure, composition and litter dynamics of a semi-deciduous forest in Mato Grosso, Brasil. *Flora* 193: 141–47.

Haber, W.A., and G.W. Frankie. 1989. A tropical hawkmoth community: Costa Rican dry forest Sphingidae. *Biotropica* 21: 155–72.

Haber, W.A., and R.D. Stevenson. 2004. Diversity, migration, and conservation of butterflies in northern Costa Rica. In *Biodiversity Conservation in Costa Rica*, ed. G.W. Frankie, A. Mata, and S.B. Vinson. 99–114. Berkeley: University of California Press.

Haffer, J. 1970. Geologic-climatic history and zoogeographic significance of the Uraba region in northwestern Colombia. *Caldasia* 10: 603–36.

——. 1982. General aspects of the Refuge Theory. In *Biological Diversification in the Tropics*, ed. G.T. Prance, 6–22. New York: Columbia University Press.

Hall, P., M.R. Chase, and K.S. Bawa. 1994. Low genetic variation but high population differentiation in a common tropical forest tree species. *Conservation Biology* 8: 471–82.

Hall, P., L.C. Orrell, and K.S. Bawa. 1994. Genetic diversity and mating system in a tropical tree, *Carapa guianensis* (Meliaceae). *American Journal of Botany* 81: 1104–11.

Hall, P., S. Walker, and K.S. Bawa. 1996. Effect of forest fragmentation on genetic diversity and mating system in a tropical tree, *Pithecellobium elegans*. *Conservation Biology* 10: 757–68.

Hall, T.A. 1999. BioEdit: A user-friendly biological sequence alignment editor and analysis program for Windows 95/98/NT. *Nucleic Acids Symposium Series* 41: 95–98.

Hallwachs, W. 1986. Agoutis (*Dasyprocta punctata*): The inheritors of guapinol (*Hymenaea courbaril*: Leguminosae). In *Frugivores and Seed Dispersal*, ed. A. Estrada and T.H. Fleming, 285–304. Dordrecht, Netherlands: W. Junk Publishers.

———. 1994. The Clumsy Dance between Agoutis and Plants: Scatterhoarding by Costa Rican Dry Forest Agoutis (*Dasyprocta punctata*: Dasyproctidae: Rodentia). PhD diss., Cornell University, Ithaca, NY.

Hamilton, M.B. 1999. Four primer pairs for the amplification of chloroplast intergenic regions with intraspecific variation. *Molecular Ecology* 8: 513–25.

Hansen, M.C., S.V. Stehman, P.V. Potapov, T. Loveland, J.R.G. Townshend, R.S. Defries, A. Pittman, F. Stolle, M. Steininger, M. Carroll, and C. DiMiceli. 2008. Humid tropical forest clearing from 2000 to 2005 quantified by using multitemporal and multiresolution remotely sensed data. *Proceedings of the National Academy of Sciences, USA* 105: 9439–44.

Hansen, R.A. 2000. Effects of habitat complexity and composition on a diverse litter microarthropod assemblage. *Ecology* 81: 1120–32.

Hansen, R.A., and D.C. Coleman. 1998. Litter complexity and composition are determinants of the diversity and species composition of oribatid mites (Acari: Oribatida) in litterbags. *Applied Soil Ecology* 9: 17–23.

Hanson, P. 2004. Biodiversity inventories in Costa Rica and their application to conservation. In *Biodiversity Conservation in Costa Rica: Learning the Lessons in a Seasonal Dry Forest*, ed. G.W. Frankie, A. Mata, and S.B. Vinson, 229–36. Berkeley: University of California Press.

Hanson, P., and I.D. Gauld, ed. 2006. Hymenoptera de la Región Neotropical. *Memoirs of the American Entomological Institute* 77.

Harcourt, A.H., and D.A. Doherty. 2005. Species-area relationships of primates in tropical forest fragments: A global analysis. *Journal of Applied Ecology* 42: 630–37.

Harmon, M.E., D.F. Whigham, J. Sexton, and I. Olmsted. 1995. Decomposition and mass of woody detritus in the dry tropical forests of the northeastern Yucatan Peninsula, Mexico. *Biotropica* 27: 305–16.

Harris, P., C. Huntingford, and P. M. Cox. 2008. Amazon Basin climate under global warming: The role of the sea surface temperature. *Philosophical Transactions of the Royal Society B* 363: 1753–59.

Harris, W. 1937. Revision of *Sciurus variegatoides*, a species of Central American squirrel. *Miscellaneous Publications, Museum of Zoology, University of Michigan* 38: 5–39.

Hart, S., G.E. Nason, D. Myrolod, and D.A. Perry. 1994. Dynamics of gross nitrogen transformations in an old-growth forest: The carbon connection. *Ecology* 75: 880–91.

Harvey, C.A., F. Alpizar, M. Chacón, and R. Madrigal. 2005. *Assessing Linkages between Agriculture and Biodiversity in Central America: Historical Overview and Future Perspectives*. San José, Costa Rica: Mesoamerican and Caribbean Region, Conservation Science Program, The Nature Conservancy.

Harvey, C.A., and W.A. Haber. 1998. Remnant trees and the conservation of biodiversity in Costa Rican pastures. *Agroforestry Systems* 44: 37–68.

Harvey, C.A., O. Komar, R. Chazdon, B.G. Ferguson, B. Finegan, D. Griffith, M. Martínez-Ramos, H. Morales, R. Nigh, L. Soto-Pinto, M. van Breugel, and M. Wishnie. 2008. Integrating agricultural landscapes with biodiversity conservation in the

Mesoamerican hotspot: Opportunities and an action agenda. *Conservation Biology* 22: 8–15.

Harvey, C.A., A. Medina, D.M. Sánchez, S. Vilchez, B. Hernández, J.C. Saenz, J.M. Maes, F. Casanoves, and F.L. Sinclair. 2006. Patterns of animal diversity in different forms of tree cover in agricultural landscapes. *Ecological Applications* 16: 1986–99.

Harvey, C.A., D. Sánchez Merlos, A. Medina, S. Vilchez, B. Hernández, and F.L. Sinclair. In preparation. Patterns of bird, bat, dung beetle and butterfly diversity associated with tree cover in Neotropical agricultural landscapes.

Harvey, C.A., N. Tucker, and A. Estrada. 2004. Live fences, isolated trees and windbreaks: Tools for conserving biodiversity in fragmented landscapes? In *Agroforestry and Biodiversity Conservation in Tropical Landscapes*, ed. G.A. Schroth, B. Fonseca, C.A. Harvey, C. Gascon, H.L. Vasconcelos, and A.M.N. Izac, 261–89. Washington, DC: Island Press.

Harvey, C.A., C. Villanueva, M. Ibrahim, R. Gómez, M. López, S. Kunth, and F.L. Sinclair. 2008. Productores, Árboles y producción ganadera en paisajes de América Central: Implicacciones para la conservación de la biodiversidad. In *Evaluacion y Conservacion de Biodiversidad en Paisajes Fragmentados de Mesoamerica*, ed. C.A. Harvey and J.C. Saénz, 197–224. Heredia, Costa Rica: INBio.

Harvey, C.A., C. Villanueva, J. Villacís, M. Chacón, D. Muñoz, M. López, M. Ibrahim, R. Gómez, R. Taylor, J. Martinez, A. Navas, J. Saenz, D. Sánchez, A. Medina, S. Vilchez, B. Hernández, A. Perez, F. Ruiz, F. López, I. Lang, and F.L. Sinclair. 2005. Contribution of live fences to the ecological integrity of agricultural landscapes. *Agriculture, Ecosystems and Environment* 111: 200–230.

Hassink, J. 1997. The capacity of soils to preserve organic C and N by their association with clay and silt particles. *Plant and Soil* 191: 77–87.

Hayward, B.J., and E.L. Cockrum. 1971. The natural history of the western long-nosed bat *Leptonycteris sanborni*. *Western New Mexico University Research in Science* 1: 75–123.

Hecht, S.B., and S.S. Saatchi. 2007. Globalization and forest resurgence: Changes in forest cover in El Salvador. *BioScience* 57: 663–72.

Heideman, P.D., and R.C.B. Utzurrum. 2003. Seasonality and synchrony of reproduction in three species of nectarivorous Philippines bats. *BMC Ecology* 3: 11, doi: 10.1186/1472-6785-3-11.

Heithaus, E.R., T.H. Fleming, and P.A. Opler. 1975. Foraging patterns and resource utilization in seven species of bats in a seasonal tropical forest. *Ecology* 56: 841–54.

Henderson, A., G. Galeano, and R. Bernal. 1995. *Field Guide to the Palms of the Americas*. Princeton, NJ: Princeton University Press.

Hensel, L.L., V. Grbic, D.A. Baumgarten, and A.B. Bleecker. 1993. Developmental and age-related processes that influence the longevity and senescence of photosynthetic tissues in arabidoposis. *Plant Cell* 5: 553–64.

Hernández-Apolinar, M., T. Valverde, and S. Purata. 2006. Demography of *Bursera glabrifolia*, a tropical tree used for folk woodcrafting in Southern Mexico: An evaluation of its management plant. *Forest Ecology and Management* 223: 139–51.

Herrera, Montalvo, L.G., K.A. Hobson, L. Mirón M., N. Ramírez P., G. Méndez C., and V. Sánchez-Cordero. 2001. Sources of protein in two species of phytophagous bats in a seasonal dry forest: Evidence from stable-isotope analysis. *Journal of Mammalogy* 82: 352–61.

Herrera M., L.G., and C. Martínez del Río. 1998. Pollen digestion by New World bats: Effects of processing time and feeding habits. *Ecology* 79: 2828–38.

Herrera M., L.G. 1997. Evidence of altitudinal movements of *Leptonycteris curasoae* (Chiroptera: Phyllostomidae) in central Mexico. *Revista Mexicana de Mastozoología* 2: 116–18.

Herrerías-Diego, Y., M. Quesada, K.E. Stoner, and J.A. Lobo. 2006. Effects of forest fragmentation on phenological patterns and reproductive success of the tropical dry forest tree *Ceiba aesculifolia*. *Conservation Biology* 20: 1111–20.

Hersch Martínez, P. 1996. *Destino común: los recolectores y su flora medicinal*. México, D.F.: Instituto Nacional de Antropología e Historia.

Heuertz, M., S. Carnevale, S. Fineschi, F. Sebastiani, J.F. Hausman, L. Paule, and G.G. Vendramin. 2006. Chloroplast DNA phylogeography of European ashes, *Fraxinus* sp. (Oleaceae): Roles of hybridization and life history traits. *Molecular Ecology* 15: 2131–40.

Hewitt, G.M. 1999. Post-glacial re-colonization of European biota. *Biological Journal of the Linnean Society* 68: 87–112.

——. 2000. The genetic legacy of the Quaternary ice ages. *Nature* 405: 907–13.

Hewitt, W.P. 1994. Hierve el Agua, Mexico: Its water and its corn-growing potential. *Latin American Antiquity* 5: 177–81.

Hidalgo-Mihart, M.G., L. Cantú-Salazar, C.A. López-González, E. Martínez-Meyer, and A. González-Romero. 2001. Coyote (*Canis latrans*) food habits in a tropical deciduous forest of western Mexico. *American Midland Naturalist* 146: 210–16.

Hoekstra, J., T. Boucher, T. Ricketts, and C. Roberts. 2005. Confronting a biome crisis: Global disparities of habitat loss and protection. *Ecology Letters* 8: 23–29.

Höfer, H., W. Hanagarth, M. Garcia, C. Martius, E. Franklin, J. Rombke, and L. Beck. 2001. Structure and function of soil fauna communities in Amazonian anthropogenic and natural ecosystems. *European Journal of Soil Biology* 37: 229–35.

Hoffman, W.A., A.C. Franco, M.Z. Moreira, and M. Haridasan. 2005. Specific leaf area explains differences in leaf traits between congeneric savanna and forest trees. *Functional Ecology* 19: 932–40.

Hoffmann, W.A., and R.B. Jackson. 2000. Vegetation-climate feedbacks in the conversion of tropical savanna to grassland. *Journal of Climate* 13: 1593–1602.

Hoffmann, W.A., W. Schroeder, and R.B. Jackson. 2003. Regional feedbacks among fire, climate, and tropical deforestation. *Journal of Geophysical Research* 108: D23, 4721, doi: 10.1029/2003JD003494.

Holbrook, N.M., J.L. Whitbeck, and H.A. Mooney. 1995. Drought responses of Neotropical dry forest trees. In *Seasonally Dry Tropical Forests*, ed. S.H. Bullock, H.A. Mooney, and E. Medina, 243–76. Cambridge: Cambridge University Press.

Holdridge, L.R. 1947. Determination of world plant formation from simple climatic data. *Science* 105: 367–68.

——. 1967. *Life Zone Ecology*. Photographic supplement prepared by Joseph A. Tosi Jr., rev. ed. San José, Costa Rica: Tropical Science Center.

Holl, K.D. 1999. Factors limiting tropical rain forest regeneration in abandoned pasture: Seed rain, seed germination, microclimate, and soil. *Biotropica* 31: 229–42.

Holzmueller, E.J., J. Shibu, and M.A. Jenkins. 2008. The relationship between fire history and an exotic fungal disease in a deciduous forest. *Oecologia* 155: 347–56.

Horvath, D.P., J.V. Anderson, W.S. Chao, and M.E. Foley. 2003. Knowing when to grow: Signals regulating bud dormancy. *Trends in Plant Science* 8: 534–40.

Houghton, R.A. 2003. Revised estimates of the annual net flux of carbon to the atmosphere from changes in land use and land management 1850–2000. *Tellus* 55B: 370–90.

Houghton, R.A., J.E. Hobbie, J.M. Melillo, B. Moore, B.J. Peterson, G.R. Shaver, and G.M. Woodwell. 1983. Changes in the carbon content of terrestrial biota and soil between 1860 and 1980: A net release of CO_2 to the atmosphere. *Ecological Monographs* 53: 235–62.

Howe, H.F. 1984. Implications of seed dispersal by animals for tropical reserve management. *Biological Conservation* 30: 261–81.

Howe, H.F., and J. Smallwood. 1982. Ecology of seed dispersal. *Annual Review of Ecology and Systematics* 13: 201–28.

Howell, D.J. 1979. Flock foraging in nectar-feeding bats: Advantages to the bats and to the host plants. *American Naturalist* 114: 23–49.

Hudson, J.W., and J.A. Rummel. 1966. Water metabolism and temperature regulation of the primitive heteromyids, *Liomys salvini* and *Liomys irrotatus*. *Ecology* 47: 346–54.

Hueck, K. 1978. *Los Bosques de Sudamérica*. Eschborn, Germany: Gesellschaft für Technische Zusammenarbeit.

Hughes, C.E. 1998a. Leucaena: *A Genetic Resources Handbook*. Tropical Forestry Paper 37. Oxford: Oxford Forestry Institute.

———. 1998b. Monograph of *Leucaena* (Leguminosae-Mimosoideae). *Systematic Botany Monographs* 55: 1–244.

Hughes, C.E., C.D. Bailey, S. Krosnick, and M.A. Luckow. 2003. Relationships among genera of the informal *Dichrostachys* and *Leucaena* groups (Mimosoideae) inferred from nuclear ribosomal ITS sequences. In *Advances in Legume Systematics*, Part 10, *Higher Level Systematics*, ed. B.B. Klitgaard and A. Bruneau, 221–38. Kew: Royal Botanic Gardens.

Hughes, R.F., J.B. Kauffman, and V.J. Jaramillo. 2000. Ecosystem-scale impacts of deforestation and land use in a humid tropical region of México. *Ecological Applications* 10: 515–27.

Hunt, J.H., R.J. Brodie, T.P. Carithers, P.Z. Goldstein, and D.H. Janzen. 1999. Dry season migration by Costa Rican lowland paper wasps to high elevation cold dormancy sites. *Biotropica* 31: 192–96.

Hutyra, L.R., J.W. Munger, C.A. Nobre, S.R. Saleska, S.A.Vieira, and S.C. Wofsy. 2005. Climatic variability and vegetation vulnerability in Amazonia. *Geophysical Research Letters* 32: L24712, doi: 10.1029/2005GL024981.

Ibáñez, R., R. Condit, G. Angehr, S. Aguilar, T. García, R. Martínez, A. Sanjur, R. Stallard, S.J. Wright, A.S. Rand, and S. Heckadon. 2002. An ecosystem report on the Panama Canal: Monitoring the status of the forest communities and the watershed. *Environmental Monitoring and Assessment* 80: 65–95.

Ibarra-Manríquez, G., and K. Oyama. 1992. Ecological correlates of reproductive traits of Mexican rain forest trees. *American Journal of Botany* 79: 383–94.

Ibarra-Manríquez, G., B. Sánchez, and L. González. 1991. Fenología de lianas y Árboles anemócoros en la selva cálido-húmeda de México. *Biotrópica* 23: 242–54.

IBGE. 2005. *Produção da Extração Vegetal e da Silvicultura 2005*. Rio de Janiero: Instituto Brasileiro de Geografia e Estatística, Distrito Federal.

Ibrahim, K.M., R.A. Nichols, and G.M. Hewitt. 1996. Spatial patterns of genetic variation generated by different forms of dispersal during range expansion. *Heredity* 77: 282–91.

Illsley, C., J. Aguilar, J. Acosta, J. García, and J. Caballero. 1998. Manchoneras y soyacahuiteras: Manejo campesino de *Brahea dulcis* (HBK) Mart. en la región de Chilapa, Guerrero. México: Memorias 7th Congreso Latinoamericanao de Botánica.

Illsley, C., J. Aguilar, J. Acosta, J. García, T. Gómez, and J. Caballero. 2001. Contribuciones al conocimiento y manejo campesino de los palmares de Brahea dulcis (HBK) Mart. en al region de Chilapa, Guerrero. In *Plantas, Cultura y Sociedad. Estudio de la Relación entre Seres Humanos y Plantas en la Albores del Siglo XXI*, ed. A. Rendón, D. Rebollar, N. Caballero, and A. Martínez, 259–386. México: UAM Iztapalapa.

Ineson, P., L.A. Levin, R. Kneib, R.O. Hall, J.M. Weslawski, R.D. Bardgett, D.A. Wardle, D.H. Wall, W.H. van der Putten, and H. Zadeh. 2004. Cascading effects of deforestation on ecosystem services across soils and freshwater sediments. In *Sustaining Biodiversity and Ecosystem Services in Soils and Sediments*, ed. D.H. Wall, 225–48. Washington, DC: Island Press.

IPCC. 2001. *Climate Change 2001: Synthesis Report*, ed. R.T. Watson and the Core Writing Team. Cambridge: Cambridge University Press.

——. 2007. *Climate Change 2007: The Physical Science Basis*. Contribution of Working Group I to the 4th Assessment Report of the Intergovernmental Panel on Climate Change, ed. S. Solomon, D. Qin, M. Manning, Z. Chen, M. Marquis, K.B. Averyt, M. Tignor, and H.L. Miller. Cambridge: Cambridge University Press.

Ireland, H., and R.T. Pennington. 1999. A revision of *Geoffroea* (Leguminosae—Papilionoideae). *Edinburgh Journal of Botany* 56: 329–47.

Iturralde-Vinent, M.A., and R.D.E. MacPhee. 1999. Paleogeography of the Caribbean region: Implications for Cenozoic biogeography. *Bulletin of the American Museum of Natural History* 238: 1–95.

Jackson, P.C., F.C. Meinzer, M. Bustamante, G. Goldstein, A. Francom, P.W. Rundel, L.Caldasm, E. Igler, and F. Causin. 1999. Partitioning of soil water among tree species in a Brazilian cerrado ecosystem. *Tree Physiology* 19: 717–24.

Jackson, R.B., H.A. Mooney, and E.D. Schulze. 1997. A global budget for fine root biomass, surface area, and nutrient contents. *Proceedings of the National Academy of Sciences, USA* 94: 7362–66.

Jaimes, I., and N. Ramírez. 1999. Breeding systems in a secondary deciduous forest in Venezuela: The importance of life form, habitat, and pollination specificity. *Plant Systematics and Evolution* 215: 23–36.

Janzen, D.H. 1967. Synchronization of sexual reproduction of trees within the dry season in Central America. *Evolution* 21: 620–37.

——. 1970. Herbivores and the number of tree species in tropical forests. *American Naturalist* 104: 501–28.

——. 1980. Specificity of seed-attacking beetles in a Costa Rican deciduous forest. *Journal of Ecology* 68: 929–52.

——. 1983. *Costa Rican natural history*. Chicago: University of Chicago Press.

——. 1986a. The future of tropical ecology. *Annual Review of Ecology and Systematics* 17: 305–24.

——. 1986b. *Guanacaste National Park: Tropical Ecological and Cultural Restoration*. San Jose, Costa Rica: UNED.

——. 1986c. Mice, big mammals, and seeds: It matters who defecates what where. In *Frugivores and Seed Dispersal*, ed. A. Estrada and T.H. Fleming, 251–71. Dordrecht: W. Junk Publishers.

——. 1987. Forest restoration in Costa Rica. *Science* 235: 15–16.

——. 1988a. Ecological characterization of a Costa Rican dry tropical forest caterpillar fauna. *Biotropica* 20: 120–35.

——. 1988b. Management of habitat fragments in a tropical dry forest: Growth. *Annals of the Missouri Botanical Garden* 75: 105–16.

——. 1988c. Tropical dry forests, the most endangered major tropical ecosystem. In *Biodiversity*, ed. E.O. Wilson, 130–37. Washington, DC: National Academies Press.

——. 1988d. Tropical ecological and biocultural restoration. *Science* 239: 243–44.

——. 1993. Caterpillar seasonality in a Costa Rican dry forest. In *Caterpillars, Ecological and Evolutionary Constraints on Foraging*, ed. N.E. Stamp and T.M. Casey, 448–77. New York: Chapman and Hall.

——. 2003. How polyphagous are Costa Rican dry forest saturniid caterpillars? In *Arthropods of Tropical Forests: Spatio-temporal Dynamics and Resource Use in the Canopy*, ed. Y. Basset, V. Novotnyh, S.E. Miller, and R.L. Kitching, 369–79. Cambridge: Cambridge University Press.

Janzen, D.H., P.J. DeVries, M.L. Higgins, and L.S. Kimsey. 1982. Seasonal and site variation in Costa Rican euglossine bees at chemical baits in lowland deciduous and evergreen forests. *Ecology* 63: 66–74.

Janzen, D.H., and W. Hallwachs. 1996. *Liomys salvini* home page (Heteromyidae, Rodentia, Mammalia). http://janzen.sas.upenn.edu/TSHP/English/Heteromyidae/Liomys_salvini_Home_Page/English_Text/Liomys_salvini_Eng.html.

——. 2005. Dynamic database for an inventory of the macrocaterpillar fauna, and its food plants and parasitoids, of Area de Conservacion Guanacaste (ACG), northwestern Costa Rica (nn-SRNP-nnnnn voucher codes). http://janzen.sas.upenn.edu.

Jaramillo, V.J., J.B. Kauffman, L. Rentería-Rodríguez, D.L. Cummings, and L.J. Ellingson. 2003. Biomass, carbon, and nitrogen pools in Mexican tropical dry forest landscapes. *Ecosystems* 6: 609–29.

Jaramillo, V.J., and R.L. Sanford Jr. 1995. Nutrient cycling in tropical deciduous forests. In *Seasonally Dry Tropical Forests*, S.H. Bullock, H.A. Mooney, and E. Medina, 346–61. Cambridge: Cambridge University Press.

Jarvis, D.I., C. Padoch, and H.D. Cooper, ed. 2007. *Managing Biodiversity in Agricultural Ecosystems*. New York: Columbia University Press.

Jesus, F.F., J.F. Wilkins, V.N. Solferini, and J. Wakeley. 2006. Expected coalescence times and segregating sites in a model of glacial cycles. *Genetics and Molecular Research* 5: 466–74.

Joergensen, R.G., and S. Scheu. 1999. Response of soil microorganisms to the addition of carbon. *Soil Biology and Biochemistry* 31: 859–66.

Johnson, D. 1972. The carnauba wax palm (*Copernicia prunifera*). 4. Economic uses. *Principes* 16: 128–31.

Johnson, N.C., and D.A. Wedin. 1997. Soil carbon, nutrients, and mycorrhizae during conversion of dry tropical forest to grassland. *Ecological Applications* 7: 171–82.

Jordano, P. 2001. Fruits and frugivory. In *Seeds: The Ecology of Regeneration in Plant Communities*. 2nd ed., ed. M. Fenner, 125–65. Wallingford, CT: CAB International.

Judas, M. 1988. The species-area relationship of European Lumbricidae (Annelida, Oligochaeta). *Oecologia* 76: 579–87.

Justiniano, M., and T. Fredericksen. 2000. Phenology of tree species in Bolivian dry forests. *Biotropica* 32: 276–81.

Kaiser, J. 2001. Bold corridor project confronts political reality. *Science* 293: 2196–99.

Kalacska, M., G.A. Sánchez-Azofeifa, J.C. Calvo-Alvarado, M. Quesada, B. Rivard, and D.H. Janzen. 2004. Species composition, similarity and diversity in three successional stages of a seasonally dry tropical forest. *Forest Ecology and Management* 200: 227–47.

Kalacska, M., G.A. Sánchez-Azofeifa, J.C. Calvo-Alvarado, B. Rivard, and M. Quesada. 2005. Effects of season and successional stage on leaf area index and spectral vegetation indices in three Mesoamerican tropical dry forests. *Biotropica* 37: 486–96.

Kalacska, M., G.A. Sánchez-Azofeifa, B. Rivard, J.C. Calvo-Alvarado, and M. Quesada. 2008. Baseline assessment for environmental services payments from satellite imagery: A case study from Costa Rica and Mexico. *Journal of Environmental Management* 88: 348–59.

Kapos, V., E. Wandelli, J.L. Camargo, and G. Ganade. 1997. Edge related change in environment and plant responses due to forest fragmentation in Central Amazonia. In *Tropical Forest Remnants: Ecology, Management and Conservation of Fragmented Communities*, ed. E.F. Lawrence and O. Bierregaard, 33–44. Chicago: University of Chicago Press.

Kauffman, J.B., D.L. Cummings, and D.E. Ward. 1998. Fire in the Brazilian Amazons. 2. Biomass, nutrient pools, and losses in cattle pastures. *Oecologia* 113: 415–27.

Kauffman, J.B., R.L. Sanford Jr., D.L. Cummings, I.H. Salcedo, and E.V.S.B. Sampaio. 1993. Biomass and nutrient dynamics associated with slash fires in Neotropical dry forests. *Ecology* 74: 140–51.

Kauffman, J.B., M.D. Steele, D.L. Cummings, and V.J. Jaramillo. 2003. Biomass dynamics associated with deforestation, fire, and conversion to cattle pastures in a Mexican tropical dry forest. *Forest Ecology and Management* 176: 1–12.

Kay, J.J., H. Rieger, M. Boyle, and G. Francis. 1999. An ecosystem approach for sustainability: Addressing the challenge of complexity. *Futures* 31: 721–42.

Keller, M., M. Bustamante, J. Gash, and P. Silva Dias, ed. 2009. *Amazonia and Global Change*, Geophysical Monograph. Series. Vol. 186. Washington, DC: AGU.

Kelm, D.H., and O.V. Helversen. 2007. How to budget metabolic energy: Torpor in a small Neotropical mammal. *Journal of Comparative Physiology, B: Biochemical, Systemic, and Environmental Physiology* 177: 667–77.

Kennard, D. 2002. Secondary forest succession in a tropical dry forest: Patterns of development across a 50-year chronosequence in lowland Bolivia. *Journal of Tropical Ecology* 18: 53–66.

Kennard, D.K., K. Gould, F.E. Putz, T.S. Frederisckons, and F. Morales. 2002. Effect of disturbance intensity on regeneration mechanisms in a Neotropical dry forest. *Forest Ecology and Management* 162: 197–208.

Kennedy, P.G., A.D. Izzo, and T.D. Bruns. 2003. There is high potential for the formation of common mycorrhizal networks between understorey and canopy trees in a mixed evergreen forest. *Journal of Ecology* 91: 1071–80.

Kerr, A.M., and A.H. Baird. 2007. Natural barriers to natural disasters. *BioScience* 57: 102–3.

Kessler, M., and N. Helme. 1999. Floristic diversity and phytogeography of the central Tuichi Valley, an isolated dry forest locality in the Bolivian Andes. *Candollea* 54: 341–66.

Khanna, P.K., R.J. Raison, and R.A. Falkiner. 1994. Chemical properties of ash derived from *Eucalyptus* litter and its effects on forest soil. *Forest Ecology and Management* 66: 107–25.

Khurana, E., and J.S. Singh. 2001. Ecology of seed and seedling growth for conservation and restoration of tropical dry forest: A review. *Environmental Conservation* 28: 39–52.

Killeen, T.J., E. Chavez, M. Peña-Claros, M. Toledo, L. Arroyo, J. Caballero, L. Correa, R. Guillen, R. Quevedo, M. Saldias, L. Soria, Y. Uslar, I. Vargas, and M. Steininger.

2006. The Chiquitano dry forest, the transition between humid and dry forest in eastern lowland Bolivia. In *Neotropical Savannas and Seasonally Dry Forests: Plant Diversity, Biogeography, and Conservation*, ed. R.T. Pennington, G. Lewis, and J.A. Ratter, 213–33. Boca Raton, FL: CRC Press.

Killeen, T.J., A. Jardim, F. Mamani, N. Rojas, and P. Saravia. 1998. Diversity, composition and structure of a tropical semideciduous forest in the Chiquitanía region of Santa Cruz, Bolivia. *Journal of Tropical Ecology* 14: 803–27.

Killingbeck, K.T. 1996. Nutrients in senesced leaves: Keys to the search for potential resorption and resorption proficiency. *Ecology* 77: 1716–27.

Kinahan, A.A., and N. Pillay. 2008. Does differential exploitation of folivory promote coexistence in an African savanna granivorous rodent community? *Journal of Mammalogy* 89: 132–37.

Kindt, R., A.J. Simons, and P. van Damme. 2004. Do farm characteristics explain differences in tree species diversity among West Kenyan farms? *Agroforestry Systems* 63: 63–74.

Kirschbaum, M.U.F., P. Bullock, J.R. Evans, K. Goulding, P.G. Jarvis, I.R. Noble, M. Rounsevell, T.D. Sharkey, M.-P. Austin, P. Brookes, S. Brown, H.K.M. Bugmann, W. Cramer, S. Díaz, H. Gitay, S.P. Hamburg, J. Harris, and J.I. Holten. 1996. Ecophysiological, ecological, and soil processes in terrestrial ecosystems: A primer on general concepts and relationships. In *Climate Change 1995: Impacts, Adaptations and Mitigation of Climate Change*, ed. R.T. Watson, M.C. Zinyowera, and R.H. Moss, 57–74. Cambridge: Cambridge University Press.

Klemens, J.A., M.L. Wieland, V.J. Flanagin, J.A. Frick, and R.G. Harper. 2003. A cross-taxa survey of organochlorine pesticide contamination in a Costa Rican wildland. *Environmental Pollution* 122: 245–51.

Klink, C.A., and R.B. Machado. 2005. Conservation of the Brazilian Cerrado. *Conservation Biology* 19: 707–13.

Klopfstein, S., M. Currat, and L. Excoffier. 2006. The fate of mutations surfing on the wave of a range expansion. *Molecular Biology and Evolution* 23: 482–90.

Korine, C., J. Speakman, and Z. Arad. 2004. Reproductive energetics of captive and free-ranging Egyptian fruit bats (*Rousettus aegyptiacus*). *Ecology* 85: 220–30.

Kowalewski, M., and G.E. Zunino. 2004. Birth seasonality in *Alouatta caraya* in northern Argentina. *International Journal of Primatology* 25: 383–400.

Kramer, E.A. 1997. Measuring landscape changes in remnant tropical dry forests. In *Tropical Forest Remnants: Ecology, Management, and Conservation of Fragmented Communities*, ed. W.F. Laurance and R.O. Bierregaard Jr., 386–99. Chicago: University of Chicago Press.

Kramer, P.J., and J.S. Boyer. 1995. *Water relations of plants and soils*. San Diego: Academic Press.

Kremen, C. 2005. Managing ecosystem services: What do we need to know about their ecology? *Ecology Letters* 8: 468–79.

Kucharik, C.J., C.C. Barford, M. El Maayar, S.C. Wofsy, R.K. Monson, and D.D. Baldocchi. 2006. A multiyear evaluation of a dynamic global vegetation model at three AmeriFlux forest sites: Vegetation structure, phenology, soil temperature, and CO_2 and H_2O vapor exchange. *Ecological Modelling* 196: 1–31.

Kummerow, J., J. Castellanos, M. Maass, and A. Larigauderie. 1990. Production of fine roots and the seasonality of their growth in a Mexican deciduous dry forest. *Vegetatio* 90: 73–80.

Kursar, T.A., B.M.J. Engelbrecht, A. Burke, M.T. Tyree, B.E. Omari, and J.P. Giraldo. 2009. Tolerance to low leaf water status of tropical tree seedlings is related to drought performance and distribution. *Functional Ecology* 23: 93–102.

Lacerda, D.R., M.D.P. Acedo, J.P.L. Filho, and M.B. Lovato. 2001. Genetic diversity and structure of natural populations of *Plathymenia reticulata* (Mimosoideae), a tropical tree from the Brazilian Cerrado. *Molecular Ecology* 10: 1143–52.

Laidlaw, R.K. 2000. Effects of habitat disturbance and protected areas on mammals of Peninsular Malaysia. *Conservation Biology* 14: 1639–48.

Lamb, D., P.D. Erskine, and J.A. Parrotta. 2005. Restoration of degraded tropical forest landscapes. *Science* 310: 1628–32.

Lambert, J.D.H., J.T. Arnason, and J.L. Gale. 1980. Leaf-litter and changing nutrient levels in a seasonally dry tropical hardwood forest, Belize, C.A. *Plant and Soil* 55: 429–33.

Lambert, J.E., and C.A. Chapman. 2005. The fate of primate-dispersed seeds: Deposition pattern, dispersal distance and implications for conservation. In *Seed Fate: Predation, Dispersal, and Seedling Establishment*, ed. P.M. Forget, J.E. Lambert, P.E. Hulme, and S.B. VanderWall, 137–50. Cambridge: CABI Publishers.

Langenheim, J.H. 1981. Terpenoids in the Leguminosae. In *Advances in Legume Systematics*, ed. R.M. Polhill and P.H. Raven, 627–55. Kew: Royal Botanic Gardens.

Lardy, L.C., M. Brossard, M.L. Lopes, and J.Y. Laurent. 2002. Carbon and phosphorus stocks of clayed Ferralsols in Cerrado native and agroecosystems, Brazil. *Agriculture, Ecosystems and Environment* 92: 147–58.

Laurance, W.F., A.K.M. Albernaz, G. Schroth, P.M. Fearnside, S. Bergen, E.M. Venticinque, and C. da Costa. 2002. Predictors of deforestation in the Brazilian Amazon. *Journal of Biogeography* 29: 737–48.

Laurance, W.F., T.E. Lovejoy, H.L. Vasconcelos, E.M. Bruna, R.K. Didham, P.C. Stouffer, C. Gascon, R.O. Bierregaard, S.G. Laurance, and E. Sampaio. 2002. Ecosystem decay of Amazonian forest fragments: A 22 year investigation. *Conservation Biology* 16: 605–18.

Laurance, W.F., H.E.M. Nascimento, S.G. Laurance, A. Andrade, J. Ribeiro, J.P. Giraldo, T.E. Lovejoy, R. Condit, J. Chave, K.E. Harms, and S. D'Angelo. 2006. Rapid decay of tree-community composition in Amazonian forest fragments. *Proceedings of the National Academy of Sciences, USA* 103: 19010–14.

LaVal, R.K. 2004a. Impact of global warming and locally changing climate on tropical cloud forest bats. *Journal of Mammalogy* 85: 237–44.

———. 2004b. An ultrasonically silent night: The tropical dry forest without bats. In *Biodiversity Conservation in Costa Rica: Learning the Lessons in a Seasonal Dry Forest*, ed. G.W. Frankie, A. Mata, and S.B. Vinson, 160–76. Berkeley: University of California Press.

Law, B.S., and M. Lean. 1999. Common blossom bats (*Syconycteris australis*) as pollinators in fragmented Australian tropical rainforest: A comparative study of two lowland tropical plants. *Biological Conservation* 91: 201–12.

Lawes, M.J., P.E. Mealin, and S.E. Piper. 2000. Patch occupancy and potential metapopulation dynamics of three forest mammals in fragmented Afromontane forest in South Africa. *Conservation Biology* 14: 1088–98.

Lawless, J. 1995. *The Illustrated Encyclopedia of Essential Oils*. London: Element Books.

Lawrence, D., and D. Foster. 2002. Changes in forest biomass, litter dynamics and soils following shifting cultivation in southern Mexico: An overview. *Interciencia* 27: 400–408.

Lawrence, D., and W.H. Schlesinger. 2001. Changes in soil phosphorus during 200 years of shifting cultivation in Indonesia. *Ecology* 82: 2769–80.

Lawton, J.H., D.E. Bignell, G.F. Bloemers, P. Eggleton, and M.E. Hodda. 1996. Carbon flux and diversity of nematodes and termites in Cameroon forest soils. *Biodiversity and Conservation* 5: 261–73.

Lawton, J.H., D.E. Bignell, B. Bolton, G.F. Bloemers, P. Eggleton, P.M. Hammond, M. Hodda, R.D. Holt, T.B. Larsen, N.A. Mawdsley, N.E. Stork, D.S. Srivastava, and A.D. Watt. 1998. Biodiversity inventories, indicator taxa and effects of habitat modification in tropical forest. *Nature* 391: 72–76.

Lebrija-Trejos, E., F. Bongers, E.A. Pérez-García, and J.A. Meave. 2008. Successional change and resilience of a very dry tropical deciduous forest following shifting agriculture. *Biotropica* 40: 422–31.

Le Corre, V., N. Machon, R.J. Petit, and A. Kremer. 1997. Colonization with long-distance seed dispersal and genetic structure of maternally inherited genes in forest trees: A simulation study. *Genetical Research Cambridge* 69: 117–25.

Lee, Y.F., and G.F. McCracken. 2005. Dietary variation of Brazilian free-tailed bats links to migratory populations of pest insects. *Journal of Mammalogy* 86: 67–76.

Leigh, E. 1999. The seasonal rhythms of fruiting and leaf flush and the regulation of animal populations. In *Tropical Forest Ecology*, ed. E. Leigh, 149–178. Cambridge: Oxford University Press.

Leigh, E., P. Davidar, C.W. Dick, J.P. Puyravaud, J. Terborgh, H. ter Steege, and S.J. Wright. 2004. Why do some tropical forests have so many species of trees? *Biotropica* 36: 447–73.

Leopold, A.C. 1951. Photoperiods in plants. *Quarterly Review of Biology* 26: 240–63.

Lessa, A.S.N., D.W. Anderson, and J.O. Moir. 1996. Fine root mineralization, soil organic matter and exchangeable cation dynamics in slash and burn agriculture in the semi-arid northeast of Brazil. *Agriculture, Ecosystems and Environment* 59: 191–202.

Levin, D.A. 1996. The evolutionary significance of pseudo-self-fertility. *American Naturalist* 148: 321–32.

Lewis, R.J., and P.M. Kappeler. 2005. Seasonality, body condition, and timing of reproduction in *Propithecus verreauxi verreauxi* in the Kirindy Forest. *American Journal of Primatology* 67: 347–64.

Lieberman, D., and M. Lieberman. 1984. The causes and consequences of synchronous flushing in a dry tropical forest. *Biotropica* 16: 193–201.

Lieberman, S., and C.F. Dock. 1982. Analysis of the leaf litter arthropod fauna of a lowland tropical evergreen forest site (La Selva, Costa Rica). *Revista de Biologia Tropical* 30: 27–34.

Light, S.F. 1933. Termites of western Mexico. *University of California Publications in Entomology* 6: 79–164.

Lilienfein, J., and W. Wilcke. 2004. Water and element input into native, agri- and silvicultural ecosystem of the Brazilian savanna. *Biogeochemistry* 67: 183–212.

Lim, P.O., H.J. Kim, and H.G. Nam. 2007. Leaf senescence. *Annual Review of Plant Biology* 58: 115–36.

Linares-Palomino, R., R.T. Pennington, and S. Bridgewater. 2003. The phytogeography of the seasonally dry tropical forests in Equatorial Pacific South America. *Candollea* 58: 473–99.

Linskens, H.F., and J.J. Jackson. 1991. *Essential Oils and Waxes*. New York: Springer-Verlag.

Little, E.L., and F.H. Wadsworth. 1964. *Common Trees of Puerto Rico and the Virgin Islands*. Agricultural Handbook 249. Washington, DC: USDA Forest Service.

Lloyd, J., M.I. Bird, L. Vellen, A.C. Miranda, E.M. Veenendaal, G. Djagbletey, S. Miranda, G. Cook, and G.D. Farquhar. 2008. Contributions of woody and herbaceous vegetation to tropical savanna ecosystem productivity: A quasi-global estimate. *Tree Physiology* 28: 451–68.

Lloyd, J., and G.D. Farquhar. 2008. Effects of rising temperatures and CO_2 on the physiology of tropical forest trees. *Philosophical Transactions of the Royal Society B* 363: 1811–17.

Lobo, J.A., M. Quesada, and K.E. Stoner. 2005. Effects of pollination by bats on the mating system of *Ceiba pentandra* (Bombacaceae) populations in two tropical life zones in Costa Rica. *American Journal of Botany* 92: 370–76.

Lobo, J.A., M. Quesada, K.E. Stoner, E.J. Fuchs, Y. Herrerías-Diego, J. Rojas, and G. Saborio. 2003. Factors affecting phenological patterns of Bombacaceous trees in seasonal forest in Costa Rica and Mexico. *American Journal of Botany* 90: 1054–63.

Lobova, T.A., C.K. Geiselman, and S.A. Mori. 2009. *Seed Dispersal by Bats in the Neotropics*. New York: New York Botanical Garden.

Lodge, D.J., W.H. McDowell, and C.P. Mcswiney. 1994. The importance of nutrient pulses in tropical forests. *Trends in Ecology and Evolution* 9: 384–87.

Longino, J.T. 2006. Ants of Costa Rica. http://www.evergreen.edu/ants/AntsofCostaRica.html.

Lopes, P.S.N., J.C. Souza, P. Rebelles, J. Mendes, and I.D. Ferreira. 2003. Caracterização do Ataque da Broca dos Frutos do Pequizeiro. *Revista Brasileira de Fruticultura* 25: 540–43.

López, A.M. 2001. Evaluación de la demanda y extracción de Madera de copal (*Bursera* spp.) para artesanía en communidades de los valles centrales de Oaxaca. MS diss., Universidad Nacional Autónoma de México.

López, M., R. Gómez, C.A. Harvey, and C. Villanueva. 2004. Caracterización del componente arbóreo en los sistemas ganaderos de Rivas, Nicaragua. *Encuentro* 36: 114–33.

López, R.P. 2003. Phytogeographical relations of the Andean dry valleys of Bolivia. *Journal of Biogeography* 30: 1659–68.

Lopezaraiza-Mikel, M.E., R.B. Hayes, M.R. Whalley, and J. Memmott. 2007. The impact of an alien plant on a native plant-pollinator network: An experimental approach. *Ecology Letters* 10: 539–50.

López-Guerrero, A. 1992. Escorrentía en pequeñas cuencas hidrológicas con selva baja caducifolia en Chamela, Jalisco. BS diss., Universidad Nacional Autónoma de México.

López-Ortiz, R., and A. Lewis. 2004. Habitat selection by *Sphaerodactylus nicholsi* (Squamata: Gekkonidae) in Cabo Rojo, Puerto Rico. *Herpetologica* 60: 438–44.

López-Tapia, D.M. 2008. Elaboración de criterios para la restauración de la Cuenca del Río Cuitzmala, Jalisco, con base en un análisis de agua. MA diss., Universidad Nacional Autónoma de México.

Loranger-Merciris, G., D. Imbert, F. Bernhard-Reversat, J. Ponge, and P. Lavelle. 2007. Soil fauna abundance and diversity in a secondary semi-evergreen forest in Guadeloupe (Lesser Antilles): Influence of soil type and dominant tree species. *Biology and Fertility of Soils* 44: 269–76.

Lorente-Adame, R.G., and L.R. Sánchez-Velásquez. 1996. Dinámica Estacional del Banco de Frutos del Teocintle *Zea diploperennis* (Gramineae). *Biotropica* 28: 267–72.

Lorenzi, H., and F.J.A. Matos. 2002. Plantas medicinais do Brasil: Nativas e exóticas. Instituto Platnarum, Nova Odessa, São Paulo.

Lott, E.J. 2002. Lista anotada de las plantas de Chamela-Cuixmala. In *Historia Natural de Chamela*, ed. F.A. Noguera, J.H. Vera-Rivera, A.N. García-Aldrete, and M. Quesada Avendaño, 137–42. Mexico City: Instituto de Biología, Universidad Nacional Autónoma de México.

Lott, E.J., and T.H. Atkinson. 2006. Mexican and Central American seasonally dry tropical forests: Chamela-Cuixmala, Jalisco, as a focal point for comparison. In *Neotropical Savannas and Seasonally Dry Forests: Plant Diversity, Biogeography, and Conservation*, ed. R.T. Pennington, G.P. Lewis, and J.A. Ratter, 315–42. Boca Raton, FL: CRC Press.

Lovegrove, B.G., G. Kortner, and F. Geiser. 1999. The energetic cost of arousal from torpor in the marsupial *Sminthopsis macroura*: Benefits of summer ambient temperature cycles. *Journal of Comparative Physiology, B: Biochemical, Systemic, and Environmental Physiology* 169: 11–18.

Lugo, A.E., E. Medina, J.C. Trejo-Torres, and E. Helmer. 2006. Botanical and ecological resilience of Antillean dry forests. In *Neotropical Savannas and Seasonally Dry Forests: Plant Diversity, Biogeography, and Conservation*, ed. R.T. Pennington, G.P. Lewis, and J.A. Ratter, 359–81. Boca Raton, FL: CRC Press.

Lugo, A.E., and P.G. Murphy. 1986. Nutrient dynamics of a Puerto Rican subtropical dry forest. *Journal of Tropical Ecology* 2: 55–72.

Lüttge, U. 1997. *Physiological Ecology of Tropical Plants*. New York: Springer-Verlag.

Lyman, C.P., J.S. Willis, A. Malan, and L.C.H. Wang. 1982. *Hibernation and Torpor in Mammals and Birds*. New York: Academic Press.

MA (Millennium Ecosystem Assessment). 2003. *Ecosystems and Human Well-being: A Framework for Assessment*. Washington, DC: Island Press.

Maass, J.M. 1995. Conversion of tropical dry forest to pasture and agriculture. In *Seasonally Dry Tropical Forests*, ed. S.H. Bullock, H.A. Mooney, and E. Medina, 399–422. Cambridge: Cambridge University Press.

Maass, J.M., P. Balvanera, A. Castillo, G.C. Daily, H.A. Mooney, P. Ehrlich, M. Quesada, A. Miranda, V.J. Jaramillo, F. García-Oliva, A. Martínez-Yrizar, H. Cotler, J. López-Blanco, J.A. Pérez-Jiménez, A. Búrquez, C. Tinoco, G. Ceballos, L. Barraza, R. Ayala, and J. Sarukhán. 2005. Ecosystem services of tropical dry forests: Insights from long-term ecological and social research on the Pacific coast of Mexico. *Ecology and Society* 10: 17. http://www.ecologyandsociety.org/vol10/iss1/art17.

Maass, J.M., V.J. Jaramillo, A. Martínez-Yrízar, F. García-Oliva, A. Pérez-Jiménez, and J. Sarukhán. 2002. Aspectos funcionales del ecosistema de selva baja caducifolia en Chamela, Jalisco. In *Historia Natural de Chamela*, ed. F.A. Noguera, J.H. Vera-Rivera, A.N. García-Aldrete, and M. Quesada Avendaño, 525–42. Mexico City: Instituto de Biología, Universidad Nacional Autónoma de México.

Maass, J.M., C. Jordan, and J. Sarukhán. 1988. Soil erosion and nutrient losses in seasonal tropical agroecosystems under various management techniques. *Journal of Applied Ecology* 25: 595–607.

Maass, J.M., A. Martínez-Yrízar, C. Patiño, and J. Sarukhán. 2002. Distribution and annual net accumulation of above-ground dead phytomass and its influence on throughfall quality in a Mexican tropical deciduous forest ecosystem. *Journal of Tropical Ecology* 18: 821–34.

Maass, J.M., J.M. Vose, W.T. Swank, and A. Martínez-Yrízar. 1995. Seasonal changes of leaf area index (LAI) on a tropical deciduous forest in west Mexico. *Forest Ecology and Management* 74: 171–80.

Machado, I.C., and A.V. Lopes. 2004. Floral traits and pollination systems in the Caatinga, a Brazilian tropical dry forest. *Annals of Botany* 94: 365–76.

Machado, I.C., A.V. Lopes, and M. Sazima. 2006. Plant sexual systems and a review of the breeding system studies in the Caatinga, a Brazilian tropical dry forest. *Annals of Botany* 97: 277–87.

Maffei, L., E. Cuellar, and A. Noss. 2004. One thousand jaguars (*Panthera onca*) in Bolivia's Chaco? Camera trapping in the Kaa-Iya National Park. *Journal of Zoology* (London) 262: 295–304.

Magri, D., G.G. Vendramin, B. Comps, I. Dupanloup, T. Geburek, D. Gömöry, M. Latałowa, T. Litt, L. Paule, J.M. Roure, I. Tantau, W.O. van der Knapp, R.J. Petit, and J.L. de Beaulieu. 2006. A new scenario for the Quaternary history of European beech populations: Paleobotanical evidence and genetic consequences. *New Phytologist* 171: 199–221.

Malhi, Y., L.E.O.C. Aragao, D. Galbraith, C. Huntingford, R.A. Fisher, P. Zelazowski, S. Sitch, C. McSweeney, and P. Meir. 2009. Exploring the likelihood and mechanism of a climate-change-induced dieback of the Amazon rainforest. *Proceedings of the National Academy of Sciences, USA* doi: 10.1073_pnas.0804619106.

Malhi, Y., T. Baker, O.L. Phillips, S. Almeida, E. Alvarez, L. Arroyo, J. Chave, C.I. Czimczik, A. Di Fiore, N. Higuchi, T.J. Killeen, S.G. Laurance, W.F. Laurance, S.L. Lewis, L.M. Mercado-Montoya, A. Monteagudo, D.A. Neill, P. Núñez-Vargas, S. Patiño, N.C.A. Pitman, C.A. Quesada, R. Salomao, J.N. Macedo-Silva, A. Torres-Lezama, R. Vásquez-Martínez, J. Terborgh, B. Vincenti, and J. Lloyd. 2004. The above-ground coarse wood productivity of 104 Neotropical forest plots. *Global Change Biology* 10: 563–91.

Malhi, Y., J.T. Roberts, R.A. Betts, T.J. Killeen, W. Li, and C.A. Nobre. 2008. Climate change, deforestation, and the fate of the Amazon. *Science* 319: 169–172.

Mandujano, S. 1999. Variation in herd size of collared peccaries in a Mexican tropical forest. *Southwestern Naturalist* 44: 199–204.

Manning, A.D., J. Fischer, and D.B. Lindenmayer. 2006. Scattered trees are keystone structures: Implications for conservation. *Biological Conservation* 132: 311–21.

Manor, R., and D. Saltz. 2004. The impact of free-roaming dogs on gazelle kid/female ratio in a fragmented area. *Biological Conservation* 119: 231–36.

Maranz, S. 2009. Tree mortality in the African Sahel indicates an anthropogenic ecosystem displaced by climate change. *Journal of Biogeography* 36: 1181–93.

Maraun, M., J.A. Salamon, K. Schneider, M. Schaefer, and S. Scheu. 2003. Oribatid mite and collembolan diversity, density and community structure in a moder beech forest (*Fagus sylvatica*): Effects of mechanical perturbations. *Soil Biology and Biochemistry* 35: 1387–94.

Maraun, M., H. Schatz, and S. Scheu. 2007. Awesome or ordinary? Global diversity patterns of oribatid mites. *Ecography* 30: 209–16.

Mares, M.A., and K.A. Ernest. 1995. Population and community ecology of small mammals in a gallery forest of central Brazil. *Journal of Mammalogy* 76: 750–68.

Marimon, B.S., E. de S. Lima, T.G. Duarte, L.C. Chieregatto, and J.A. Ratter. 2006. Observations on the vegetation of northeastern Mato Grosso, Brazil. 4. An analysis of the cerrado-Amazonian forest ecotone. *Edinborough Journal of Botany* 63: 323–41.

Mari Mutt, J.A., P.F. Bellinger, and F. Janssens. 1996–2001. Checklist of the Collembola: Supplement of the Catalog of the Neotropical Collembola—May 1996–2001. http://www.collembola.org/publicat/neotrcat.htm.

Marin, D., and E. Medina. 1981. Duración foliar, contenido de nutrientes, y esclerofilia en árboles de un bosque muy seco tropical. *Acta Cientifica Venezolana* 32: 508–14.

Marod, D., U. Kutintara, H. Tanaka, and T. Nakashuka. 2002. The effects of drought and fire on seed and seedling dynamics in a tropical seasonal forest in Thailand. *Plant Ecology* 161: 41–57.

Marquis, R.J. 1988. Phenological variation in the Neotropical understory shrub *Piper arieianum*: Causes and consequences. *Ecology* 69: 1552–65.

Martínez, M. 1969. *Las Plantas Medicinales de México*. México, D.F: Botas.

Martínez-Ballesté, A., C. Martorell, and J. Caballero. 2006. Cultural or ecological sustainability? The effect of cultural change on sabal palm management among the lowland Maya of Mexico. *Ecology and Society* 11: 27.

Martínez-Yrízar, A. 1995. Biomass distribution and primary productivity of tropical dry forests. In *Seasonally Dry Tropical Forests*, ed. S.H. Bullock, H.A. Mooney, and E. Medina, 326–45. Cambridge: Cambridge University Press.

Martínez-Yrízar, A., A. Búrquez, and J.M. Maass. 2000. Structure and functioning of tropical deciduous forest in western Mexico. In *The Tropical Deciduous Forest of Alamos: Biodiversity of a Threatened Ecosystem in Mexico*, ed. R.H. Robichaux and D.A. Yetman, 19–35. Tucson: University of Arizona.

Martínez-Yrízar, A., J.M. Maass, L.A. Pérez-Jiménez, and J. Sarukhán. 1996. Net primary productivity of a tropical deciduous forest ecosystem in western Mexico. *Journal of Tropical Ecology* 12: 169–75.

Martínez-Yrízar, A., and J. Sarukhán. 1990. Litterfall patterns in a tropical deciduous forest in Mexico over a five-year period. *Journal of Tropical Ecology* 6: 433–44.

Martínez-Yrízar, A., J. Sarukhán, A. Pérez-Jiménez, E. Rincón, J.M. Maass, A. Solís-Magallanes, and L. Cervantes. 1992. Above-ground phytomass of a tropical deciduous forest on the coast of Jalisco, Mexico. *Journal of Tropical Ecology* 8: 87–96.

Martini, A.M.Z., R.A.F. Lima, G.A.D.C. Franco, and R.R. Rodrigues. 2008. The need for full inventories of tree modes of disturbance to improve forest dynamics comprehension: An example from a semideciduous forest in Brazil. *Forest Ecology and Management* 255: 1479–88.

Martins, K., L.J. Chaves, G.S.C. Buso, and P.Y. Kageyama. 2006. Mating system and fine-scale spatial genetic structure of *Solanum lycocarpum* St. Hil. (Solanaceae) in the Brazilian Cerrado. *Conservation Genetics* 7: 957–69.

Mas, J.F., A. Velazquez, J.L. Palacio-Prieto, G. Bocco, A. Peralta, and J. Prado. 2002. Assessing forest resources in Mexico: Wall-to-wall land use/cover mapping. *Photogrammetric Engineering and Remote Sensing* 68: 966–68.

Mascia Vieira, D.L., V. Vinícius de Lima, A. Cássio Sevilha, and A. Scariot. 2008. Consequences of dry season seed dispersal on seedling establishment of dry forest trees: Should we store seeds until the rains? *Forest Ecology and Management* 256: 471–81.

Masera, O., A. Ghillardi, R. Drigo, and M. Trossero. 2006. Wisdom: A GIS-based supply demand mapping tool for woodfuel management. *Biomass and Bioenergy* 30: 618–37.

Masera, O., M. Ordoñez, and R. Dirzo. 1997. Carbon emissions from Mexican forests: Current situation and long term scenarios. *Climatic Change* 35: 265–95.

Mastache, F., G. Alba, and E.N. Moret. 1982. *El Trabajo de la Palma en la Región de la Montaña, Guerrero*. México: Universidad Autónoma de Guerrero.

Matos, F.J.A. 1999. *Plantas da Medicina Popular do Nordeste*. Fortaleza, Brazil: Editora UFC.

Mayfield, M.M., and G.C. Daily. 2005. Countryside biogeography of Neotropical herbaceous and shrubby plants. *Ecological Applications* 15: 423–39.

Mayle, F.E. 2004. Assessment of the Neotropical dry forest hypothesis in the light of palaeoecological and vegetation model simulations. *Journal of Quaternary Science* 19: 713–20.

——. 2006. The Late Quaternary biogeographic history of South American seasonally dry tropical forests: Insights from palaeoecological data. In *Neotropical Savannas and Seasonally Dry Forests: Plant Diversity, Biogeography, and Conservation*, ed. R.T. Pennington, G. Lewis, and J.A. Ratter, 395–416. Boca Raton, FL: CRC Press.

Mayle, F.E., D.J. Beerling, W.D. Gosling, and M.B. Bush. 2004. Responses of Amazonian ecosystems to climatic and atmospheric carbon dioxide changes since the last glacial maximum. *Philosophical Transactions of the Royal Society B* 29: 499–514.

McCrary, J.K., B. Walsh, and A.L. Hammett. 2005. Species, sources, seasonality and sustainability of fuelwood commercialization in Masaya, Nicaragua. *Forest Ecology and Management* 205: 299–309.

McCune, B., and M.J. Mefford. 1999. *Multivariate Analysis of Ecological Data, Version 4.17*. Gleneden Beach, OR: MjM Software Design.

McDonald, M.A., and K.P. McLaren. 2003. Coppice regrowth in a disturbed tropical dry limestone forest in Jamaica. *Forest Ecology and Management* 180: 98–111.

McDowell, N., W.T. Pockman, C.D. Allen, D.D. Breshears, N. Cobb, T. Kolve, J. Plaut, J. Sperry, A. West, D.G. Williams, and E.A. Yepez. 2008. Mechanisms of plant survival and mortality during drought: Why do some plants survive while others succumb to drought? *New Phytologist* 178: 719–39.

McFadden, K.W., R.N. Sambrotto, R.A. Medellín, and M.E. Gompper. 2006. Feeding habits of endangered pygmy raccoons (*Procyon pygmaeus*) based on stable isotope and fecal analyses. *Journal of Mammalogy* 87: 501–9.

McGuire, K.L. 2006. Common ectomycorrhizal networks maintain monodominace in a tropical rain forest. *Ecology* 88: 567–74.

McLaren, K., and M. McDonald. 2005. Seasonal patterns of flowering and fruiting in a dry tropical forest in Jamaica. *Biotropica* 37: 584–90.

McNab, B.K., and P. Morrison. 1963. Body temperature and metabolism in subspecies of *Peromyscus* from arid and mesic environments. *Ecological Monographs* 33: 68–82.

McNeely, J.A., and G. Schroth. 2006. Agroforestry and biodiversity conservation: Traditional practices, present dynamics, and lessons for the future. *Biodiversity and Conservation* 15: 549–54.

Mead, R.A. 1993. Embryonic diapause in vertebrates. *Journal of Experimental Zoology* 266: 629–41.

Medellín, R.A., M. Equihua, and M.A. Amín. 2000. Bat diversity and abundance as indicators of disturbance in Neotropical rainforests. *Conservation Biology* 14: 1666–75.

Medellín, R.A., and O. Gaona. 1999. Seed dispersal by bats and birds in forest and disturbed habitats of Chiapas, Mexico. *Biotropica* 31: 478–85.

Medina, A., C.A. Harvey, D. Sánchez, S. Vílchez, and B. Hernández. 2007. Bat diversity and movement in an agricultural landscape in Matiguás, Nicaragua. *Biotropica* 39: 120–28.

Medina, E. 1995. Diversity of life forms of higher plants in Neotropical dry forests. In *Seasonally Dry Tropical Forests*, ed. S.H. Bullock, H.A. Mooney, and E. Medina, 221–38. Cambridge: Cambridge University Press.

Medina, E., and M. Zewler. 1972. Soil respiration in tropical plant communities. In *Tropical Ecology, with Emphasis on Organic Matter Production*, ed. P. Golley and F.B. Golley, 245–67. Athens: University of Georgia.

Medina, E., and E. Cuevas. 1990. Propiedades fotosintéticas y eficiencia de uso de agua de plantas leñosas del bosque deciduo de Guánica: Consideraciones generales y resultados preliminares. *Acta Científica* (Puerto Rico) 4: 25–36.

Meir, P., P. Cox, and J. Grace. 2006. The influence of terrestrial ecosystems on climate. *Trends in Ecology and Evolution* 21: 254–60.

Meir, P., J. Grace, and A.C. Miranda. 2001. Leaf respiration in two tropical rain forests: Constraints on physiology by phosphorus, nitrogen and temperature. *Functional Ecology* 15: 378–87.

Meir, P., D.B. Metcalfe, A.C.L. Costa, and R.A. Fisher. 2008. The fate of assimilated carbon during drought: Impacts on respiration in Amazon rainforests. *Philosophical Transactions of the Royal Society B* 363: 1849–55.

Mejia-Recamier, E., and J.G. Palacios-Vargas. 2007. Three new species of *Neoscirula* (Prostigmata: Cunaxidae) from a tropical dry forest in Jalisco, Mexico. *Zootaxa* 1545: 17–31.

Melo, F.P.L., B. Rodríguez-Herrera, R.L. Chazdon, R. Medellin, and G. Ceballos. 2009. Small tent-roosting bats promote dispersal of large-seeded plants in a Neotropical forest. *Biotropica* 41: 737–43.

Mendes Pontes, A.R. 1999. Environmental determinants of primate abundance in Maracá Island, Roraima, Brazilian Amazonia. *Journal of Zoology* (London) 247: 189–99.

——. 2004. Ecology of a community of mammals in a seasonally dry forest in Roraima, Brazilian Amazon. *Mammalian Biology* 69: 319–36.

Mendes Pontes, A.R., and D.J. Chivers. 2007. Peccary movements as determinants of the movements of large cats in Brazilian Amazonia. *Journal of Zoology* (London) 273: 257–65.

Mercado, L.M., N. Belluin, S. Sitch, O. Boucher, C. Huntingford, M. Wild, and P.M. Cox. 2009. Impact of changes in diffuse radiation on the global land carbon sink. *Nature* 458: 1014–17.

Mertens, B., D. Kaimowitz, A. Puntodewo, J. Vanclay, and P. Mendez. 2004. Modeling deforestation at distinct geographic scales and time periods in Santa Cruz, Bolivia. *International Regional Science Review* 27: 271–96.

Metcalfe, D.B., P. Meir, L. Aragão, A.C.L. da Costa, A.P. Braga, P.H.L. Gonçalves, J. de Athaydes Silva Junior, S.S. de Almeida, L.A. Dawson, and Y. Malhi. 2008. The effects of water availability on root growth and morphology in an Amazon rainforest. *Plant and Soil* 311: 189–99.

Michon, G., H. De Foresta, P. Levang, and F. Verdeaux. 2007. Domestic forests: A new paradigm for integrating local communities' forestry into tropical forest science. *Ecology and Society* 12: 1. http://www.ecologyandsociety.org/vol12/iss2/art1/.

Middleton, B.A., E. Sanchez-Rojas, B. Suedmeyer, and A. Michels. 1997. Fire in a tropical dry forest of Central America: A natural part of the disturbance regime? *Biotropica* 29: 515–17.

Miles, L., A.C. Newton, R.S. DeFries, C. Ravilious, I. May, S. Blyth, V. Kapos, and J.E. Gordon. 2006. A global overview of the conservation status of tropical dry forests. *Journal of Biogeography* 33: 491–505.

Miller, A., C. Schlagnhaufer, M. Spalding, and S. Rodermel. 2000. Carbohydrate regulation of leaf development: Prolongation of leaf senescence in Rubisco antisense mutants of tobacco. *Photosynthesis Research* 63: 1–8.

Miller, K., E. Chang, and N. Johnson. 2001. *Defining Common Ground for the Mesoamerican Biological Corridor*. Washington, DC: World Resources Institute.

Miller, P.M., and J.B. Kaufman. 1998. Effect of slash and burn agriculture on species abundance and composition of a tropical deciduous forest. *Forest Ecology and Management* 103: 191–201.

Milne, B. 1999. Motivation and benefits of complex systems approaches in ecology. *Ecosystems* 1: 449–56.

Miranda, A. 2002. Diversidad, historia natural, ecología y conservación de los mamíferos de Chamela. In *Historia Natural del Bosque Caducifolia de Chamela*, ed. F.A. Noguera, M. Quesada, J. Vega, and A. García-Aldrete, 359–77. México City: Instituto de Biología, Universidad Nacional Autónoma de México.

Miranda, E.J., G.L. Vourlitis, N. Priante, P.C. Priante, J.H. Campelo Jr., G.S. Suli, C.L. Fritzen, F. de Almeida, and S. Shiraiwa. 2005. Seasonal variation in the leaf gas exchange of tropical forest trees in the rain forest-savanna transition of the southern Amazon Basin. *Journal of Tropical Ecology* 21: 451–60.

Miranda, F. 1947. Estudios sobre la vegetación de México. V. Rasgos de la vegetación de la cuenca del Río Balsas. *Revista de la Sociedad Mexicana de Historia Natural* 8: 95–114.

Mitchell, A.W., K. Secoy, N. Mardas, M. Trivedi, and R. Howard. 2008. *Forests Now in the Fight against Climate Change*. Oxford: Global Canopy Programme.

Montaño, N.M., F. García-Oliva, and V.J. Jaramillo. 2007. Dissolved organic carbon affects soil microbial activity and nitrogen dynamics in a Mexican tropical deciduous forest. *Plant and Soil* 295: 265–77.

Monteiro, J.M., C.F. Almeida, U.P. Albuquergue, R.F.P. Lucena, A.T.N. Florentino, and R.L.C. Oliveira. 2006. Use and traditional management of *Anadenanthera colubrine* (Vell.) Brenan in the semi-arid region of northeastern Brazil. *Journal of Ethnobiology and Ethnomedicine* 2: 6–13.

Montes-Hernández, S., and L.E. Eguiarte. 2002. Genetic structure and indirect estimates of gene flow in three taxa of Cucurbita (Cucurbitaceae) in western Mexico. *American Journal of Botany* 89: 1156–63.

Moody, A., and C. Woodcock. 1995. The influence of scale and the spatial characteristics of landscapes on land-cover mapping using remote sensing. *Landscape Ecology* 10: 363–79.

Mooney, H.A., S.H. Bullock, and E. Medina. 1995. Introduction. In *Seasonally Dry Tropical Forests*, ed. S.H. Bullock, H.A. Mooney, and E. Medina, 1–8. Cambridge: Cambridge University Press.

Moorcroft, P.R. 2006. How close are we to a predictive science of the biosphere? *Trends in Ecology and Evolution* 21: 401–7.

Moore, J.E., and R.K. Swihart. 2007. Importance of fragmentation-tolerant species as seed dispersers in disturbed landscapes. *Oecologia* 151: 663–74.

Moreira, M.Z., F.G. Scholz, S.J. Bucci, L.S. Sternberg, G. Goldstein, F.C. Meinzer, and A.C. Franco. 2003. Hydraulic lift in a Neotropical savanna. *Functional Ecology* 17: 573–81.

Morellato, L.P.C. 1992. Nutrient cycling in two south-east Brazilian forests. 1. Litterfall and litter standing crop. *Journal of Tropical Ecology* 8: 205–15.

Moreno, E. 1998. Variación espacial y temporal de la conductancia estomática y del potencial hídrico foliar de una selva baja caducifolia en Chamela, Jalisco, México. BS diss., Universidad Michoacana de San Nicolás de Hidalgo.

Moura, A.C.D.A. 2007. Primate group size and abundance in the Caatinga dry forest, northeastern Brazil. *International Journal of Primatology* 28: 1279–97.

Muchoney, D., J. Borak, H. Chi, M. Friedl, S. Gopal, J. Hodges, N. Morrow, and A. Strahler. 2000. Application of the MODIS global supervised classification model to vegetation and land cover mapping of Central America. *International Journal of Remote Sensing* 21: 1115–38.

Mulkey, S.S., S.J. Wright, and A.P. Smith. 1991. Drought acclimation of an understory shrub (*Psychotria limonensis*; Rubiaceae) in a seasonally dry tropical forest in Panama. *American Journal of Botany* 78: 579–87.

Muñoz, D., C.A. Harvey, F.L. Sinclair, J. Mora, and M. Ibrahim. 2003. Conocimiento local de la cobertura arbórea en sistemas de producción ganadera en dos localidades de Costa Rica. *Agroforestería en las Américas* 10: 61–68.

Muñoz, J.D. 1990. Anacardiaceae. In *Flora del Paraguay*, ed. R. Spichiger and L. Ramella. Genève: Conservatoire et Jardin Botaniques, Ville Genève and Missouri Botanical Garden.

Murali, K.S., and R. Sukumar. 1994. Reproductive phenology of a tropical dry forest in Mudumalai, southern India. *Journal of Ecology* 82: 759–67.

Murawski, D.A., and J.L. Hamrick. 1992. Mating system and phenology of *Ceiba pentandra* (Bombacaceae) in central Panama. *Journal of Heredity* 83: 401–4.

Murphy, P.G., and A.E. Lugo. 1986a. Ecology of tropical dry forest. *Annual Review of Ecology and Systematics* 17: 67–88.

———. 1986b. Structure and biomass of a subtropical dry forest in Puerto Rico. *Biotropica* 18: 89–96.

———. 1995. Dry forests of Central America and the Caribbean. In *Seasonally Dry Tropical Forests*, ed. S.H. Bullock, H.A. Mooney, and E. Medina, 9–34. Cambridge: Cambridge University Press.

Murren, C.J. 2002. Effects of habitat fragmentation on pollination: Pollinators, pollinia viability and reproductive success. *Journal of Ecology* 90: 100–107.

———. 2003. Spatial and demographic population genetic structure in *Catasetum viridiflavum* across a human-disturbed habitat. *Journal of Evolutionary Biology* 16: 333–42.

Murty, D., M.U.F. Kirschbaum, R.E. McMurtrie, and H. McGilvray. 2002. Does conversion of forest to agricultural land change soil carbon and nitrogen? A review of the literature. *Global Change Biology* 8: 105–23.

Muscarella, R., and T. Fleming. 2007. The role of frugivorous bats in tropical forest succession. *Biological Reviews* 82: 573–90.

Mutere, F.A. 1967. The breeding biology of equatorial vertebrates: Reproduction in the fruit bat, *Eidolon helvum*, at latitude 0°20' N. *Journal of Zoology* (London) 153: 153–61.

Naciri, Y., S. Caetano, R.T. Pennington, D. Prado, and R. Spichiger. 2006. Population genetics and inference of ecosystem history: An example using two Neotropical seasonally dry forest species. In *Neotropical Savannas and Dry Forests: Plant Diversity, Biogeography, and Conservation*, ed. R.T. Pennington, G.P. Lewis, and J.A. Ratter, 417–32. Boca Raton, FL: CRC Press.

Naciri-Graven, Y., S. Caetano, D.E. Prado, R.T. Pennington, and R. Spichiger. 2005. Development and characterization of 11 microsatellite markers in a widespread Neotropical seasonally dry forest tree species, *Geoffroea spinosa* Jacq. (Leguminosae). *Molecular Ecology Notes* 5: 542–45.

Naranjo, L.G. 1992. Estructura de la avifauna en un área ganadera en el Valle del Cauca, Colombia. *Caldasia* 17: 55–66.

Nardoto, G.B., M.M.C. Bustamante, A.S. Pinto, and C.A. Klink. 2006. Nutrient use efficiency at ecosystem and species level in savanna areas of central Brazil and impacts of fire. *Journal of Tropical Ecology* 22: 191–201.

Nason, J.D., and J.L. Hamrick. 1997. Reproductive and genetic consequences of forest fragmentation: Two case studies of Neotropical canopy trees. *Journal of Heredity* 88: 264–76.

Nava-Mendoza, M., L. Galicia, and F. García-Oliva. 2000. Efecto de dos especies de árboles remanentes y de un pasto en la capacidad amortiguadora del pH del suelo en un ecosistema tropical estacional. *Boletín de la Sociedad Botánica de México* 67: 17–24.

Neal, B.R. 1984. Relationship between feeding habits, climate and reproduction of small mammals in Meru National Park, Kenya. *African Journal of Ecology* 22: 195–205.

Nei, M. 1987. *Molecular Evolutionary Genetics*. New York: Colombia University Press.

Neill, C., J.M. Melillo, P.A. Steudler, C.C. Cerri, J.F.L. Moraes, M.C. Piccolo, and M. Brito. 1997. Soil carbon and nitrogen stocks following forest clearing for pasture in the southwestern Brazilian Amazon. *Ecological Applications* 7: 1216–25.

Nepstad, D., G. Carvalho, A.C. Barros, A. Alencar, J.P. Capobianco, J. Bishop, P. Moutinho, P. Lefebvre, U.L. Silva Jr., and E. Prins. 2001. Road paving, fire regime feedbacks, and the future of Amazon forests. *Forest Ecology and Management* 154: 395–407.

Nepstad, D.C., A. Verissimo, A. Alencar, C. Nobre, E. Lima, P. Lefebvre, P. Schlesinger, C. Potter, P. Moutinho, E. Mendoza, M. Cochrane, and V. Brooks. 1999. Large-scale impoverishment of Amazonian forests by logging and fire. *Nature* 398: 505–8.

New, M., M. Hulme, and P. Jones. 2000. Representing twentieth-century space-time climate variability. Part II: Development of 1901–96 monthly grids of terrestrial surface climate. *Journal of Climate* 13: 2217–38.

Nicolas, V., and M. Colyn. 2003. Seasonal variations in population and community structure of small rodents in a tropical forest of Gabon. *Canadian Journal of Zoology* 81: 1034–46.

Nilsen, E.T., M.R. Sharifi, P.W. Rundel, I.N. Forseth, and J.R. Ehleringer. 1990. Water relations of stem succulent trees in north-central Baja California. *Oecologia* 82: 299–303.

Noble, I.R., A.M. Gill, and G.A.V. Bary. 1980. McArthur's fire danger meters expressed as equations. *Journal of Ecology* 5: 201–3.

Noblick, L.R. 1986. *Palmeiras das caatingas da Bahia e as potencialidades econômicas*. Simpósio sobre a Caatinga e sua Exploração Racional. Brasilia: EMBRAPA.

Nobre, C.A., P.J. Sellers, and J. Shukla. 1991. Amazonian deforestation and regional climate change. *Journal of Climate* 4: 957–88.

Noguera, F., S. Zaragoza-Caballero, J.A. Chemsak, A. Rodríguez-Palafox, E. Ramírez, E. González-Soriano, and R. Ayala. 2002. Diversity of the family Cerambycidae (Coleoptera) of the tropical dry forest of Mexico. 1. Sierra de Huautla, Morelos. *Annals of the Entomological Society of America* 95: 617–27.

Noodén, L.D., and A.C. Leopold. 1988. *Senescence and Aging in Plants*. London: Academic Press.

Norscia, I., V. Carrai, and S.M. Borgognini-Tarli. 2006. Influence of dry season and food quality and quantity on behavior and feeding strategy of *Propithecus verreauxi* in Kirindy, Madagascar. *International Journal of Primatology* 27: 1001–22.

Novick, R.R., C.W. Dick, M.R. Lemes, C. Navarro, A. Caccone, and E. Bermingham. 2003. Genetic structure of Mesoamerican populations of big-leaf mahogany (*Swietenia macrophylla*) inferred from microsatellite analysis. *Molecular Ecology* 12: 2885–93.

Noyes, J.S. 2000. Encyrtidae of Costa Rica (Hymenoptera: Chalcidoidea). 1. The subfamily Tetracneminae, parasitoids of mealybugs (Homoptera: Pseudococcidae). *Memoirs of the American Entomological Institute* 62: 1–355.

Núñez-Iturri, G., O. Olsson, and F.H. Howe. 2008. Hunting reduces recruitment of primate dispersed trees in Amazonian Peru. *Biological Conservation* 141: 1536–46.

Núñez Pérez, R. 2006. Area de actividad, patrones de actividad y movimiento del jaguar (*Panthera onca*) y del puma (*Puma concolor*) en la Reserva de la Biosfera Chamela-Cuixmala, Jalisco. MS diss., Universidad Nacional Autónoma de México.

Olivares, E., and E. Medina. 1992. Water and nutrient relations of woody perennials from tropical dry forests. *Journal of Vegetation Science* 3: 383–92.

Oliveira, A.R., R.A. Norton, and G.J. Des-Moraes. 2005. Edaphic and plant inhabiting oribatid mites (Acari: Oribatida) from Cerrado and Mata Atlantica ecosystems in the state of Sao Paulo, southeast Brazil. *Zootaxa* 1049: 49–68.

Oliveira, P.E., and P.E. Gibbs. 2000. Reproductive biology of woody plants in a cerrado community of central Brazil. *Flora* 195: 311–29.

Oliveira, P.E.A.M., and A.G. Moreira. 1992. Anemocoria em espécies de Cerrado e Mata de Galeria de Brazília, DF. *Revista Brasileira de Botânica* 15: 163–74.

Oliveira, R.S., L. Bezerra, E.A. Davidson, F. Pinto, C.A. Klink, D.C. Nepstad, and A. Moriera. 2005. Deep root function in soil water dynamics in cerrado savannas of central Brazil. *Functional Ecology* 19: 574–81.

Olivera, O.G. 1998. Estudio fitoquímico del Cuachalalate (*Amphipterygium adstringens*). MS diss., Colegio de Postgraduados, Montecillo, Estado de México.

Olson, D., and E. Dinerstein. 2002. The Global 200: Priority ecoregions for global conservation. *Annals of the Missouri Botanical Garden* 89: 199–224.

Olson, D.M., E. Dinerstein, E.D. Wikramanayake, N.D. Burgess, G.V.N. Powell, E.C. Underwood, J.A. D'amico, I. Itoua, H.E. Strand, J.C. Morrison, C.J. Loucks, T.F. Allnutt, T.H. Ricketts, Y. Kura, J.F. Lamoreux, W.W. Wettengel, P. Hedao, and K.R. Kassem. 2001. Terrestrial ecoregions of the world: A new map of life on Earth. *BioScience* 51: 933–38.

Ono, K., Y. Nishi, A. Watanabe, and I. Terashima. 2001. Possible mechanisms of adaptive leaf senescence. *Plant Biology* 3: 234–43.

Ono, K., and A. Watanabe. 1997. Levels of endogenous sugars, transcripts of rbcS and rbcL, and of RuBisCo protein in senescing sunflower leaves. *Plant and Cell Physiology* 38: 1032–38.

Opler, P.A., G.W. Frankie, and H.G. Baker. 1976. Rainfall as a factor in the release, timing, and synchronization of anthesis by tropical tress and shrubs. *Journal of Biogeography* 3: 231–36.

———. 1980. Comparative phenological studies of treelet and shrub species in tropical wet and dry forests in the lowlands of Costa Rica. *Journal of Ecology* 68: 167–88.

Ortega, A.T. 1995. El desarrollo socioeconómico de Jalisco. Perspectivas de recursos naturales. *Revista Universidad de Guadalajara* 41–48.

Ortíz, T. 2001. Estructura arbórea en sitios perturbados y caracterizados por la presencia de Mimosa arenosa (Willd.) Poir. var. leiocarpa (D.C.) Barneby, en el bosque tropical seco de la Costa de Jalisco, México. BS diss., Universidad Nacional Autónoma de México.

Ostner, J., P.M. Kappeler, and M. Heistermann. 2002. Seasonal variation and social correlates of androgen excretion in male redfronted lemurs (*Eulemur fulvus rufus*). *Behavioral Ecology and Sociobiology* 52: 485–95.

Ostrom, E., J. Burger, C.B. Field, R.B. Norgaard, and D. Policansky. 1999. Revisiting the commons: Local lessons, global challenges. *Science* 284: 278–82.

Otterstrom, S.M., M.W. Schwartz, and I. Velázquez-Rocha. 2006. Responses to fire in selected tropical dry forest trees. *Biotropica* 38: 592–98.

Pagiola, S., P.J. Agostini, J. Gobbi, C. de Haan, M. Ibrahim, E. Murgueitio, E. Ramírez, M. Rosales, and J.P. Ruiz. 2005. Paying for biodiversity conservation services: Experience in Colombia, Costa Rica, and Nicaragua. *Mountain Research and Development* 25: 206–11.

Palacios-Vargas, J.G., G. Castaño-Meneses, J.A. Gómes-Anaya, A. Martínez-Yrízar, E. Mejia-Recamier, and J. Martínez-Sánchez. 2007. Litter and soil arthropod diversity and density in a tropical dry forest ecosystem in western Mexico. *Biodiversity and Conservation* 16: 3703–17.

Pallardy, S.G., and J.L. Rhoads. 1993. Morphological adaptations to drought in seedlings of deciduous angiosperms. *Canadian Journal of Forest Research* 23: 1766–74.

Pariona, W., T.S. Fredericksen, and J.C. Licona. 2003. Natural regeneration and liberation of timber species in logging gaps in two Bolivian tropical forests. *Forest Ecology and Management* 181: 313–22.

Parra-Tabla, V., C.F.S. Vargas, S. Magaña-Rueda, and J. Navarro. 2000. Female and male pollination success of *Oncidium ascendens* Lindey (Orchidaceae) in two contrasting habitat patches: Forest vs. agricultural field. *Biological Conservation* 94: 335–40.

Paul, E.A., and F.E. Clark. 1989. *Soil Microbiology and Biochemistry*. San Diego: Academic Press.

Pearson, D.L., and A.P. Vogler. 2001. *Tiger Beetles: The Evolution, Ecology, and Diversity of the Cicindelids*. Ithaca, NY: Cornell University Press.

Pennington, R.T., M. Lavin, and A. Oliveira-Filho. 2009. Woody plant diversity, evolution, and ecology in the tropics: Perspectives from seasonally dry tropical forests. *Annual Review of Ecology, Evolution and Systematics* 40: 437–57.

Pennington, R.T., M. Lavin, D.E. Prado, C.A. Pendry, S. Pell, and C.A. Butterworth. 2004. Historical climate change and speciation: Neotropical seasonally dry forest plants show patterns of both Tertiary and Quaternary diversification. *Philosophical Transactions of the Royal Society B* 359: 515–38.

Pennington, R.T., G.P. Lewis, and J.A. Ratter. 2006. An overview of the plant diversity, biogeography and conservation of Neotropical savannas and seasonally dry forests. In *Neotropical Savannas and Seasonally Dry Forests: Plant Diversity, Biogeography, and Conservation*, ed. R.T. Pennington, G.P. Lewis, and J.A. Ratter, 1–29. Boca Raton, FL: CRC Press.

Pennington, R.T., D.E. Prado, and C.A. Pendry. 2000. Neotropical seasonally dry forests and Quaternary vegetation changes. *Journal of Biogeography* 27: 261–73.

Pennington, R.T., J.E. Richardson, and Q.C.B. Cronk. 2004. Plant phylogeny and the origin of major biomes: Introduction and synthesis. *Philosophical Transactions of the Royal Society B* 359: 1455–65.

Pennington, R.T., J.E. Richardson, and M. Lavin. 2006. Insights into the historical construction of species-rich biomes from dated plant phylogenies, phylogenetic community structure and neutral ecological theory. *New Phytologist* 172: 605–16.

Pennington, T.D., and J. Sarukhán. 1998. *Arboles tropicales de México: manual para la identificación de las principales especies*. México, D.F.: Universidad Nacional Autónomo de México.

Peres, C.A. 1997. Effects of habitat quality and hunting pressure on arboreal folivore densities in Neotropical forests: A case study of howler monkeys (*Alouatta* spp.). *Folia Primatologica* 68: 199–222.

Peres, C.A., and I.R. Lake. 2003. Extent of nontimber resource extraction in tropical forests: Accessibility to game vertebrates by hunters in the Amazon basin. *Conservation Biology* 17: 521–35.

Perfecto, I., R.A. Rice, R. Greenberg, and M.E. van der Voort. 1996. Shade coffee: A disappearing refuge for biodiversity. *BioScience* 46: 598–608.

Perfecto, I., and J. Vandermeer. 2002. Quality of agroecological matrix in a tropical montane landscape: Ants in coffee plantations in southern Mexico. *Conservation Biology* 16: 174–82.

Perret, M. 1992. Environmental and social determinants of sexual function in the male lesser mouse lemur (*Microcebus murinus*). *Folia Primatologica* 59: 1–25.

Perret, M., F. Aujard, and G. Vannier. 1998. Influence of daylength on metabolic rate and daily water loss in the male prosimian primate *Microcebus murinus*. *Comparative Biochemistry and Physiology, A: Comparative Physiology* 119: 981–89.

Peters, C.M. 1994. *Sustainable Harvest of Non-timber Plant Resources in Tropical Moist Forests: An Ecological Primer*. Washington, DC: Biodiversity Support Program.

Peters, C.M., S.E. Purata, M. Chibnick, B.J. Brosi, A.M. López, and M. Ambrosio. 2003. The life and times of *Bursera glabrifolia* (HBK) Engl. in Mexico: A parable for ethnobotany. *Economic Botany* 57: 431–41.

Petit, L.J., and D.R. Petit. 2003. Evaluating the importance of human-modified lands for Neotropical bird conservation. *Conservation Biology* 17: 687–94.

Petit, R.J., E. Pineau, B. Demesure, R. Bacilieri, A. Ducousso, and A. Kremer. 1997. Chloroplast DNA footprints of postglacial recolonization by oaks. *Proceedings of the National Academy of Sciences, USA* 94: 9996–10001.

Pfaff, A.S.P., S. Kerr, F.R. Hughes, S. Li, G.A. Sanchez-Azofeifa, D. Shimel, J. Tosi, and V. Watson. 2000. The Kyoto Protocol and payments for tropical forest: An interdisciplinary method for estimating carbon-offset supply and increasing the feasibility of a carbon market under the CDM. *Ecological Economics* 35: 2003–221.

Phillips, O.L., and 65 coauthors. 2009. Drought sensitivity of the Amazon rainforest. *Science* 323: 1344–47.

Pickett, S.T.A., M. Cadenasso, and J.M. Grove. 2005. Biocomplexity in coupled natural-human systems: A multidimensional framework. *Ecosystems* 8: 225–32.

Pickett, S.T.A., S.L. Collins, and J. Armesto. 1987. A hierarchical consideration of causes and mechanisms of succession. *Vegetatio* 69: 109–14.

Pimienta Barrios, E., and P. Nobel. 1994. Pitaya (*Stenocereus* spp., Cactaceae): An ancient and modern fruit crop of Mexico. *Economic Botany* 48: 76–83.

Piperno, D.R. 2006. Quaternary environmental history and agricultural impact on vegetation in Central America. *Annals of the Missouri Botanical Garden* 93: 274–96.

Pires, J.M. 1974. Tipos de vegetação da Amazônia. *Brasil Florestal* 17: 48–58.

Pires, J.M., and G.T. Prance. 1985. The vegetation types of the Brazilian Amazon. In *Amazonia: Key Environments*, ed. G.T. Prance and T.E. Lovejoy, 109–45. Oxford: Pergamon Press.

Pitman, N.C.A., J. Terborgh, M.R. Silman, and P. Nuñez. 1999. Tree species distributions in an upper Amazonian forest. *Ecology* 80: 2651–61.

Pizo, M., and P. Oliveira. 2000. The use of fruits and seeds by ants in the Atlantic forest of southeast Brazil. *Biotropica* 32: 851–61.

Portillo-Quintero, C., and G.A. Sánchez-Azofeifa. 2009. Extent and conservation of tropical dry forest in the Americas. *Biological Conservation* 143 (1): 144–55.

Pourtau, N., M. Mares, S. Purdy, N. Quentin, A. Ruel, and A. Wingler. 2004. Interactions of abscisic acid and sugar signalling in the regulation of leaf senescence. *Planta* 219: 765–72.

Power, A. 1996. Arthropod diversity in forest patches and agroecosystems of tropical landscapes. In *Forest Patches in Tropical Landscapes*, ed. J. Schelhas and R. Greenberg, 91–110. Washington, DC: Island Press.

Powers, J.S. 2004. Changes in soil carbon and nitrogen after contrasting land-use transitions in northeastern Costa Rica. *Ecosystems* 7: 134–46.

Powers, J.S., J.M. Becknell, J. Irving, and D. Pérez-Aviles. 2009. Diversity and structure of regenerating tropical dry forests in Costa Rica: Geographic patterns and environmental drivers. *Forest Ecology and Management* 258: 959–70.

Prado, D.E. 2000. Seasonally dry forests of tropical South America: From forgotten ecosystems to a new phytogeographic unit. *Edinburgh Journal of Botany* 57: 437–61.

———. 2003. As caatingas da América do Sul. In *Ecologia e Conservação da Caatinga*, ed. I.R. Leal, M.Tabarelli, and J.M. Cardoso da Silva, 3–73. Recife, Brasil: Editora Universitária, Universidade Federal de Pernambuco.

Prado, D.E., and P.E. Gibbs. 1993. Patterns of species distributions in the dry seasonal forests of South America. *Annals of the Missouri Botanical Garden* 80: 902–27.

Prance, G.T. 1972. Chrysobalancaceae. *Flora Neotropica Monograph* 9: 85–86.

———. 1982. Forest refuges: Evidence from woody angiosperms. In *Biological Diversification in the Tropics*, ed. G.T. Prance, 137–59. New York: Columbia University Press.

Provan, J., and K.D. Bennett. 2008. Phylogeographic insights into cryptic refugia. *Trends in Ecology and Evolution* 23: 564–71.

Quesada, C.A., J. Lloyd, M. Schwarz, S. Patiño, T.R. Baker, C. Czimczik, N.M. Fyllas, L. Martinelli, G.B. Nardoto, J. Schmerler, A.J.B. Santos, M.G. Hodnett, R. Herrera, F.J. Luizão, A. Arneth, G. Lloyd, N. Dezzeo, I. Hilke, I. Kuhlmann, M. Raessler, W.A. Brand, H. Geilmann, J.O. Moraes Filho, F.P. Carvalho, R.M. Araujo Filho, J.E. Chaves, O.F. Cruz Junior, T.P. Pimentel, and R. Paiva. 2009. The chemical and physical properties of Amazon forest soils in relation to their genesis. *Biogeosciences Discussions* 6: 3923–92.

Quesada, M., E.J. Fuchs, and J.A. Lobo. 2001. Pollen load size, reproductive success, and progeny kinship of naturally pollinated flowers of the tropical dry forest tree *Pachira quinata* (Bombacaceae). *American Journal of Botany* 88: 2113–18.

Quesada, M., G.A. Sánchez-Azofeifa, M. Alvarez-Añorve, K.E. Stoner, L. Avila-Cabadilla, J. Calvo-Alvarado, A. Castillo, M.M. Espírito-Santo, M. Fagundes, G.W. Fernandes, J. Gamon, M. Lopezaraiza-Mikel, D. Lawrence, L.P. Cerdeira Morellato, J.S. Powers, F. de S. Neves, V. Rosas-Guerrero, R. Sayago, and G. Sanchez Montoya. 2009. Succession and management of tropical dry forests in the Americas: Review and new perspectives. *Forest Ecology and Management* 258: 1014–24.

Quesada, M., and K.E. Stoner. 2004. Threats to the conservation of the tropical dry forest in Costa Rica. In *Biodiversity Conservation in Costa Rica: Learning the Lessons in a*

Seasonal Dry Forest, ed. G.W. Frankie, A. Mata, and S.B. Vinson, 266–80. Berkeley: University of California Press.

Quesada, M., K.E. Stoner, J.A. Lobo, Y. Herrerías-Diego, C. Palacios-Guevara, M.A. Munguía-Rosas, K.A. O.-Salazar, and V. Rosas-Guerrero. 2004. Effects of forest fragmentation on pollinator activity and consequences for plant reproductive success and mating patterns in bat-pollinated bombacaceous trees. *Biotropica* 36: 131–38.

Quesada, M., K.E. Stoner, V. Rosas-Guerrero, C. Palacios-Guevara, and J.A. Lobo. 2003. Effects of habitat disruption on the activity of nectarivorous bats (Chiroptera: Phyllostomidae) in a dry tropical forest: Implications for the reproductive success of the Neotropical tree *Ceiba grandiflora*. *Oecologia* 135: 400–406.

Quigley, M.F., and W.J. Platt. 2003. Composition and structure of seasonally deciduous forests in the Americas. *Ecological Monographs* 73: 87–106.

Quirino, B.F., Y.S. Noh, E. Himelblau, and R.M. Amasino. 2000. Molecular aspects of leaf senescence. *Trends in Plant Science* 5: 278–82.

Rabinowitz, A. 1990. Notes on the behavior and movements of leopard cats, *Felis bengalensis*, in a dry tropical forest mosaic in Thailand. *Biotropica* 22: 397–403.

———. 1991. Behavior and movements of sympatric civet species in Huai-Kha-Khaeng Wildlife Sanctuary, Thailand. *Journal of Zoology* (London) 223: 281–98.

Racey, P.A. 1982. Ecology of bat reproduction. In *Ecology of Bats*, ed. T.H. Kunz, 57–104. New York: Plenum Press.

Racey, P.A., and A.C. Entwistle. 2000. Life-history and reproductive strategies of bats. In *Reproductive Biology of Bats*, ed. E.G. Crichton and P.H. Krutzsch, 364–401. London: Academic Press.

Raich, J.W., and W.H. Schlesinger. 1992. The global carbon dioxide flux in soil respiration and its relationship to vegetation and climate. *Tellus* 44: 81–99.

Raison, R.J. 1979. Modification of the soil environment by vegetation fires, with particular reference to nitrogen transformations: A review. *Plant and Soil* 51: 73–108.

Ramakrishnan, P.S., and O.P. Toky. 1981. Soil nutrient status of hill agro-ecosystem and recovery pattern after slash-and-burn agriculture (Jhum) in north-eastern India. *Plant and Soil* 60: 41–64.

Ramankutty, N., H.K. Gibbs, F. Achard, R. DeFries, J.A. Foley, and R.A. Houghton. 2007. Challenges to estimating carbon emissions from tropical deforestation. *Global Change Biology* 13: 51–66.

Ramírez, M., and M.L. Enríquez. 2003. Riqueza y diversidad de hormigas en sistemas silvopastoriles del Valle del Cauca, Colombia. *Livestock Research for Rural Development* 15.

Ramos, A.C.S., J.P. Lemos-Filho, R.A. Ribeiro, F.R. Santos, and M.B. Lovato. 2007. Phylogeography of the tree *Hymenaea stigonocarpa* (Fabaceae: Caesalpinioideae) and the influence of Quaternary climate changes in the Brazilian cerrado. *Annals of Botany* 100: 1219–28.

Ramos, F.N., and F.A.M. Santos. 2006. Floral visitors and pollination of *Psychotria tenuinervis* (Rubiaceae): Distance from the anthropogenic and natural edges of an Atlantic forest fragment. *Biotropica* 38: 383–89.

Randrianambinina, B., S. Mbotizafy, S. Rasoloharijaona, R.O. Ravoahangimalala, and E. Zimmermann. 2007. Seasonality in reproduction of *Lepilemur edwardsi*. *International Journal Primatology* 28: 783–90.

Ranganathan, J., and G.C. Daily. 2008. La biogeografía del paisaje rural: Oportunidad de conservación para paisajes de Mesoamérica manejados por humanos. In *Evalu-*

ación y conservación de biodiversidad en paisajes fragmentados de Mesoamérica, ed. C.A. Harvey and J.C. Saenz, 15–30. Heredia, Costa Rica: INBio.

Rasweiler, J.J., and N.K. Badwaik. 1997. Delayed development in the short-tailed fruit bat, *Carollia perspicillata*. *Journal of Reproduction and Fertility* 109: 7–20.

Ratcliffe, B.C. 2003. The dynastine scarab beetles of Costa Rica and Panama (Coleoptera: Scarabaeidae: Dynastinae). *Bulletin of the University of Nebraska State Museum* 16: 1–506.

Ratter, J.A. 1992. Transitions between cerrado and forest vegetation in Brazil. In *Neotropical Savannas and Seasonally Dry Forests: Plant Diversity, Biogeography, and Conservation*, ed. R.T. Pennington, G.P. Lewis, and J.A. Ratter, 417–29. Boca Raton, FL: CRC Press.

Ratter, J.A., G.P. Askew, R.F. Montgomery, and D.R. Gifford. 1978. Observations of the vegetation of northeastern Mato Grosso. 2. Forests and soils of the Rio Suiá-Missu area. *Philosophical Transactions of the Royal Society B* 203: 191–208.

Ratter, J.A., S. Bridgewater, and J.F. Ribeiro. 2003. Analysis of the Brazilian cerrado vegetation. 3: Comparison of the woody vegetation of 376 areas. *Edinburgh Journal of Botany* 60: 57–109.

Ratter, J.A., J.F. Ribeiro, and S. Bridgewater. 1997. The Brazilian Cerrado vegetation and threats to its biodiversity. *Annals of Botany* 80: 223–30.

Ratter, J.A., P.W. Richards, G. Argent, and D.R. Gifford. 1973. Observations on the vegetation of northeastern Mato Grosso. 1. The woody vegetation types of the Zavantina-Cachimbo expedition area. *Philosophical Transactions of the Royal Society B* 266: 449–92.

Ray, G.J., and B.J. Brown. 1995. Restoring Caribbean dry forest: Evaluation of tree propagation techniques. *Restoration Ecology* 3: 86–94.

Read, L., and D. Lawrence. 2003. Litter nutrient dynamics during succession in dry tropical forests of the Yucatan: Regional and seasonal effects. *Ecosystems* 6: 747–61.

Redford, K.H., and J.G. Robinson. 1991. Subsistence and commercial uses of wildlife in Latin America. In *Neotropical Wildlife Use and Conservation*, ed. J.G. Robinson and K.H. Redford, 6–23. Chicago: University of Chicago Press.

Reeder, D.M., K.M. Helgen, and D.E. Wilson. 2007. Global trends and biases in new mammal species discoveries. *Occasional Papers, The Museum, Texas Tech University* 269: 1–35.

Reich, P.B. 1995. Phenology of tropical forests: Patterns, causes, and consequences. *Canadian Journal of Botany* 73: 164–74.

Reich, P.B., and R. Borchert. 1982. Phenology and ecophysiology of the tropical tree, *Tabebuia neochrysantha* (Bignoniaceae). *Ecology* 63: 294–99.

——. 1984. Water-stress and tree phenology in a tropical dry forest in the lowlands of Costa Rica. *Journal of Ecology* 72: 61–74.

Renner, S.S. 2005. Relaxed molecular clocks for dating historical plant dispersal events. *Trends in Plant Science* 10: 550–58.

Rentería, L.Y., V.J. Jaramillo, A. Martínez-Yrízar, and A. Pérez-Jiménez. 2005. Nitrogen and phosphorus resorption in tree species of a Mexican tropical dry forest. *Trees* 19: 431–41.

Reyna-Hurtado, R., and G.W. Tanner. 2005. Habitat preferences of ungulates in hunted and nonhunted areas in the Calakmul forest, Campeche, Mexico. *Biotropica* 37: 676–85.

Rice, R.A., and R. Greenberg. 2004. Silvopastoral systems: Ecological and socioeconomic benefits and migratory bird conservation. In *Agroforestry and Biodiversity*

Conservation in Tropical Landscapes, ed. G.A. Schroth, B. Fonseca, C.A. Harvey, C. Gascon, H.L. Vasconcelos, and A.M.N. Izac, 453–72. Washington, DC: Island Press.

Richardson, J.E., R.T. Pennington, T.D. Pennington, and P.M. Hollingsworth. 2001. Rapid diversification of a species-rich genus of Neotropical rain forest trees. *Science* 293: 2242–45.

Richardson, T.E., A. Hrincevich, T.H. Kao, and A.G. Stephenson. 1990. Preliminary studies into age dependent breakdown of self-incompatibility in *Campanula rapunculoides*: Seed set, pollen tube growth, and molecular data. *Plant Cell Incompatibility Newsletter* 22: 41–47.

Richer, R.A. 2008. Leaf phenology and carbon dynamics in six leguminous trees. *African Journal of Ecology* 46: 88–95.

Ricketts, T.H., G.C. Daily, P.R. Ehrlich, and C.D. Michener. 2004. Economic value of tropical forest to coffee production. *Proceedings of the National Academy of Sciences, USA* 101: 12579–82.

Rivera, G., S. Elliott, L.S. Caldas, G. Nicolossi, V.T.R. Coradin, and R. Borchert. 2002. Increasing day-length induces spring flushing of tropical dry forest trees in the absence of rain. *Trees—Structure and Function* 16: 445–56.

Rivero, K., D.I. Rumiz, and A.B. Taber. 2005. Differential habitat use by two sympatric brocket deer species (*Mazama americana* and *M. gouazoubira*) in a seasonal Chiquitano forest of Bolivia. *Mammalia* 69: 169–83.

Robichaux, R.H., and D.A. Yetman. 2000. *The Tropical Deciduous Forest of Alamos*. Tucson: University of Arizona Press.

Rocha, O.J., and G. Aguilar. 2001. Reproductive biology of the dry forest tree *Enterolobium cyclocarpum* (Guanacaste) in Costa Rica. *American Journal of Botany* 88: 1607–14.

Rodríguez, J.P., J.M. Nassar, K.M. Rodríguez-Clark, I. Zager, C.A. Portillo-Quintero, F. Carrasquel, and S. Zambrano. 2008. Tropical dry forests in Venezuela: Assessing status, threats and future prospects. *Environmental Conservation* 35: 311–18.

Rojas-Jiménez, K., N.M. Holbrook, and M.V. Gutiérrez-Soto. 2007. Dry-season leaf flushing of *Enterolobium cyclocarpum* (ear-pod tree): Above- and belowground phenology and water relations. *Tree Physiology* 27: 1561–68.

Rojas-Martínez, A., A. Valiente-Banuet, M. del Coro Arizmendi, A. Alcántara-Eguren, and H.T Arita. 1999. Seasonal distribution of the long-nosed bat (*Leptonycteris curasoae*) in North America: Does a generalized migration pattern really exist? *Journal of Biogeography* 26: 1065–77.

Rolland, F., E. Baena-Gonzalez, and J. Sheen. 2006. Sugar sensing and signaling in plants: Conserved and novel mechanisms. *Annual Review of Plant Biology* 57: 675–709.

Romero-Duque, L.P., V.J. Jaramillo, and J.A. Pérez-Jiménez. 2007. Structure and diversity of secondary tropical dry forests in Mexico differing in their prior land-use history. *Forest Ecology and Management* 253: 38–47.

Rose, T.J., Z. Rengel, Q. Ma, and J.W. Bowden. 2008. Hydraulic lift by canola plants aids P and K uptake from dry topsoil. *Australian Journal of Agricultural Research* 59: 38–45.

Rosenberg, M.S., D.C. Adams, and J. Gurevitch. 2000. *MetaWin: Statistical Sofware for Meta-analysis, Version 2.0*. Sunderland, MA: Sinauer Associates.

Rossetto, M., C.L. Gross, R. Jones, and J. Hunter. 2004. The impact of clonality on an endangered tree (*Elaeocarpus williamsianus*) in a fragmented rainforest. *Biological Conservation* 117: 33–39.

Rowell, C.H.F. 1998. The grasshoppers of Costa Rica: A survey of the parameters influencing their conservation and survival. *Journal of Insect Conservation* 2: 225–34.

Ruiz, J., M.C. Fandiño, and R.L. Chazdon. 2005. Vegetation structure, composition, and species richness across a 56-year chronosequence of dry tropical forest on Providencia Island, Colombia. *Biotropica* 37: 520–30.

Ruíz-Alemán, F., R. Gómez, and C.A. Harvey. 2005. Caracterización del componente arbóreo en los sistemas ganaderos de Matiguás, Nicaragua. Managua, Nicaragua: Tropitécnica, Nitlapán, UCA.

Rusell, J.D., A.R. Fraser, J.R. Watson, and J.W. Parsons. 1974. Thermal decomposition of protein in soil organic matter. *Geoderma* 11: 63–66.

Rzedowski, J. 1978. *Vegetación de Mexico*. México, D.F.: Editorial Limusa.

Sabogal, C. 1992. Regeneration of tropical dry forests in Central America, with examples from Nicaragua. *Journal of Vegetation Science* 3: 407–16.

Sack, L., and N.M. Holbrook. 2006. Leaf hydraulics. *Annual Review of Plant Biology* 57: 361–81.

Sakai, S., K. Momose, T. Yumoto, T. Nagamitsu, H. Nagamasu, A.A. Hamid, and T. Nakasiiizuka. 1999. Plant reproductive phenology over four years including an episode of general flowering in a lowland Dipterocarp forest, Sarwak, Malaysia. *American Journal of Botany* 86: 1414–36.

Sala, O.E., F.S. Chapin III, J.J. Armesto, E. Berlow, J. Bloomfield, R. Dirzo, E. Huber-Sanwald, L.F. Huenneke, R. Jackson, A. Kinzig, R. Leemans, D. Lodge, H.A. Mooney, M. Oesterheld, N.L. Poff, M.T. Sykes, B.H. Walker, M. Walker, and D.H. Wall. 2000. Global biodiversity scenarios for the year 2100. *Science* 287: 1770–74.

Salamon, J.A., M. Schaefer, J. Alphei, B. Schmid, and S. Scheu. 2004. Effects of plant diversity on Collembola in an experimental grassland ecosystem. *Oikos* 106: 51–60.

Salazar, L.F., C.A. Nobre, and M.D. Oyama. 2007. Climate change consequences on the biome distribution in tropical South America. *Geophysical Research Letters* 34: L09708.

Salcedo, I.H., H. Tiessen, and E.V.S.B. Sampaio. 1997. Nutrient availability in soil samples from shifting cultivation sites in the semi-arid Caatinga of NE Brazil. *Agriculture, Ecosystems and Environment* 65: 177–86.

Salgueiro, F., F. Durvalina, J.F. Caldas, M. Margis-Pinheiro, and R. Margis. 2004. Even population differentiation for maternal and biparental gene markers in *Eugenia uniflora*, a widely distributed species from the Brazilian coastal Atlantic rain forest. *Diversity and Distributions* 10: 201–10.

Sampaio, A.B., K.D. Holl, and A. Scariot. 2007. Does restoration enhance regeneration of seasonal deciduous forests in pastures in central Brazil? *Restoration Ecology* 15: 462–71.

Sampaio, E.V.S.B. 2002. Uso das plantas da caatinga. In *Vegetação e Flora da Caatinga*, ed. E.V.S.B. Sampaio, A.M. Giulietti, J. Virgínio, and C.L. Gamarra-Rojas, 49–90. Recife: Centro Orestino de Informação sobre Plantas.

Sampaio, G., C. Nobre, M.H. Costa, P. Satyamurty, B.S. Soares, M. Cardoso. 2007. Regional climate change over eastern Amazonia caused by pasture and soybean cropland expansion. *Geophysical Research Letters* 34: L17709.

Sánchez-Azofeifa, G.A. 2000. Land use/cover change in Costa Rica: A geographic perspective. In *Quantifying Sustainable Development*, ed. C.A. Hall, C. Leon-Perez, and G. Leclerc, 473–501. New York: Academic Press.

Sánchez-Azofeifa, G.A., G.C. Daily, S.P.A. Pfaff, C. Busch. 2003. Integrity and isolation of Costa Rica's national parks and biological reserves: Examining the dynamics of land cover change. *Biological Conservation* 109: 123–35.

Sánchez-Azofeifa, G.A., M.E. Kalaczka, M. Quesada, K.E. Stoner, J.A. Lobo, and P. Arroyo-Mora. 2003. Tropical dry climates. In *Phenology: An Integrative Environmental Science*, ed. M.D. Schwartz, 121–37. Dordrecht, Netherlands: Klewer Academic Publishers.

Sánchez-Azofeifa, G.A., M. Quesada, P. Cuevas-Reyes, A. Castillo, and G. Sánchez-Montoya. 2009. Land cover and conservation in the area of influence of the Chamela-Cuixmala Biosphere Reserve, Mexico. *Forest Ecology and Management* 258: 907–12.

Sánchez-Azofeifa, G.A., M. Quesada, J.P. Rodriguez, J.M. Nassar, K.E. Stoner, A. Castillo, T. Garvin, E.L. Zent, J.C. Calvo-Alvarado, M.E.R. Kalacska, L. Fajardo, J.A. Gamon, and P. Cuevas-Reyes. 2005. Research priorities for Neotropical dry forests. *Biotropica* 37: 477–85.

Sandell, M. 1990. The evolution of seasonal delayed implantation. *Quarterly Review of Biology* 65: 23–42.

Santiago-Valentin, E., and R.G. Olmstead. 2004. Historical biogeography of Caribbean plants: Introduction to current knowledge and possibilities from a phylogenetic perspective. *Taxon* 53: 299–319.

Santos, C.A.F. 1999. In-situ evaluation of fruit yield and estimation of repeatability coefficient for major fruit traits of umbu tree, *Spondias tuberosa* (Anacardiaceae), in the semi-arid region of Brazil. *Genetic Resources and Crop Evolution* 46: 455–60.

Sarmiento, G. 1972. Ecological and floristic convergences between seasonal plant formations of tropical and subtropical South America. *Journal of Ecology* 60: 367–410.

———. 1975. The dry plant formations of South America and their floristic connections. *Journal of Biogeography* 2: 233–51.

———. 1992. A conceptual model relating environmental factors and vegetation formations in the lowlands of tropical South America. In *Nature and Dynamics of Forest-Savanna Boundaries*, ed. P. Furley, J. Ratter, and J. Proctor, 583–601. London: Chapman and Hall.

Sassi, P.L., C.E. Borghi, and F. Bozinovic. 2007. Spatial and seasonal plasticity in digestive morphology of cavies (*Microcavia australis*) inhabiting habitats with different plant qualities. *Journal of Mammalogy* 88: 165–72.

Saynes, V., C. Hidalgo, J.D. Etchevers, and J. Campo. 2005. Soil C and N dynamics in primary and secondary seasonally dry tropical forests in Mexico. *Applied Soil Ecology* 29: 282–89.

Saysel, A.K., and Y. Barlas. 2001. A dynamic model of salinization on irrigated lands. *Ecological Modelling* 139: 177–99.

SCBD (Secretary of the Convention on Biological Diversity). 2001. *Global Biodiversity Outlook*. Montreal: Secretary of the Convention on Biological Diversity.

Schaal, B.A., D.A. Hayworth, K.M. Olsen, J.T. Rauscher, and W.A. Smith. 1998. Phylogeographic studies in plants: Problems and prospects. *Molecular Ecology* 7: 465–74.

Schemske, D.W., and C.C. Horvitz. 1984. Variation among floral visitors in pollination ability: A precondition for mutualism specialization. *Science* 225: 519–21.

Schmid, J. 2000. Torpor in the tropics: The case of the gray mouse lemur (*Microcebus murinus*). *Basic and Applied Ecology* 1: 133–39.

Schmidt, I.K., A. Michelsen, and S. Jonasson. 1997. Effects of labile soil carbon on nutrient partitioning between an arctic graminoid and microbes. *Oecologia* 112: 557–62.

Schmidt-Nielsen, B., and K. Schmidt-Nielsen. 1950. Evaporative water loss in desert rodents in their natural habitat. *Ecology* 31: 75–85.

Schmidt-Nielsen, K., B. Schmidt-Nielsen, A. Brokaw, and H. Schneiderman. 1948. Water conservation in desert rodents. *Journal of Cellular and Comparative Physiology* 32: 331–60.

Scholz, F.G., S.J. Bucci, G. Goldstein, F.C. Meinzer, and A.C. Franco. 2002. Hydraulic redistribution of soil water by Neotropical savanna trees. *Tree Physiology* 22: 603–12.

Schroth, G.A., G.A.B. da Fonseca, C.A. Harvey, C. Gascon, H.L. Vasconcelos, and A.M.N. Izac, ed. 2004. *Agroforestry and Biodiversity Conservation in Tropical Landscapes*. Washington, DC: Island Press.

Schwendenmann, L., and E. Veldkamp. 2006. Long-term CO_2 production from deeply weathered soils of a tropical rain forest: Evidence for a potential positive feedback to climate warming. *Global Change Biology* 12: 1878–93.

Schwendenmann, L., E. Veldkamp, T. Brenes, J.J. O'Brien, and J. Mackensen. 2003. Spatial and temporal variation in soil CO_2 efflux in an old-growth Neotropical rain forest, La Selva, Costa Rica. *Biogeochemistry* 64: 111–28.

Seastedt, T.R. 1984. Microarthropods of burned and unburned tallgrass prairie. *Journal of the Kansas Entomological Society* 57: 468–76.

Segura, G., P. Balvanera, E. Durán, and A. Peréz. 2003. Tree community structure and stem mortality along a water availability gradient in a Mexican tropical dry forest. *Plant Ecology* 169: 259–71.

Sekercioglu, C.H., S.R. Loarie, F. Oviedo Brenes, P.R. Ehrlich, and G.C. Daily. 2007. Persistence of forest birds in the Costa Rican agricultural countryside. *Conservation Biology* 21: 482–94.

SEMARNAT. 2006. *Estrategia de Educación Ambiental para la Sustentabilidad en México*. México, D.F.: Secretaría de Medio Ambiente y Recursos Naturales (SEMARNAT).

Sewell, M.M., C.R. Parks, and M.W. Chase. 1996. Intraspecific chloroplast DNA variation and biogeography of North American *Liriodendron* L. (Magnoliaceae). *Evolution* 50: 1147–54.

Sheng, C., and H. Tiessen. 2000. Carbon turnover and carbon-13 natural abundance in organo-mineral fractions of a tropical dry forest soil under cultivation. *Soil Science Society of America Journal* 64: 2149–55.

Sibanda, H., and S. Young. 1989. The effect of humus acids and soil heating on the availability of phosphate in oxide-rich tropical soils. In *Mineral Nutrients in Tropical Forest and Savanna Ecosystems*, ed. J. Proctor, 71–83. Oxford: Blackwell Scientific Publications.

Silva, A.C.O., and U.S.P. Albuquerque. 2005. Woody medicinal plants of the caatinga in the state of Pernambuco (northeast Brazil). *Acta Botanica Brasilica* 19: 17–26.

Silva, A.G., R.R. Guedes-Bruni, and M.P.M. Lima. 1997. Sistemas sexuais e recursos florais do componente arbustivo-arbóreo em mata preservada na reserva ecológica de Macaé de Cima. In *Serra de Macaé de Cima: Diversidade Florística e Conservacao em Mata Atlantica*, orgs. H.C. Lima and R.R. Guedes-Bruni, 187–211. Rio de Janeiro: Jardin Botanico.

Silva, M.A. 1986. *Plantas Úteias da Caatinga. Simpósio Sobre Caatinga e sua Exploração Racional*. Brasilia: EMBRAPA.

Silva, S., and H. Tassara. 1996. *Frutas no Brasil*. São Paolo: Empresa das Artes.

Simpson, B.B., and J.L. Neff. 1985. Plants, their pollinating bees, and the Great American Interchange. In *The Great American Biotic Interchange*, ed. F.G. Stehli and S.D. Webb, 427–52. New York: Plenum Publishing.

Singh, J.S., A.S. Raghubanshi, R.S. Singh, and S.C. Srivastava. 1989. Microbial biomass acts as a source of plant nutrients in dry tropical forest and savanna. *Nature* 338: 499–500.

Singh, J.S., L. Singh, and C.B. Pandey. 1991. Savannization of dry tropical forest increases carbon flux relative to storage. *Current Science* 61: 477–80.

Singh, K.P. 1989. Mineral nutrients in tropical dry deciduous forest and savanna ecosystems in India. In *Mineral Nutrients in Tropical Forest and Savanna Ecosystems*, ed. J. Proctor, 153–68. Oxford: Blackwell Scientific Publications.

Sitch, S., B. Smith, I.C. Prentice, A. Arneth, A. Bondeau, W. Cramer, J.O. Kaplan, S. Levis, W. Lucht, M.T. Sykes, K. Thronicke, and S. Venevsky. 2003. Evaluation of ecosystem dynamics, plant geography and terrestrial carbon cycling in the LPJ dynamic global vegetation model. *Global Change Biology* 9: 161–85.

Six, J., H. Bossuyt, S. Degryze, and K. Denef. 2004. A history of research on the link between (micro)aggregates, soil biota, and soil organic matter dynamics. *Soil and Tillage Research* 79: 7–31.

Six, J., R.T. Conant, E.A. Paul, and K. Paustian. 2002. Stabilization mechanisms of soil organic matter: Implication for C-saturation of soils. *Plant and Soil* 241: 155–76.

Šklíba, J., R. Šumbera, W.N. Chitaukali, and H. Burda. 2007. Determinants of daily activity patterns in a free-living Afrotropical solitary subterranean rodent. *Journal of Mammalogy* 88: 1009–16.

Skole, D., and C. Tucker. 1993. Tropical deforestation and habitat fragmentation in the Amazon: Satellite data from 1978 to 1988. *Science* 206: 1905–10.

Soares-Filho, B.S., D.C. Nepstad, L.M. Curran, G.C. Cerqueira, R.A. Garcia, C.A. Ramos, E. Voll, A. McDonald, P. Lefebvre, and P. Schlesinger. 2006. Modelling conservation in the Amazon basin. *Nature* 440: 520–23.

Sobrado, M.A. 1993. Trade-off between water transport efficiency and leaf life-span in a tropical dry forest. *Oecologia* 96: 19–23.

Sobrevila, C., and M.T.K. Arroyo. 1982. Breeding systems in a montane tropical cloud forest in Venezuela. *Plant Systematics and Evolution* 140: 19–37.

Solares, A.F. 1995. Capacidad de regeneración de la corteza y evaluación fitoquímica antes y despues descortezamiento en Cuachalalate. MS diss., Programa Forestal, Colegio de Postgraduadoas, Montecillo, Estado de México.

Somanathan, H., and R.M. Borges. 2000. Influence of exploitation on population structure, plant spacing and reproductive success in dioecious tree species within a fragmented cloud forest in India. *Biological Conservation* 94: 243–56.

Sorensen, T.C., and L.M. Fedigan. 2000. Distribution of three monkey species along a gradient of regenerating tropical dry forest. *Biological Conservation* 92: 227–40.

Sotta, E.D., E. Veldkamp, L. Schwendenmann, B.R. Guimarães, R.K. Paixão, M.L.P. Ruivo, A.C.L. Costa, and P. Meir. 2007. Effects of an induced drought on soil CO_2 efflux and soil CO_2 production in an eastern Amazonian rainforest, Brazil. *Global Change Biology* 13: 2218–29.

Southwood, T.R.E., V.K. Brown, and P.M. Reader. 1979. The relationship of plant and insect diversities in succession. *Biological Journal of the Linnean Society* 12: 327–48.

Spector, J.M., D. Christensen, A. Sioutine, and D. McCormac. 2001. Models and simulations for learning in complex domains: Using causal loop diagrams for assessment and evaluation. *Computers in Human Behavior* 17: 517–45.

Spichiger, R., C. Calenge, and B. Bise. 2004. Geographical zonation in the Neotropics of tree species characteristic of the Paraguay-Paraná basin. *Journal of Biogeography* 31: 1489–1501.

Spichiger, R., R. Palese, A. Chautems, and L. Ramella. 1995. Origin, affinities and diversity hot spots of the Paraguayan dendofloras. *Candollea* 50: 515–37.

Spichiger, R., L. Ramella, R. Palese, and F. Mereles. 1991. Proposición de leyenda para la cartografía de las formaciones vegetales del Chaco Paraguayo. Contribución al estudio de la flora y de la vegetación del Chaco. *Candollea* 46: 541–64.

Srivastava, L.M. 2002. *Plant Growth and Development: Hormones and Environment*. Amsterdam: Academic Press/Elsevier Science.

Srivastava, S.C., and J.S. Singh. 1989. Effect of cultivation on microbial carbon and nitrogen in dry tropical forest soil. *Biology and Fertility of Soils* 8: 343–48.

St. John, M.G., D.H. Wall, and V. Behan-Pelletier. 2006. Does plant species co-occurrence influence soil mite diversity? *Ecology* 87: 625–33.

St. John, M.G., D.H. Wall, and H.W. Hunt. 2006. Are soil mite assemblages structured by the identity of native and invasive alien grasses? *Ecology* 87: 1314–24.

Standley, P.C. 1923. Trees and shrubs of Mexico. *Contributions of the United States National Herbarium* 23: 542–52.

Steele, M.A., and J.L. Koprowski. 2001. *North American Tree Squirrels*. Washington, DC: Smithsonian Institution Press.

Steele, M.D. 2000. Biomass and nutrient dynamics associated with deforestation, biomass burning, and conversion to pasture in a tropical dry forest in Mexico. MS diss., Oregon State University.

Steininger, M.K. 2000. Secondary forest structure and biomass following short and extended land-use in central and southern Amazonia. *Journal of Tropical Ecology* 16: 689–708.

Steininger, M.K., C.J. Tucker, P. Ersts, J. Killeen, Z. Villegas, and S.B. Hecht. 2001. Clearance and fragmentation of tropical deciduous forest in the Tierras Bajas, Santa Cruz, Bolivia. *Conservation Biology* 15: 856–66.

Stephenson, A.G. 1982. When does outcrossing occur in a mass-flowering plant? *Evolution* 36: 762–67.

Stiles, F.G. 1975. Ecology, flowering phenology, and hummingbird pollination of some Costa Rican *Heliconia* species. *Ecology* 56: 285–301.

——. 1977. Coadapted competitors: The flowering seasons of hummingbird-pollinated plants in a tropical forest. *Science* 198: 1170–78.

Stone, A.I. 2007. Responses of squirrel monkeys to seasonal changes in food availability in an eastern Amazonian forest. *American Journal of Primatology* 69: 142–57.

Stoner, K.E. 2001. Differential habitat use and reproductive patterns of frugivorous bats in tropical dry forest of northwestern Costa Rica. *Canadian Journal of Zoology* 79: 1626–33.

——. 2002. Murciélagos nectarívoros y frugívoros del bosque caducifolio de la Reserva de la Biosfera Chamela-Cuixmala. In *Historia natural del bosque caducifolia de Chamela*, ed. F.A. Noguera, M. Quesada, J. Vega, and A. García-Aldrete, 379–95. México City: Instituto de Biología, Universidad Nacional Autónoma de México.

Stoner, K.E., M. Quesada, V. Rosas-Guerrero, and J.A. Lobo. 2002. Effects of forest fragmentation on the (*Musonycteris harrisoni*) Colima long-nosed bat foraging in tropical dry forest in Jalisco, Mexico. *Biotropica* 34: 462–67.

Stoner, K.E., P. Riba-Hernández, K. Vulinec, and J.E. Lambert. 2007. The role of mammals in creating and modifying seedshadows in tropical forests and some possible consequences of their elimination. *Biotropica* 39: 316–27.

Stoner, K.E., A.O. Salazar, R. Fernández, and M. Quesada. 2003. Population dynamics, reproduction, and diet of the lesser long-nosed bat (*Leptonycteris curasoae*) in Jalisco, Mexico: Implications for conservation. *Biodiversity and Conservation* 12: 357–73.

Stoner, K.E., and R.M. Timm. 2004. Tropical dry-forest mammals of Palo Verde: Ecology and conservation in a changing landscape. In *Biodiversity Conservation in Costa Rica: Learning the Lessons in a Seasonal Dry Forest*, ed. G.W. Frankie, A. Mata, and S.B. Vinson, 48–66. Berkeley: University of California Press.

Stoner, K.E., K. Vulinec, S.J. Wright, and C.A. Peres. 2007. Hunting and plant community dynamics in tropical forests: A synthesis and future directions. *Biotropica* 39: 385–92.

Stott, P. 1990. Stability and stress in the savanna forests of mainland South-East Asia. *Journal of Biogeography* 17: 373–83.

Swift, M.J., O.W. Heal, and J.M. Anderson. 1979. *Decomposition in Terrestrial Ecosystems*. Oxford: Blackwell.

Taberlet, P., L. Gielly, G. Pautou, and J. Bouvet. 1991. Universal primers for amplification of three non-coding regions of chloroplast DNA. *Molecular Biology* 17: 1105–9.

Taiz, L., and E. Zeiger. 2006. *Plant Physiology*, 4th ed. Sunderland, MA: Sinauer Associates.

Tajima, F. 1983. Evolutionary relationship of DNA sequences in finite populations. *Genetics* 105: 437–60.

Tallak Nilsen, E., and W.H. Muller. 1981. Phenology of the drought-deciduous shrub *Lotus scoparius*: Climatic controls and adaptive significance. *Ecological Monographs* 51: 323–41.

Tang, J., and D.D. Baldocchi. 2005. Spatial-temporal variation in soil respiration in an oak-grass savanna ecosystem in California and its partitioning into autotrophic and heterotrophic components. *Biogeochemistry* 73: 183–207.

Tardieu, F., and T. Simonneau. 1998. Variability among species of stomatal control under fluctuating soil water status and evaporative demand: Modelling isohydric and anisohydric behaviours. *Journal of Experimental Botany* 49: 419–32.

Teixeira, F.C.P., F. Reinert, N.G. Rumjanek, and R.M. Boddey. 2006. Quantification of the contribution of biological nitrogen fixation to *Cratylia mollis* using the ^{15}N natural abundance technique in the semi-arid Caatinga region of Brazil. *Soil Biology and Biochemistry* 38: 1989–93.

Temperton, V., R.J. Hobbs, T. Nuttle, and S. Halle. 2004. *Assembly Rules and Restoration Ecology: Bridging the Gap between Theory and Practice*. Washington, DC: Island Press.

Theunis, L., M. Gilbert, Y. Roisin, and M. Leponce. 2005. Spatial structure of litter-dwelling ant distribution in a subtropical dry forest. *Insectes Sociaux* 52: 366–77.

Thorne, B.L., M.I. Haverty, and D.H. Bensing. 1996. Associations between termites and bromeliads in two dry tropical habitats. *Biotropica* 28: 781–85.

Thorne, B.L., M.I. Haverty, and M.S. Collins. 1994. Taxonomy and biogeography of *Nasutitermes acajutlae* and *N. nigriceps* (Isoptera: Termitidae) in the Caribbean and Central America. *Annals of the Entomological Society of America* 87: 762–70.

Tiessen, H., I.H. Salcedo, and E.V.S.B. Sampaio. 1992. Nutrient and soil organic matter dynamics under shifting cultivation in semi-arid northeastern Brazil. *Agriculture, Ecosystems and Environment* 38: 139–51.

Tiessen, H., and M.C.D. Santos. 1989. Variability of C, N and P contents of a tropical semiarid soil as affected by soil genesis, erosion and land clearing. *Plant and Soil* 119: 337–41.

Tiessen, H., and J.W.B. Stewart. 1983. Carbon and nitrogen in the light fraction of a forest soil: Vertical distribution and seasonal patterns. *Soil Science* 135: 79–87.

Timm, R.M., and R.K. LaVal. 2000. Mammals. In *Monteverde: Ecology and conservation of a tropical cloud forest*, ed. N.M. Nadkarni and N.T. Wheelwright, 223–44. New York: Oxford University Press.

Timm, R.M., and S.E. Lewis. 1991. Tent construction and use by *Uroderma bilobatum* in coconut palms (*Cocos nucifera*) in Costa Rica. *Bulletin of the American Museum of Natural History* 206: 251–60.

Timm, R.M., D. Lieberman, M. Lieberman, and D. McClearn. 2009. Mammals of Cabo Blanco: History, diversity, and conservation after 45 years of regrowth of a Costa Rican dry forest. *Forest Ecology and Management* 258: 997–1013.

Timm, R.M., and D.K. McClearn. 2007. The bat fauna of Costa Rica's Reserva Natural Absoluta Cabo Blanco and its implications for bat conservation. *University of California Publications in Zoology* 134: 303–52.

Timmermann, A. 1999. Detecting the nonstationary response of ENSO to greenhouse warming. *Journal of the Atmospheric Sciences* 56: 2313–25.

Tole, L. 2001. Jamaica's disappearing forests: Physical and human aspects. *Environmental Management* 28: 455–67.

Toledo, V.M. 1992. Bio-economic cost. In *The Conversion of Tropical Forest to Pasture in Latin America*, ed. T. Downing, S. Hecht, and H. Pearson, 63–71. New York: Westview Press.

——. 2001. Indigenous peoples and biodiversity. In *Encyclopedia of Biodiversity*, ed. S. Levin. San Diego: Academic Press.

Torres, J.A., and G. González. 2005. Wood decomposition of *Cyrilla racemiflora* (Cyrillaceae) in Puerto Rican dry and wet forests: A 13-year case study. *Biotropica* 37: 452–56.

Tosi, J.A., and R.F. Voertman. 1964. Some environmental factors in the economic development of the tropics. *Economic Geography* 40: 189–205.

Tracy, R.L., and G.E. Walsberg. 2001. Intraspecific variation in water loss in a desert rodent, *Dipodomys merriami*. *Ecology* 82: 1130–37.

Trejo, I., and R. Dirzo. 2000. Deforestation of seasonally dry tropical forest: A national and local analysis in Mexico. *Biological Conservation* 94: 133–42.

——. 2002. Floristic diversity of Mexican seasonally dry tropical forests. *Biodiversity and Conservation* 11: 2063–84.

Trumbore, S. 2006. Carbon respired by terrestrial ecosystems: Recent progress and challenges. *Global Change Biology* 12: 141–53.

Trumbore, S., E.A. Davidson, P. Barbosa de Carmargo, D.E. Nepstad, and L.A. Martinelli. 1995. Belowground cycling of carbon in forest and pasture of Eastern Amazonia. *Global Biogeochemistry Cycles* 9: 515–28.

TSBF. 2007. The Small Business Forum. http://www.tsbf.org/csm_bgbd.htm.

Tschapka, M., E.B. Sperr, L.A. Caballero-Martínez, and R.A. Medellín. 2008. Diet and cranial morphology of *Musonycteris harrisoni*, a highly specialized nectar-feeding bat in western Mexico. *Journal of Mammalogy* 89: 924–32.

Turner B.L., II, R.E. Kasperson, P.A. Matson, J.J. McCarthy, R.W. Corell, L. Christensen, N. Eckley, J.X. Kasperson, A. Luers, M.L. Martello, C. Polsky, A. Pulsipher, and A. Schiller. 2003. A framework for vulnerability analysis in sustainability science. *Proceedings of the National Academy of Sciences, USA* 100: 8074–79.

Turner B.L., II, P.A. Matson, J.J. McCarthy, R.W. Corell, L. Christensen, N. Eckley, G.K. Hovelsrud-Broda, J.X. Kasperson, R.E. Kasperson, A. Luers, M.L. Martello, R.S. Mathiesen, R. Naylor, C. Polsky, A. Pulsipher, A. Schiller, H. Selin, and N. Tyler. 2003. Illustrating the coupled human-environment system for vulnerability analysis: Three case studies. *Proceedings of the National Academy of Sciences, USA* 100: 8080–85.

Tyree, M.T., and J.S. Sperry. 1988. Do woody plants operate near the point of catastrophic xylem disfunction caused by dynamic water stress: Answers from a model. *Plant Physiology* 88: 574–80.

Urquiza-Haas, T., P.M. Dolman, and C.A. Peres. 2007. Regional scale variation in forest structure and biomass in the Yucatan Peninsula, Mexico: Effects of forest disturbance. *Forest Ecology and Management* 247: 80–90.

Valdespino, P., R. Romualdo, L. Cadenazzi, and J. Campo. 2009. Phosphorus cycling in primary and secondary seasonally dry tropical forests in Mexico. *Annals of Forest Science* 66: 107, doi: 10.1051/forest: 2008075.

Valenzuela, D. 1998. Natural history of the white-nosed coati, *Nasua narica*, in a tropical dry forest of western México. *Revista Mexicana de Mastozoología* 3: 26–44.

——. 2002. *Nasua narica* (Merriam 1902). Tejon, coati. In *Historia Natural de Chamela*, ed. F.A. Noguera, J.H. Vera-Rivera, A.N. García-Aldrete, and M. Quesada Avendaño, 151–53. Mexico City: Instituto de Biología, Universidad Nacional Autónoma de México.

Valenzuela, D., and G. Ceballos. 2000. Habitat selection, home range, and activity of the white-nosed coati (*Nasua narica*) in a Mexican tropical dry forest. *Journal of Mammalogy* 81: 810–19.

Vallejo, M.A., and F.J. Oveido. 1994. *Características Botánicas, Usos y Distribución de los Principales Arboles y Arbustos con Potencial Forrajero de América Central*. Vol. 2. Serie Técnica, Informe Técnica No. 236. Costa Rica: Centro Agronómico Tropical de Investigación y Ensenañza (CATIE).

van Aarde, R.J., T.P. Jackson, and S.M. Ferreira. 2006. Conservation science and elephant management in southern Africa. *South African Journal of Science* 102: 385–88.

Van der Hammen, T. 1972. Changes in vegetation and climate in the Amazon basin and surrounding areas during the Pleistocene. *Geologie en Mijnbouw* 51: 641–43.

van der Hammen, T., and M.L. Absy. 1974. The Pleistocene changes of vegetation and climate in tropical South America. *Journal of Biogeography* 1: 3–26.

van der Heijden, G.M.F., and O.L. Phillips. 2008. What controls liana success in Neotropical forests? *Global Ecology and Biogeography* 17: 372–83.

Vandermeer, J., and I. Perfecto. 2005. The future of farming and conservation. *Science* 308: 1257–1258.

——. 2007. The agricultural matrix and a future paradigm for conservation. *Conservation Biology* 21: 274–77.

van der Merwe, M., and R.L. Stirnemann. 2007. Reproduction of the banana bat, *Neoromicia nanus*, in Mpumalanga Province, South Africa, with a discussion on sperm storage and latitudinal effects on reproductive strategies. *South African Journal of Wildlife Research* 37: 53–60.

van der Putten, W.H., J.M. Anderson, R.D. Bardgett, V. Behan-Pelletier, D.E. Bignell, G.G. Brown, V.K. Brown, L. Brussaard, H.W. Hunt, P. Ineson, T.H. Jones, P. Lavelle, E.A. Paul, M. St. John, D.A. Wardle, T. Wojtowicz, and D.H. Wall. 2004.

The sustainable delivery of goods and services provided by soil biota. In *Sustaining Biodiversity and Ecosystem Services in Soil and Sediments*, ed. D.H. Wall, 15–43. Washington, DC: Island Press.

van Gestel, M., R. Merckx, and K. Vlassak. 1993. Microbial biomass responses to soil drying and rewetting: The fate of fast- and slow-growing microorganisms in soils from different climates. *Soil Biology and Biochemistry* 25: 109–23.

van Schaik, C., J.W. Terborgh, and J. Wright 1993. The phenology of tropical forests: Adaptive significance and consequences for primary consumers. *Annual Review of Ecology and Systematics* 24: 353–77.

van Vuuren, D.P., P.L. Lucas, and H. Hilderink. 2007. Downscaling drivers of global environmental change: Enabling use of global SRES scenarios at the national and grid levels. *Global Environmental Change* 17: 114–30.

Varella, R.F., M.M.C. Bustamante, A.S. Pinto, K.W. Kisselle, R.V. Santos, R.A. Burke, R.G. Zepp, and L.T. Viana. 2004. Soil fluxes of CO_2, CO, NO, and N_2O from an old pasture and from native savanna in Brazil. *Ecological Applications* 14: S221–S231.

Vargas, R., and E.B. Allen. 2008. Diel patterns of soil respiration in a tropical forest after Hurricane Wilma. *Journal of Geophysical Research* 113: G03021, doi: 10.1029/2007JG000620.

Vargas, R., M.F. Allen, and E.B. Allen. 2008. Biomass and carbon accumulation in a fire chronosequence of a seasonally dry tropical forest. *Global Change Biology* 14: 109–24.

———. 2009. Effects of vegetation thinning on above and belowground carbon in a seasonally dry topical forest in Mexico. *Biotropica* 41: 302–11.

Vargas-Salazar, E. 1993. Anacardiaceae. In *Guia de Arboles de Bolivia*, ed. T. Killeen, E. Garcia, and S. Beck, 93–97. La Paz, Bolivia: Herbario Nacional de Bolivia, Missouri Botanical Garden.

Vaughan, C., and M. Rodríguez. 1986. Comparación de los hábitos alimentarios del coyote (*Canis latrans*) en dos localidades en Costa Rica. *Vida Silvestre Neotropical* 1: 6–11.

Vázquez, R.M.C. 1991. Tendencias en el proceso de domesticación del papaloquelite (*Porophyllum ruderale* [Jacq.] Cass. subsp. *macrocephalum* [DC] R.R. Johnson Asteraceae). MS diss., Universidad Nacional Autónomo de México.

Veiga, V.F., L. Zunino, J.B. Calixto, M.L. Patitucci, and A.C. Pinto. 2001. The *Copaifera* genus. *Quimica Nova* 25: 272–86.

Verboom, B., and H. Huitema. 1997. The importance of linear landscape elements for the pipistrelle *Pipistrellus pipistrellus* and the serotine bat *Eptesicus serotinus*. *Landscape Ecology* 12: 117–25.

Victor, P. 1990. Plantas medicinais: Comparação da flora de quarto municipios de Pernambco. PhD diss., Universidade Federal de Pernambuco.

Vieira, D.L.M., A.B. Sampaio, and K.D. Holl. 2006. Tropical dry-forest regeneration from root suckers in Central Brazil. *Journal of Tropical Ecology* 22: 353–57.

Vieira, D.L.M, and A. Scariot. 2006. Principles of natural regeneration of tropical dry forests for restoration. *Restoration Ecology* 14: 11–20.

Vieira, E.M., and D. Port. 2007. Niche overlap and resource partitioning between two sympatric fox species in southern Brazil. *Journal of Zoology* (London) 272: 57–63.

Viera, R.F. 1999. Conservation of medicinal and aromatic plants in Brazil. In *New Crops and New Uses: Biodiversity and Agricultural Sustainability*, ed. J. Janick, 152–59. Alexandria, VA: ASHS Press.

Viglizzo, E.F., and F.C. Frank. 2006. Land-use options for del plata basin in South America: Tradeoffs analysis based on ecosystem service provision. *Ecological Economics* 57: 140–51.

Vílchez, S.J., C.A. Harvey, D. Sánchez, A. Medina, B. Hernández, and R. Taylor. 2008. La diversidad y composición de aves en un agropaisaje de Nicaragua. In *Evaluación y Conservación de la Biodiversidad en Agropaisajes de Mesoamérica*, ed. C.A. Harvey and J.C. Sáenz, 547–76. Costa Rica: Editorial EUNA.

Villanueva, C., M. Ibrahim, C.A. Harvey, F.L. Sinclair, and D. Muñoz. 2003. Estudio de las decisiones claves que influyen sobre la cobertura arbórea en fincas ganaderas de Cañas, Costa Rica. *Agroforestería en las Américas* 10: 69–77.

Villers-Ruiz, L., and I. Trejo-Vázquez. 1997. Assessment of the vulnerability of forest ecosystems to climate change in Mexico. *Climate Research* 9: 87–93.

Vitousek, P.M. 1984. Litterfall, nutrient cycling, and nutrient limitation in tropical forests. *Ecology* 65: 285–98.

Vizcaíno, M. 1983. Patrones temporales y espaciales de producción de hojarasca en una selva baja caducifolia en la costa de Jalisco, México. MS diss., Universidad Nacional Autónoma de México.

Vogler, D.W., C. Das, and A.G. Stephenson. 1998. Phenotypic plasticity in the expression of self incompatibility in *Campanula rapunculoides*. *Heredity* 81: 546–55.

Von Helversen, O., and Y. Winter. 2003. Glossophagine bats and their flowers: Costs and benefits for plants and pollination. In *Bat Ecology*, ed. T.H. Kunz and M.B. Fenton, 346–97. Chicago: University of Chicago Press.

Vose, J.M., and J.M. Maass. 1999. A comparative analysis of hydrologic responses of tropical deciduous and temperate deciduous watershed ecosystems to climatic change. In *USDA Forest Service Proceedings RMRS-P-12 of North American Science Symposium: Toward a Unified Framework for Inventorying and Monitoring Forest Ecosystem Resources (November 2–6, 1998), Guadalajara, Mexico*, ed. C. Aguirre-Bravo and C. Rodríguez-Franco, 292–98. Fort Collins, CO: U.S. Department of Agriculture.

Vourlitis, G.L., J. de Souza Nogueira, N.P. Filho, W. Hoeger, F. Raiter, M.S. Biudes, J.C. Arruda, V.B. Capistrano, J.L. Brito de Faria, and F. de Almeida Lobo. 2005. The sensitivity of diel CO_2 and H_2O vapor exchange of a tropical transitional forest to seasonal variation in meteorology and water availability. *Earth Interactions* 9: 1–23.

Vourlitis, G.L., N. Priante, M.M.S. Hayashi, J.S. Nogueira, F. Raiter, W. Hoeel, and J.H. Campelo. 2004. Effects of meteorological variations on the CO_2 exchange of a Brazilian transitional tropical forest. *Ecological Applications* 14: S89–S100.

Vreugdenhil, D., J. Meerman, A. Meyrat, L. Diego Gómez, and D.J. Graham. 2002. Map of the Ecosystems of Central America: Final Report. Washington, DC: World Bank.

Walker, J., R.J. Raison, and P.K. Khanna. 1986. Fire. In *Australian Soils: The Human Impact*, ed. J. Russell and R. Isbell, 185–216. Brisbane: University of Queensland Press.

Wall, D.H. 2004. *Sustaining Biodiversity and Ecosystem Services in Soils and Sediments*. Washington, DC: Island Press.

Wall, D.H., G. Adams, and A.N. Parsons. 1999. Soil biodiversity. In *Global Biodiversity in a Changing Environment*, ed. F.S. Chapin III, O.E. Sala, and E. Huber-Sannwald, 376. New York: Springer.

Wall, D.H., A.H. Fitter, and E.A. Paul. 2005. Developing new perspectives from advances in soil biodiversity research. In *Biological Diversity and Function in Soils*, ed. R.D. Bardgett, M.B. Usher, and D.W. Hopkins, 3–27. Cambridge: Cambridge University Press.

Wall, D.H., and J.C. Moore. 1999. Interactions underground: Soil biodiversity, mutualism, and ecosystem processes. *BioScience* 49: 109–17.

Wall, D.H., and R.A. Virginia. 2000. The world beneath our feet: Soil biodiversity and ecosystem functioning. In *Nature and Human Society: The Quest for a Sustainable World*, ed. P.R. Raven and T. Williams, 225–41. Washington, DC: National Academy of Sciences and National Research Council.

Wallace, R.B. 2005. Seasonal variations in diet and foraging behavior of *Ateles chamek* in a southern Amazonian tropical forest. *International Journal of Primatology* 26: 1053–75.

Wallwork, J.A. 1976. Fauna of decaying wood, rocks and trees. In *The Distribution and Diversity of Soil Fauna*, ed. J.A. Wallwork, 243–73. London: Academic Press.

Walsh, S.J. 2008. Biocomplexity in coupled human-natural systems: The study of population and environment interactions. *Geoforum* 39: 773–75.

Walter, D.E., and D.T. Kaplan. 1991. Observation on the *Colescirus simplex* (Acarina: Prostigmata), a predatory mite colonizing greenhouse cultures of rootknot nematodes (*Meloidogyne* spp.) and review of feeding behavior in the Cunaxidae. *Experimental and Applied Acarology* 12: 47–59.

Wang, J., and S.A. Christopher. 2006. Mesoscale modeling of Central American smoke transport to the United States. 2. Smoke radiative impact on regional surface energy budget and boundary layer evolution. *Journal of Geophysical Research—Atmospheres* 111: D14S92, doi: 10.1029/2005JD006720.

Wardle, D.A. 2002. *Communities and Ecosystems: Linking Aboveground and Belowground Components*. Princeton, NJ: Princeton University Press.

Wardle, D.A., V.K. Brown, V. Behan-Pelletier, M. St. John, T. Wojtowicz, L. Brussaard, H.W. Hunt, E.A. Paul, and D.H. Wall. 2004. Vulnerability to global change of ecosystem goods and services driven by soil biota. In *Sustaining Biodiversity and Ecosystem Services in Soil and Sediments*, ed. D.H. Wall, 101–36. Washington, DC: Island Press.

Warnecke, L., J.M. Turner, and F. Geiser. 2008. Torpor and basking in a small arid zone marsupial. *Naturwissenschaften* 95: 73–78.

Weberbauer, A. 1914. Die Vegetationsgliederung des nördlichen Peru um 5° südl. Br. (Departmento Piura und Provincia Jaen des Departamento Cajamarca). *Botanische Jahrbücher und Systematik* 50: 72–94.

———. 1945. *El Mundo Vegetal de los Andes Peruanos Estudio*. Lima, Peru: Editorial Lume.

Weeks, A., D.C. Daly, and B.B. Simpson. 2005. Phylogenetic relationships and historical biogeography of the Burseraceae based on nuclear and chloroplast sequence data. *Molecular Phylogenetics and Evolution* 35: 85–101.

Werth, D., and R. Avissar. 2002. The local and global effects of Amazon deforestation. *Journal of Geophysical Research—Atmospheres* 107, doi: 10.1029/2001JD000717.

West, A.G., K.R. Hultine, J.S. Sperry, S.E. Bush, and J.R. Ehleringer. 2008. Transpiration and hydraulic strategies in a pinyon-juniper woodland. *Ecological Applications* 18: 911–27.

Whigham, D.F., I. Olmsted, E. Cabrera-Cano, and M.E. Harmon. 1991. The impact of Hurricane Gilbert on trees, litterfall, and woody debris in a dry tropical forest in the northeastern Yucatan peninsula. *Biotropica* 23: 434–41.

Whigham, D.F., P. Zugasty Towle, E.F. Cabrera, J. O'Neill, and E. Ley. 1990. The effect of variation in precipitation on basal area growth and litter production in a tropical dry forest in Mexico. *Tropical Ecology* 31: 23–34.

White, A., M.G.R. Cannell, and A.D. Friend. 1999. Climate change impacts on ecosystems and the terrestrial carbon sink: A new assessment. *Global Environmental Change* 9: S21–S30.

White, G.M., D.H. Boshier, and W. Powell. 1999. Genetic variation within a fragmented population of *Swietenia humilis* Zucc. *Molecular Ecology* 8: 1899–1910.

Wick, B., H. Tiessen, and R.S.C. Menezes. 2000. Land quality changes following the conversion of the natural vegetation into silvo-pastoral systems in semi-arid NE Brazil. *Plant and Soil* 222: 59–70.

Wikander, T. 1984. Mecanismos de dispersión de diasporas de una selva deciduas en Venezuela. *Biotropica* 16: 276–83.

Wilcock, C., and R. Neiland. 2002. Pollination failure in plants: Why it happens and when it matters. *Trends in Plant Science* 7: 270–77.

Williams, H.E., and C. Vaughan. 2001. White-faced monkey (*Cebus capucinus*) ecology and management in Neotropical agricultural landscapes during the dry season. *Revista de Biologia Tropical* 49: 1199–1206.

Williams, M., Y. Malhi, A.D. Nobre, E.B. Rastetter, J. Grace, and M.G.P. Pereira. 1998. Seasonal variation in net carbon exchange and evapotranspiration in a Brazilian rain forest: A modelling analysis. *Plant Cell and Environment* 21: 953–68.

Williams, R.J., B.A. Myers, W.J. Muller, G.A. Duff, and D. Eamus. 1997. Leaf phenology of woody species in a north Australian tropical savanna. *Ecology* 78: 2542–58.

Williams-Linera, G. 1997. Phenology of deciduous and broad leaf evergreen tree species in a Mexican tropical lower montane forest. *Global Ecology and Biogeography Letters* 6: 115–27.

Wilson, D.E. 1979. Reproductive patterns. In *Biology of Bats of the New World Family Phyllostomidae, Part III*, ed. R.J. Baker, J.K. Jones Jr., and D.C. Carter, 1–441. Lubbock, TX: Special Publications, Museum, Texas Tech University.

Wilson, D.E., C.O. Handley Jr., and A.L. Gardner. 1991. Reproduction of Barro Colorado Island. In *Demography and Natural History of the Common Fruit Bat*, Artibeus jamaicensis, *on Barro Colorado Island, Panamá*, ed. C.O. Handley Jr., D.E. Wilson, and A.L. Gardner, 43–52. Washington, DC: Smithsonian Institution Press.

Wingler, A., M. Mares, and N. Pourtau. 2004. Spatial patterns and metabolic regulation of photosynthetic parameters during leaf senescence. *New Phytologist* 161: 781–89.

Winter, Y., and O. von Helversen. 2001. Bats as pollinators: Foraging energetics and floral adaptations. In *Cognitive Ecology of Pollination: Animal Behavior and Floral Evolution*, ed. L. Chittka and J.D. Thomson, 148–70. Cambridge: Cambridge University Press.

Woo, H.R., and C.H. Goh, J.H. Park, B.T. de la Serve, J.H. Kim, Y.I. Park, and H.G. Nam. 2002. Extended leaf longevity in the ore4-1 mutant of *Arabidopsis* with a reduced expression of a plastid ribosomal protein gene. *Plant Journal* 31: 331–40.

Wood, H.C., and A. Osol. 1943. *The Dispensatory of the USA*. 23rd ed. Philadelphia: J.B. Lippincott.

Wood, J.R.I. 2006. Inter-Andean dry valleys of Bolivia: Floristic affinities and patterns of endemism—Insights from Acanthaceae, Asclepiadaceae and Labiatae. In *Neotropical Savannas and Seasonally Dry Forests: Plant Diversity, Biogeography, and Conservation*, ed. R.T. Pennington, G. Lewis, and J.A. Ratter, 235–56. Boca Raton, FL: CRC Press.

Wright, I.J., P.B. Reich, O.K. Atkin, C.H. Lusk, M.G. Tjoelker, and M. Westoby. 2006. Irradiance, temperature and rainfall influence leaf dark respiration in woody plants: Evidence from comparisons across 20 sites. *New Phytologist* 169: 309–19.

Wright, J.C., G.A. Sánchez-Azofeifa, C. Portillo-Quintero, and D. Davies. 2007. Poverty and corruption compromises tropical forest reserves. *Ecological Applications* 17: 1259–66.

Wright, S.J. 1991. Seasonal drought and the phenology of understory shrubs in a tropical moist forest. *Ecology* 72: 1643–57.

———. 1996. Phenological responses to seasonality in tropical forest plants. In *Tropical Forest Plant Ecophysiology*, ed. S.S. Mulkey, R.L. Chazdon, and A.P. Smith, 444–60. New York: Chapman and Hall.

Wright, S.J., C. Carrasco, O. Calderón, and S. Paton. 1999. The El Niño Southern Oscillation, variable fruit production, and famine in a tropical forest. *Ecology* 80: 1632–47.

Wright, S.J., and F.H. Cornejo. 1990. Seasonal drought and leaf fall in a tropical forest. *Ecology* 71: 1165–75.

Wright, S.J., A. Hernández, and R. Condit. 2007. The bushmeat harvest alters seed banks by favoring large seeds, and seeds dispersed by bats, birds, and wind. *Biotropica* 39: 363–71.

Wright, S.J., and H.C. Muller-Landau. 2006. The uncertain future of tropical forest species. *Biotropica* 38: 443–45.

Wright, S.J., and C.P. van Schaik. 1994. Light and the phenology of tropical trees. *American Naturalist* 143: 192–99.

Wunder, S. 2007. The efficiency of payments for environmental services in tropical conservation. *Conservation Biology* 21: 48–58.

Wunderle, J.M., and S.C. Latta. 1996. Avian abundance in sun and shade coffee plantations and remnant pine forest in the Cordillera Central, Dominican Republic. *Ornitología Neotropical* 7: 19–34.

Yamada, A., T. Inoue, D. Wiwatwitaya, M. Ohkuma, T. Kudo, and A. Sugimoto. 2006. Nitrogen fixation by termites in tropical forests, Thailand. *Ecosystems* 9: 75–83.

Yetman, D.A., T.R. Van Devender, R.A. López Estudillo, and A.L.R. Guerrero. 2000. Monte Mojino: Maya people and trees in southern Sonora. In *The Tropical Deciduous Forest of Alamos: Biodiversity of a Threatened Ecosystem in Mexico*, ed. R.H. Robichaux and D.A. Yetman, 102–51. Tucson: University of Arizona Press.

York, H.A. 2007. Interspecific ecological differentiation in the short-tailed fruit bats (Chiroptera: Phyllostomidae: *Carollia*). PhD diss., University of Kansas.

Young, A.G., T. Boyle, and T. Brown. 1996. The population genetic consequences of habitat fragmentation for plants. *Trends in Ecology and Evolution* 11: 413–18.

Young, O.R., F. Berkhout, G. Gallopin, M. Janssen, E. Ostrom, and S. van der Leeuw. 2006. The globalization of socio-ecological systems: An agenda for scientific research. *Journal of Global Environmental Change* 16: 304–16.

Zahawi, R.A. 2005. Establishment and growth of living fence species: An overlooked tool for the restoration of degraded areas in the tropics. *Restoration Ecology* 13: 92–102.

Zak, M.R., M. Cabido, and J.G. Hodgson. 2004. Do subtropical seasonal forests in the Gran Chaco, Argentina, have a future? *Biological Conservation* 120: 589–98.

Zarco, A. 2001. Dinámica y distribución espacial y temporal del C y N en el suelo en un ecosistema tropical estacional: Un enfoque de paisaje. BS diss., Universidad Nacional Autónoma de México.

Zarin, D.J., J.R.R. Alavalapati, F.E. Putz, and M. Schmink. 2004. *Working Forests in the Neotropics: Conservation through Sustainable Management*. New York: Columbia University Press.

Zimmerman, J.K., J.B. Pascarella, and T.M. Aide. 2000. Barriers to forest regeneration in an abandoned pasture in Puerto Rico. *Restoration Ecology* 8: 350–60.

Zimmerman, J.K., D.W. Roubik, and J.D. Ackerman. 1989. Asynchronous phenologies of a Neotropical orchid and its euglossine bee pollinator. *Ecology* 70: 1192–95.

Zimmerman, J.K., S.J. Wright, O. Calderon, M. Aponte Pagan, and S. Paton. 2007. Flowering and fruiting phenologies of seasonal and aseasonal Neotropical forests: The role of annual changes in irradiance. *Journal of Tropical Ecology* 23: 231–51.

Zimmermann, M., P. Meir, M.I. Bird, Y. Malhi, and A.J.Q. Ccahuana. 2009a. Climate dependence of heterotrophic soil respiration from a soil translocation experiment along a 3000 m altitudinal tropical forest gradient. *European Journal of Soil Science* 60: 895–906.

———. 2009b. Litter contribution to diurnal and annual soil respiration in a tropical montane cloud forest. *Soil Biology and Biochemistry* doi: 10.1016/j.soilbio.2009.02.023.

Zunino, G.E., V. Gonzalez, M.M. Kowalewski, and S.P. Bravo. 2001. *Alouatta caraya*: Relations among habitat, density and social organization. *Primate Report* 61: 37–46.

CONTRIBUTORS

RAMIRO AGUILAR obtained his BS in Biology and his PhD in Biological Sciences at the Facultad de Ciencias Exactas Físicas y Naturales, Universidad Nacional de Córdoba (UNC), Argentina. He conducted postdoctoral research at the Centro de Investigaciones en Ecosistemas, Universidad Nacional Autónoma de Mexico (UNAM), where he studied the effect of forest fragmentation on the genetic structure of tropical trees. He is a researcher at the Institute of Research in Plant Biology at UNC. His research interests are the ecology, evolution, and conservation of plants.

LORENA ASHWORTH obtained her BS in Biology and her PhD in Biological Sciences at the Facultad de Ciencias Exactas Físicas y Naturales, Universidad Nacional de Córdoba (UNC), Argentina. She conducted postdoctoral research at the Centro de Investigaciones en Ecosistemas, UNAM and FCEFYN, UNC. She is a researcher at the Institute of Research in Plant Biology at UNC with a research interest centered in plant ecology, evolution, and conservation, with am emphasis on plant-pollinator interactions.

LUIS DANIEL AVILA CABADILLA is a PhD student at UNAM, with an undergraduate biology degree from Universidad de La Habana, Cuba. His research focuses on determining the response of vertebrate (lizard, bird, rodent, bat) populations and communities to habitat modification, as well as the importance of these organisms for ecosystem function and recovery. His research approach is population genetics, community ecology, seed dispersal ecology, and landscape ecology, with a focus on the Neotropics.

PATRICIA BALVANERA studies the links between biodiversity, the functioning of ecosystems, and the benefits societies derive from them. Through interdisciplinary collaborations in Mexico and abroad, she is tackling these questions at local, regional, national, and global scales. She obtained a BS at Universidad Autónoma Metropolitana and an MS and a PhD in Ecology from UNAM. She has been working in the tropical dry forest of Chamela since the early 1990s. She is a professor at the Center for Ecosystem Research, UNAM, where she teaches community ecology and ecosystem services.

ANA BURGOS is a research associate at the Environmental Geography Research Center (CIGA) of UNAM, Campus Morelia. She got her PhD at the Center for

Ecosystem Research of UNAM. She currently leads a research group working on land use and water issues among local communities in tropical deciduous forests of the Zicuirán-Infiernillo Biosphere Reserve, in Michoacán, Mexico.

SOFIA CAETANO is a postdoc at the Conservatory and Botanical Garden of Geneva, working on a DNA barcode project that focuses on population genetics. She worked on this study as part of her PhD, which dealt with the phylogeography and population genetics of seasonally dry tropical forest trees. Her main research interests include population genetics and phylogeography, using gene sequencing and genotyping and the application of these methodologies for the reconstruction of historical and present distribution patterns of plants species.

ALICIA CASTILLO holds a bachelor's degree in biology from UNAM, a master's degree in museologic studies from the University of Leicester, UK, and a PhD in Environmental Education from the University of Reading, UK. Her lines of research are science communication and the analysis of the social dimensions of ecosystem management. She is a professor at the Center for Ecosystem Research, UNAM; she teaches undergraduate and graduate courses and mentors students at those levels.

JEANNINE CAVENDER-BARES is associate professor in the Department of Ecology, Evolution and Behavior, University of Minnesota. Her work aims at understanding how plant functional traits link evolutionary history to current ecological processes, with consequences for ecosystem function. Her research integrates levels of organization from cells to ecosystems, with the aim of understanding the origins and organization of plant biodiversity. Much of her work focuses on oaks of Central American SDTFs. Before joining the Minnesota faculty, she studied at Cornell (BA), Yale (MES), and Harvard (PhD) and was a postdoc at the Smithsonian Environmental Research Center and at CEFE-CNRS, Montpellier, France.

GERARDO CEBALLOS is professor at Instituto de Ecología, UNAM. He obtained a BS in Biology at Universidad Metropolitana, Mexico City, an MS at the University of Wales, and a PhD at the University of New Mexico. An expert in the ecology and conservation of terrestrial mammals, he has written extensively on this topic. His work has received much recognition, including the Presidential Award for Conservation in Mexico. He currently teaches conservation biology and mammalian ecology.

ROBIN CHAZDON is a professor in the Ecology and Evolutionary Biology Department at the University of Connecticut. Her research interests include biodiversity conservation and restoration of tropical forests, tropical second-growth forest dynamics, and biodiversity conservation in agricultural landscapes. An authority on the ecology and regeneration of tropical forests, she leads a long-term vegetation dynamics study of secondary forests in northeastern Costa Rica. She served as editor in chief of *Biotropica*, was president of the Association for Tropical Biology and Conservation, and was a member of the governing board of the Ecological Society of America.

NICHOLAS J. DEACON is an adjunct assistant professor of biology at Hamline University and received his PhD from the University of Minnesota Plant Biological Sciences Program. He is interested in native plant communities and how they respond to anthropogenically altered landscapes. His research experience includes work on molecular genetic variation in *Quercus oleoides* of Guanacaste, Costa Rica, tests for evidence of limited pollen dispersal, and local population adaptation. Nick's background with conservation organizations has motivated him to pursue projects that address the protection and maintenance of biodiversity.

RODOLFO DIRZO obtained a BS in Biology from the State University of Morelos, Mexico, and his MS and PhD from the University of Wales, UK. He has been a professor at UNAM and director of the Los Tuxtlas Tropical Research Station. Currently he is a professor in the Biology Department at Stanford University. He has worked extensively in the tropics of Latin America, focusing on the evolutionary ecology of species interactions and conservation biology. He teaches in Mexico and other Latin American countries and in the United States.

MÓNICA FLORES-HIDALGO is a graduate student in the Environmental Sciences Master's Program of the Centro de Investigaciones en Ecosistemas, UNAM. Her field of interest is land-bird ecology. She has focused her work on understanding patterns of habitat use by birds in the subtropical scrub of Mexico's Central Plateau and in different successional stages of the tropical dry forest of the Chamela-Cuixmala Reserve.

FELIPE GARCÍA OLIVA has been a research scientist at Centro de Investigaciones en Ecosistemas, UNAM, since 1995 and chairs the Science Steering Group of the Mexican Carbon Program. His research interests include soil biogeochemistry, carbon dynamics in terrestrial ecosystems, and links between soil bacteria biodiversity and soil nutrient dynamics. He has taught graduate and undergraduate courses on soil biogeochemistry at UNAM since 1999. He received his BS in Geography and his PhD in Ecology at UNAM and conducted postdoctoral research at Colorado State University.

JUAN PABLO GIRALDO is a PhD candidate in the Department of Organismic and Evolutionary Biology at Harvard University. His research interests focus on the plant vascular system as a signaling system controlling plant development and phenology. Before his graduate studies at Harvard, he worked on plant physiological ecology as an intern at the Smithsonian Tropical Research Institute in Panama. He received his BS in Biology and Physics from Universidad de Los Andes in Bogotá, Colombia.

GRIZELLE GONZÁLEZ's research examines the effects of soil organisms on ecosystem processes in tropical, temperate, and arctic ecosystems. She is a research ecologist at the International Institute of Tropical Forestry of the U.S. Department of Agriculture, Forest Service, Puerto Rico, and adjunct faculty of the Department of Biology, University of Puerto Rico, Rio Piedras Campus.

PAUL HANSON, originally from rural Minnesota, obtained BA and MS degrees from the University of Minnesota and a PhD in Entomology from Oregon State Univesity. Currently he is full professor at the University of Costa Rica, where he has worked since 1987. He teaches insect systematics, biological control, and social insects. His principal research involves the biology and systematics of Hymenoptera, especially parasitoids, and he has coedited a book on the Hymenoptera of Costa Rica and another, in Spanish, on the Hymenoptera of the Neotropical region.

CELIA HARVEY is an ecologist with expertise in conservation biology, climate change, tropical agroforestry, and landscape ecology. She holds a BS in Biological Sciences from Stanford University and a PhD in Ecology from Cornell University. She currently serves as a vice president at Conservation International, leading the Department of Global Change and Ecosystem Services. Previously, she was a professor at the Centro Agronómico Tropical de Investigación y Enseñanza in Turrialba, Costa Rica. Her research focuses on understanding patterns of biodiversity within agricultural landscapes and examining how different agricultural and management practices influence biodiversity conservation and ecosystem services provision.

YVONNE HERRERÍAS-DIEGO received her BS in Biology and her PhD in Biological Sciences at UNAM. Her general area of research is in plant ecology, particularly focusing on the ecology, ecological genetics, and conservation biology of woody plants. Specifically, she is interested in the reproductive success, gene flow, and population genetics of tropical plants and the effects of forest fragmentation on biotic interactions, mainly pollination and seed predation. Her publications focus on the effects of forest fragmentation on tropical plants.

N. MICHELE HOLBROOK is professor of biology and Charles Bullard Professor of Forestry in the Department of Organismic and Evolutionary Biology at Harvard University. Her research in whole-plant physiology focuses on the mechanisms that maintain the vascular integrity needed for the transport of water and carbohydrates.

VÍCTOR J. JARAMILLO is a research scientist at the Centro de Investigaciones en Ecosistemas, UNAM. He has a BS in Biology from UNAM and an MS and PhD in Range Science from Colorado State University. His research focuses on the biogeochemistry (C, N, P) of tropical forests and the consequences of land use change, with an emphasis on tropical dry forests.

JEFFREY A. KLEMENS is an independent biologist who has worked in the Area de Conservación Guanacaste (ACG) since 1998. He is the founder of Investigadores ACG, a nonprofit that works to link scientific research to conservation, education, and biological development in northwest Costa Rica.

REYNALDO LINARES-PALOMINO has broad interests in the ecology and biogeography of Neotropical ecosystems. At regional to continental geographical scales, his focus is on understanding the origin, evolution, and current relationships of dry tropical ecosystems. At smaller scales his studies focus on the plant diversity and community ecology of Peruvian Andean ecosystems. He holds a BS in Biology from La Molina University (Lima, Peru), an MS in Plant Biodiversity and Taxonomy from the University of Edinburgh (UK), and a PhD in Biological Diversity and Ecology from the University of Göttingen (Germany).

JORGE A. LOBO obtained his BS in Biology at the Escuela de Biología, Universidad de Costa Rica (UCR), and his MS and PhD in Genetics at the Departamento de Genética, Universidad de Sao Paulo, Brazil. He is currently professor of genetics at UCR's Escuela de Biología. He has conducted research in ecology, genetics, and evolution of tropical insects and plants in Brazil, Costa Rica, and Mexico. Specifically, he has conducted studies on population genetics of Africanized bees in the Americas. He studies pollination, reproduction, and population genetics of tropical trees and has published extensively on these topics.

MANUEL MAASS is a research scientist at the Center for Ecosystem Research, UNAM, Campus Morelia. He got his PhD at the Institute of Ecology of the University of Georgia. Since the mid-1980s he has been studying the structure and function of the tropical deciduous forest ecosystem at the Chamela Biological Station, on the west coast of Mexico. He is currently the cochair of the Mexican Long Term Ecological Research Network (Mex-LTER).

MARÍA JOSÉ MARTÍNEZ-HARMS holds a bachelor's degree in Natural Resource Management from the University of Chile and a master's degree in Environmental Sciences from UNAM. She has developed a methodology for the qualitative assessment of the ecosystem services provided by the protected areas of western Patagonia. She has worked in environmental restoration of river systems and wetlands impacted by productive activities in Chile and developed a methodology to quantify and map ecosystem services in a Mexican watershed.

MIGUEL MARTÍNEZ-RAMOS has a PhD in Ecology from UNAM. He is a professor at the Center for Ecosystem Research at UNAM, where he conducts research on basic and applied population and community ecology in tropical forest ecosystems of Mexico, including studies on secondary succession, forest regeneration, and restoration ecology. He has published extensively on these topics and is an active advisor and mentor of graduate and undergraduate students. He has been Bullard Fellow at Harvard University, president of the Association for Tropical Biology and Conservation, and president of the Botanical Society of Mexico.

ANGELINA MARTÍNEZ-YRÍZAR is a research ecologist at the Instituto de Ecologia, UNAM (Unidad Hermosillo). She has a PhD in Ecology from Cambridge University, UK. Her interests are focused on the study of the structure and function of arid and semiarid ecosystems in Mexico. Most of her reseach is centered on the Sonoran Desert and Mexican tropical dry forests, where she conducts projects on biomass, primary productivity, decomposition, dieback, and drought. She also studies the ecology of nontimber forest products extraction and forest recovery. She is coordinator of the Chamela Group of the Mex-LTER network.

PATRICK MEIR is a reader in ecology at the School of Geosciences, at the University of Edinburgh, UK. He has a BA in Biology from the University of Oxford, UK, and a PhD in Ecological Science from the University of Edinburgh. His research interests focus primarily on forest ecosystem processes and their response to climate. He maintains a central research theme examining the environmental physiology of plants and soil organisms in tropical forest ecosystems, their roles in the cycles of water and carbon, and their influence on vegetation-atmosphere interactions.

HAROLD A. MOONEY is a professor at Stanford University. His research has examined the physiological ecology of plants across a variety of ecosystems, from the tropics to the Arctic. Professor Mooney has served as chairman of the U.S. National Research Council Committee on Ecosystem Management for Sustainable Marine Fisheries, as coordinator of the UN Global Biodiversity Assessment, and as one of the primary organizers of the Millennium Ecosystem Assessment. He has been president of the Ecological Society of America and received numerous recognitions for his work in ecology and biodiversity science. He chairs the international biodiversity science program, DIVERSITAS.

YAMAMA NACIRI is the research officer of the Conservatory and Botanical Garden of Geneva, where she also heads the Phylogeny and Molecular Genetics Unit. She teaches courses on population genetics and biodiversity at the University of Geneva. Her research focuses on the population genetics of different organisms, especially plants, with a particular interest in phylogeography and conservation biology. Over the last few years she has been involved in reconstructing the history of seasonally dry Neotropical forests and studying the way these forests have reacted to past environmental change.

ARY T. OLIVEIRA-FILHO is a senior professor at the Institute of Biological Sciences of the Federal University of Minas Gerais, Brazil. He holds an undergraduate degree in forestry from the University of Viçosa and a master's degree and PhD from the University of Campinas, Brazil. His main fields of interest are the ecology and phytogeography of vegetation, the structure and dynamics of plant communities and populations, and biodiversity science. An active teacher and mentor, he has published extensively in the fields of his expertise.

R. TOBY PENNINGTON leads the Tropical Diversity Programme at the Royal Botanic Garden Edinburgh, UK. He has a BA and D.Phil. in Botany from the University of Oxford, UK. His research covers taxonomy, phylogenetics, and biogeography, with particular emphasis on the legume family and on Latin American forests. Much of his recent work has focused on floristic inventory, biogeography, and diversification in the seasonally dry tropical forests of Latin America.

CHARLES M. PETERS is the Kate E. Tode Curator of Botany in the Institute of Economic Botany, New York Botanical Garden. His research focuses on the ecology, use, and management of tropical forest resources, most of which is done in collaboration with local community groups. He has conducted field research or community forestry projects in the Peruvian Amazon, Papua New Guinea, West Kalimantan (Indonesia), Mexico, Sri Lanka, India, Nepal, Uganda, and Cameroon. He is an associate professor of Tropical Ecology at Yale University and an adjunct research scientist at Columbia University.

CARLOS PORTILLO-QUINTERO obtained his PhD in Earth and Atmospheric Sciences at the University of Alberta. His research focuses on the design and implementation of field survey methods to identify land cover types for satellite imagery supervised classification of vegetation, particularly seasonally dry tropical forests of the Neotropics.

MAURICIO QUESADA obtained his BS in Biology at the Universidad de Costa Rica and his MS and PhD in Ecology at the Department of Biology, Pennsylvania State University. His research focuses on pollination and plant reproductive systems, plant genetics, tree ecophysiology, and conservation biology. He is currently a research professor at the Centro de Investigaciones en Ecosistemas, UNAM. His research program focuses on seasonally dry tropical forests of Mesoamerica. In addition to his research, he is an active teacher and advisor of undergraduate and graduate students.

FERNANDO ROSAS obtained his BS in Biology and his PhD in Ecology at UNAM. He was a postdoctoral fellow at the Centro de Investigaciones en Ecosistemas, UNAM, and is currently a postdoc at the University of California, Los Angeles. He has conducted research on the evolutionary biology of reproductive systems of flowering plants. At present he is conducting studies regarding the effects of forest fragmentation on the patterns of gene flow as well as on the genetic diversity and structure of trees in tropical dry forests.

VÍCTOR ROSAS-GUERRERO obtained his BS in Biology at the Universidad Autónoma Metropolitana, Mexico, and his MS in Biological Sciences at UNAM, where he is currently a PhD student. His research is focused on ecology, evolution, and conservation biology of animal-pollinated plants. He has worked on the effects of forest fragmentation and natural succession on plant-animal interactions, evolution of floral traits, and pollination ecology. He has taught courses in pollination ecology.

G. ARTURO SÁNCHEZ-AZOFEIFA's research is related to the study of impacts of land use/cover change on biodiversity loss and habitat fragmentation in tropical dry forest environments. He received an undergraduate degree in Civil Engineering from the University of Costa Rica, an MS in Hydrology from the University of New Hampshire, and a PhD in Earth Sciences from the University of New Hampshire. He is currently a professor in the Department of Earth and Atmospheric Sciences at the University of Alberta, Canada. He is also the director and founding member of TROPI-DRY and director of the Center for Earth Observation Sciences at the University of Alberta.

GUMERSINDO SÁNCHEZ-MONTOYA obtained his BS in Biology at the Facultad de Ciencias, UNAM. He is currently a research assistant at the Centro de Investigaciones en Ecosistemas, UNAM. He has conducted research on pollination biology, plant reproduction, and plant population genetics. He has experience in the application of molecular biology techniques to the field of ecology and evolution of plants. He has served as instructor in courses of general biology, plant reproductive biology, field ecology, and molecular techniques applied to ecology.

ROBERT L. SANFORD JR. is University Distinguished Scholar and Professor of Biological Sciences at the University of Denver, Colorado, as well as a visiting scientist in the Natural Resource Ecology Laboratory at Colorado State University at Fort Collins, Colorado. His expertise is terrestrial biogeochemistry and forest ecology; he has worked in tropical ecosystems since 1976.

ROBERTO SAYAGO obtained his BS in Biology from the Universidad Michoacana de San Nicolás de Hidalgo, where he has taught courses on ecology and biogeography. He obtained his MS in Biological Sciences at the Centro de Investigaciones en Ecosistemas, UNAM. His work examines the use of bird and woody plant diversity for the selection of priority conservation areas. He is currently a PhD student at UNAM, looking at the effects of forest fragmentation on epiphyte diversity and reproduction in seasonally dry tropical forests of Mexico.

JORGE E. SCHONDUBE is a functional ecologist who studies organism-level physiology to understand ecological end evolutionary processes and patterns. To do this, his lab integrates information from molecular biology with ecology, using physiological and behavioral approaches. His research systems include wildlife in human-modified landscapes, digestive and metabolic physiology of nectar- and fruit-eating animals, and urban settings. He studied biology at the University of Guadalajara and got his PhD in Evolutionary Ecology from the University of Arizona. Presently he is a professor at UNAM's Center for Ecosystem Research.

BREANA SIMMONS surveys soil animal communities and investigates the effects of belowground biodiversity on large-scale ecological processes in managed lands and extreme environments. Dr. Simmons is an assistant professor of biology at East Georgia College, where she teaches biology and microbiology.

KATHRYN E. STONER received her BS and MA from the University of Michigan and her PhD from the University of Kansas. She served as codirector of the Palo Verde Biological Station of the Organization for Tropical Studies, Costa Rica. She has served as a subject editor for *Biotropica* and has worked on tropical forests in Costa Rica, Mexico, and Brazil. Her interests include the effect of forest fragmentation on primates and bats and consequences on forest regeneration, the evolution of color vision in primates, and bat-plant pollination interactions. She is currently a researcher for the Centro de Investigaciones en Ecosistemas, UNAM.

ROBERT M. TIMM first visited the tropical dry forest in 1974 when he was a graduate student at the University of Minnesota. He has conducted research and taught field biology courses in the tropics regularly since then. His research includes systematics and ecology of Neotropical mammals and host-parasite interactions. He has described a number of new species of mammals and their parasites. Dry forests have been one of the focal regions for his research. In recent years he has focused on conservation issues. He is a faculty member at the University of Kansas.

DIANA H. WALL is an ecologist and environmental scientist actively engaged in global research on the contribution of belowground invertebrate biodiversity to soil sustainability. She is a University Distinguished Professor, professor of biology, and director of the School of Global Environmental Sustainability at Colorado State University, Fort Collins. She has served as president of the Ecological Society of America.

HILLARY S. YOUNG is a postdoctoral researcher at Stanford University. She is interested in the indirect effects of anthropogenic disturbances on tropical plant-animal interactions and ecological communities. She received a BA from Princeton University, a Masters in Environmental Management from Yale University, and a PhD from Stanford University. She has conducted research in Panama, Ecuador, the Palmyra Atoll, Northern California, and Alaska, and she is currently doing research on the links between defaunation/land use change and the risks of disease in humans in African savannas.

INDEX

Note: Throughout the index "SDTF" refers to seasonally dry tropical forests.

Island Press | Board of Directors